Coastal and Estuarine Studies

Coastal and Estuarine Studies

formerly Lecture Notes on Coastal and Estuarine Studies

34

S.J. Neshyba Ch.N.K. Mooers
R.L. Smith R.T. Barber (Eds.)

Poleward Flows Along Eastern Ocean Boundaries

Springer-Verlag
New York Berlin Heidelberg London Paris Tokyo Hong Kong

Editors

S.J. Neshyba
College of Oceanography, Oregon State University
Corvallis, OR 97331-5503, USA

Ch.N.K. Mooers
Institute for Naval Oceanography
Stennis Space Center, Mississippi 39529-5005, USA

R.L. Smith
College of Oceanography, Oregon State University
Corvallis, OR 97331-5503, USA

R.T. Barber
Monterey Bay Aquarium Research Institute, 160 Central Avenue
Pacific Grove, CA 93950, USA

ISBN 0-387-97175-0 Springer-Verlag New York Berlin Heidelberg
ISBN 3-540-97175-0 Springer-Verlag Berlin Heidelberg New York

Printed in Germany

Printing and binding: Druckhaus Beltz, Hemsbach/Bergstr.
2837/3140-543210 – Printed on acid-free paper

PROLOGUE

This is the first volume to deal exclusively with an ubiquitous feature of all the major ocean basins--the poleward-flowing surface currents and undercurrents along the continental shelf and slope of their eastern boundaries. The volume emanates from the First International Workshop held on this topic, and which was hosted by the Naval Postgraduate School, Monterey, California, in December 1986.

Organized initially by Steve Neshyba, Robert Smith, Dick Barber, and Chris Mooers, the list of those interested in the workshop topic grew to more than 90 investigators, worldwide, who were then directly involved in or had previously carried out studies of eastern ocean boundary flows. We had come to realize, ca. 1984, that such poleward flows were being observed and described, and that hypotheses were being generated to account for their cause, by many investigators within the context of descriptive theoretical and physical oceanography of regional phenomena.

It was timely to convene the scientific community involved for the broad purpose of comparing the "anatomy" of eastern boundary flow concepts, of deciding how to integrate theoretical and modeling tools more effectively into future eastern boundary studies, and of recommending future sampling recipes from which more wholesome descriptions could come.

And it was also high time to elevate the study of eastern boundary flow processes to its rightful importance alongside that of intense western boundary currents and of their concomitant rings and eddy dynamics. Indeed, it was the discovery of features along eastern boundaries that we now call squirts and jets (made possible by the technology of satellite ocean surveillance) that led us to conclude how the complexity of eastern flow regimes might be organized into an analytical framework, just as the study of the dynamics of western boundary regimes has been given over to its broad-scale description and theory. Eastern ocean zones formerly labeled simply as "transition zones" for lack of adequate tools of analysis, now become recognized as dynamic ocean process regimes worthy of study in their own analytical domain.

And it was the development of the scientific understanding of Pacific basin-wide equatorial waves, as processes integral to the large-scale ocean-atmosphere coupling, that gave birth to explanations of events like El Niño, and that also elevated eastern ocean boundary processes to their proper place in the dynamics of total ocean circulation.

Just as eastern boundary flow processes are important to the physical understanding of ocean circulation, they are equally important to the study of aphysical phenomena. It happens that eastern ocean boundary regimes of both the Atlantic and the Pacific also contain the four major zones of coastal upwelling driven by gyre-dominated wind systems: Baja California-California-Oregon-Washington in the NE Pacific; Peru to the central Chilean coast in the SE Pacific; the Senegal-Mauritania region in the NE Atlantic; and the Angola-Namibia coastal states in the SE Atlantic. Major upwelling zones of the Indian Ocean are driven by monsoon-dominated wind systems.

These are zones of major fisheries, particularly of the clupeoid species from which the major part of world fishmeal production derives: herring, anchovy and menhaden. One of the central problems, posed to appropriate management of these fisheries, is to define the physical constraints to the spawning-egg-larval-juvenile-adult life cycles of members of their biotic communities. Accompanying the physical-biological aspect of these biomes is the chemistry of abiotic cycling on which the biota depend. We recognized early in our design of the workshop agenda how the several disciplines might contribute important insights to one another, to the common goal of understanding how eastern ocean boundary regimes function.

On a larger scale, there is the question of how eastern ocean boundary flows participate in the basin-wide distribution of heat, salt, nutrients, and water, i.e., the "budget" oceanography studies that are also key aspects of global circulation models. An integral part of this same question is whether or not the poleward flow is continuous in transport from low to high latitudes, whether or not it is continuous in time, how it is linked to the intensity and time scale of gyre circulation, and how it is linked to the coastal upwelling process.

All such questions came to the fore during this workshop. We did not find answers to them all. What we did accomplish, however, was a first global scale recommendation for the methodology of proceeding with both the field work of observations in situ and for the refining of theory and models which will eventually need to assimilate these data. This volume presents both of these achievements.

ACKNOWLEDGEMENTS

The authors express appreciation to all those persons whose efforts have made this volume a reality.

Steve Neshyba deserves special recognition among the editors for "staying the course" from inception of the Workshop on Poleward Flowing Undercurrents on Eastern Boundaries held in Monterey December 1986, to its completion in this text.

The Naval Postgraduate School (NPS) provided excellent facilities for the workshop, and this within ten kilometers of an Eastern Boundary Poleward Undercurrent, the (presumed) California Undercurrent off Monterey Bay. Special thanks to Chris Mooers, then the Oceanography Chairman at NPS, for arranging workshop details at the NPS.

Special thanks also to Bob Smith for assembling the initial list of participants, from whose replies to inquiry we eventually compiled a mailing group of over 70 scientists the world over, and to Florence Beyer for her patient mailing effort. More than a third of us gathered in Monterey.

Chris Mooers mobilized support for assembly of the camera-ready copy through the Institute for Naval Oceanography (INO). A very special thanks goes to Ms. Lydia Harper, INO, who performed exceptionally in coordinating the final version with the authors and editors, as well as with Ms. Sabine Kessler, technical editor of Lecture Notes on Coastal and Estuarine Studies.

We thank the special efforts of Richard Barber in providing a biological and chemical continuity to the workshop deliberations of otherwise physics-oriented oceanographers.

We express appreciation also to the College of Oceanography of Oregon State University for sustaining the costs of paper work and mailings and telecommunication needed to bring this international effort to success.

The National Science Foundation supported travel for one international participant to the workshop, and has supported the research of several of the participants which resulted in their contributions to this volume.

This volume is a contribution of the Institute for Naval Oceanography, which is sponsored by the Navy and administered by the Office of the Chief of Naval Research. Any opinion, findings and conclusions or recommendations expressed in this publication are those of the authors and do not necessarily reflect the views of the Chief of Naval Research.

Lastly, we give our best thanks to all colleagues who participated in the reviews of papers.

Table of Contents

PART I:

WORKSHOP SUMMARIES

WORKSHOP SUMMARY: POLEWARD FLOW-OBSERVATIONAL AND THEORETICAL ISSUES

Christopher N.K. Mooers
Institute for Naval Oceanography
Stennis Space Center, MS 39529-5005

I will summarize from my own perspective what I think the essence of the workshop has been, to help set the stage for discussion of scientific futures. With that, let me fill in some of the historical background leading to this workshop.

Some of us, who over the past 10 to 30 years of association with the coastal upwelling studies which "bumped" into poleward undercurrents, have long wondered about their greater importance, but not a great deal has been done about the topic in general. I always thought the topic deserved more attention on its own. Beginning a half-dozen years ago, through the Eastern Pacific Oceanic Conference (EPOC) where western North American physical and other oceanographers gather annually for informal discussions, the undercurrent continually "popped up" as a topic for discussion. And we kept hoping that someone, like Bob Smith or Dudley Chelton or Andrew Willmott, who were "designated hitters" for awhile, would produce an organized research effort on poleward undercurrents.

The opportunity for this workshop arose about two years ago when Steve Neshyba proposed holding a topical session on poleward flows in the context of a Fall AGU meeting. Both Joe Reid and I were then involved as officers of the Ocean Sciences Section of the AGU, and we supported the idea. As it was, the date was a bit late for scheduling it that year, and the event was rescheduled as a poster session for the following year, the 1986 Fall AGU. In the sequence of events that followed, I offered the facilities of the Naval Postgraduate School for an expanded workshop on the topic just prior to the AGU meeting, the better to organize a more comprehensive discussion than is possible within the AGU context, but still designed to interact with the AGU. Bob Smith joined the organizers, drawing up a first list of invited participants who were then encouraged to suggest others with demonstrated interest in the poleward undercurrent topic. The list grew and here we are.

The 1986 AGU meeting next week has special interest to our group because of its sessions celebrating the centennial of scientific investigations of the Equatorial Undercurrent—this first cousin of the poleward undercurrents of interest to us is to be given special treatment in recognition of 100 years of its somewhat systematic study.

Along the way to this workshop, we discussed how the results might be published. It is always good to hold a forum like this, to examine old ideas and old results, and put forth new ideas and new results. But it is my opinion that we have at times lost the benefits of the progress made in the workshop mode by failing to establish a permanent record. In addition to the papers we all will submit to refereed journals, we plan also to compile our workshop papers, notes, and discussions as a volume of the Springer-Verlag series of Lecture Notes on Coastal and Estuarine Studies. One of the most important ingredients of this volume will be our collective thoughts on scientific futures.

I have, therefore, organized my talk accordingly. By taking a systematic approach, I want to bemuse you a bit with some fairly fresh results that will remind you that my research group here at the Naval Postgraduate School does not understand everything about poleward flows in eastern boundary current regimes! Just at the time that we thought we had begun to understand the regime off Northern California, through the NPS/Harvard OPTOMA Program's studies conducted over a period of four-and-a-half years, (cf. Rienecker and Mooers, 1989a) we encountered a big surprise. I want to go back in time, to show some highlights, though at the risk of being very superficial. You will hear more about these later—at the moment I wish to paint a picture for you.

During 1986 we acquired a series of quasi-synoptic maps off Northern California, as a "swan song" burst of activity in the last phase of the OPTOMA Program. Over about a 270 kilometer square domain off Northern California, the surface dynamic topography relative to 450 db in late March/early April indicated the development of a cyclonic coastal upwelling center, just south of Point Arena, Figure 1a, at the beginning of the upwelling season. By late April/end of May, it had separated from the coast, Figure 1b. In mid-July, instead of sampling a regular grid of some 200 stations, as done for the other maps shown here, a cool anomaly moving offshore in association with a jet system was tracked over a more limited domain, Figure 1c. The cool anomaly and jet system flow between an anticyclone and a cyclonic belt extending offshore from near Point Arena; it extends offshore, until it turns sharply to the south in a convergence with an offshore-to-onshore-oriented front and flow.

The shocking result appears in the surface dynamic height field from the next cruise acquired two weeks later, Figure 1d, which covers a

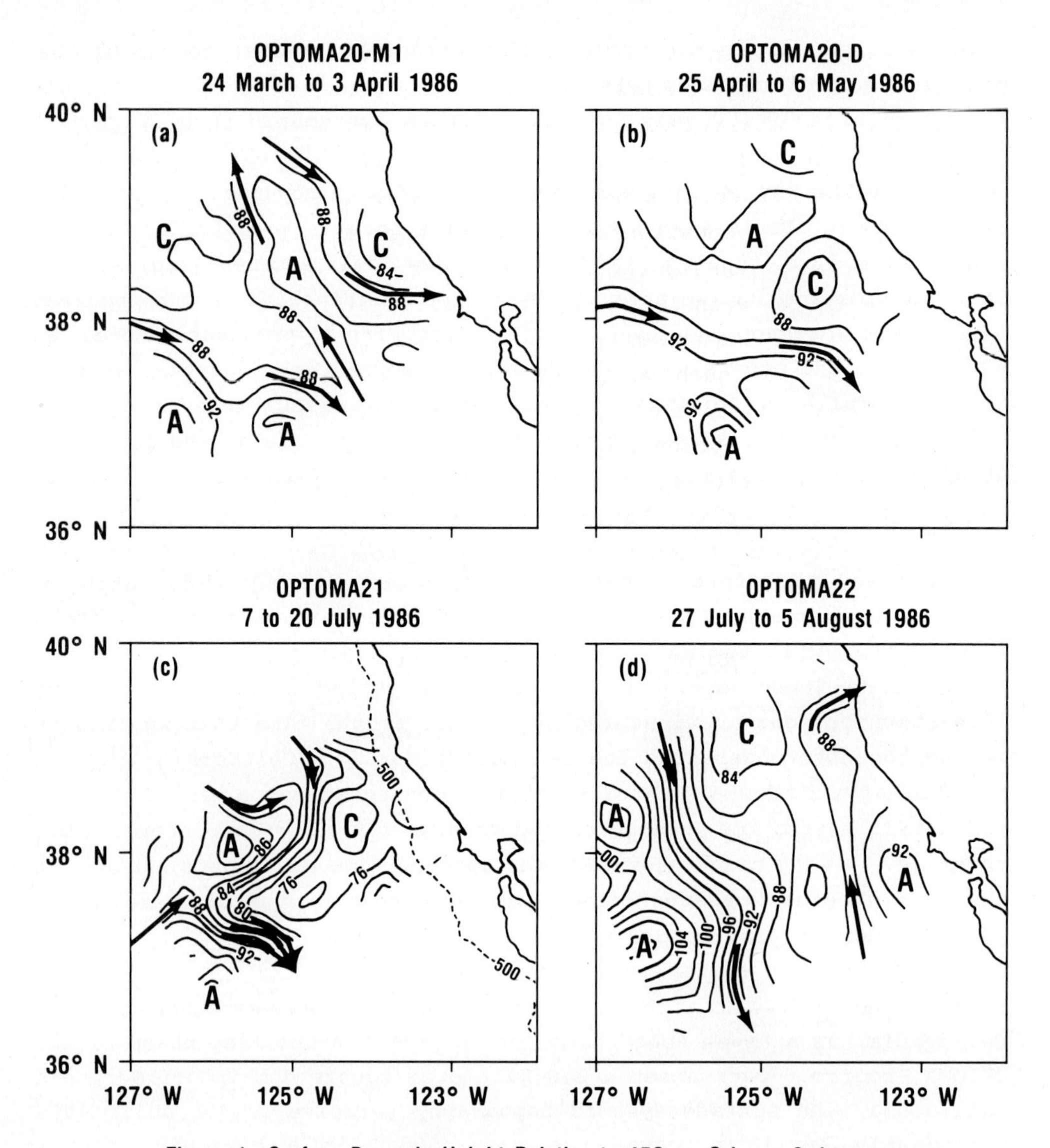

Figure 1. Surface Dynamic Height Relative to 450 m; C.I. = 2 dyn. cm.

somewhat larger domain of 320 kilometers square. There is no offshore flowing jet remaining--basically, all flow is coast-parallel within 200 kilometers of the coast. There has been a massive change in the flow regime, and we do not know why--it seems to be associated with an enhanced poleward flow over the continental slope. While the wind field fluctuated during this period, the fluctuations were not of such magnitude as to provide an obvious explanation for the reversed coastal circulation and onset of strong poleward flow in late July, the latter disrupting the upwelling center off Point arena as well as the offshore-flowing jet. We had never before encountered such abrupt changes. (N.B. Subsequent analyses suggested that relaxation of the local equatorward winds and an increase in the large scale cyclonic wind stress and may have played a role, Rienecker and Mooers (1989b)). The moral of this story is that we ought to keep "looking," that there are probably more surprises linked to poleward flows to be found in these eastern boundary regimes.

WHAT I THINK I SAW AND HEARD DURING THESE PAST TWO DAYS

With the above preface, I turn now to summarize what I have gathered from our workshop. My notes are organized with these topic headings:

1. Occurrence of Poleward Flow

2. Scientific Issues

3. Observational Implications

4. Theoretical Implications

5. Overall Implications

OCCURRENCE--THE GEOGRAPHY OF POLEWARD (PUC) FLOW REGIMES

Spatial. The set of coarsely-defined pigeon-holes into which we have characterized the spatial domains of poleward flows has consisted of these elements:

- There is evidence for flow, or the dynamical equivalent of poleward undercurrents, along all eastern continental margins, plus the Mid-Atlantic Ridge, and perhaps extending into polar regions.

- They clearly occur from tropical to polar latitudes, but we do not imply continuity across this broad latitudinal range.

- They can occur nearshore, typically as near-surface, buoyancy-driven poleward flows. The most common experience and interest relates to flows on the outer shelves and upper slope. Examples are given in this workshop by Jane Huyer et al., [Poleward Flow in the California Current System], Howard Freeland [Observations of the Low-Frequency Circulation Off the West Coast of British Columbia], and others. One must wonder if there are not also flows over the deep slope; we know little of this because no one has yet made the definitive measurements off the eastern boundaries. Some measurements have recently been made on the deep slope off Iberia by Des Barton [The Poleward Undercurrent on the Eastern Boundary of the Subtropical North Atlantic], and Greville Nelson reports some data taken on the deep slope in the region of the Benguela Current. The need to extend our studies of poleward flows to the boundary regions of the deep slope, and even farther offshore, has arisen repeatedly as an issue in our discussions. It is not yet clear whether the poleward flows observed over the deep offshore margins are topographically trapped or not.

Most of the data described here relate to subsurface flows. But the poleward flow regime clearly incorporates surface flows as well; these probably include surface manifestations of coastally trapped wave modes and surfaced undercurrents. Flow that is trapped topographically to the bottom, has also been described by Greg Holloway [Topographic Stress in Coastal Circulation Dynamics].

Temporal. From the temporal view, we have compared much data on poleward flows from margins around the world ocean, mostly on the basis of seasonal variation.

- Overall, our data bases are "short" of the extended measurements needed to focus on the mean poleward flow characteristics; this remark applies across the entire shelf, near-shore and deep-slope, as well as to the extended offshore zones. Once such data are acquired, we shall then also have solid descriptions of seasonal cycles in poleward flows throughout the coastal ocean.

- The problem of variability definition extends to interannual periods as well. Wherever eastern boundary flow data cover two years or more, we see evidence of interannual variability.

Surprisingly, not a great deal has been said in this workshop of variation on the time scale of El Niño events.

- At the other end of the temporal scale, there is much evidence of variability in poleward flow on the storm scales associated with atmospheric forcing. The corresponding coastal ocean response occurs at periods from several-days-to-a-few-weeks.

- We have heard a few reports describing oscillations in coastal flows with periods of the order of one to two months. This particular range of time scales in ocean motion is currently receiving much attention in studies of equatorial flows in connection with tropical meteorology. This point highlights another rationale for more extensive monitoring plans in eastern boundary regimes; i.e., the potential for "connecting" poleward flows on eastern boundaries with flow patterns in equatorial current systems at these time scales. Here we find also the phenomena of coastal trapped waves.

Quantitative Characteristics. The following are quantitative descriptions derived from our discussions:

- Typical along-shore speeds of poleward flows range between 10 to 30 cm/sec; I estimate that the range includes more than 50% of the measured values.

- In spatial terms the flows are found in the depth range 100 to 300 m, over a zone extending 10 to 30 km (or more) offshore.

- If we translate these data into transports, these range from 1 to 2 Sverdrups; again, such characterization applies to the central half of the transport distribution, with values easily extended to 0.5 Sv greater or less. What our workshop has defined clearly is that we do not know transport numbers at all well, and this deficiency could be one of the more important results of our discussions.

- Des Barton has pointed out that the core of the undercurrent off Northwest Africa appears to sink as it moves poleward; this apparent behavior is based on rather short-term observations. I do not know if anyone else has so organized their data as to draw analogous views of other poleward flow regimes, from which one might extract such behavior as a general characteristic. Jane Huyer reports no obvious trend of this type in her data; Steve

Neshyba states that the Gunther Undercurrent data does not show such a trend either. Nevertheless, the depth-latitude behavior of the core of poleward undercurrents, and its variation in time, is a factor which we ought to watch for in the future, as a possible clue to the governing dynamics. Core-depth variation on a local scale can yield clues to topography-connected dynamic terms as well.

SCIENTIFIC ISSUES IN POLEWARD FLOW STUDIES

I break this topic into four categories: kinematics, dynamics, energetics, and roles.

Kinematics. Here there is a rich lode of features to observe:

- The continuity of the poleward flow features, both in space and time, is receiving more and more attention, and rightfully so. This characteristic is a rich one to study, because even though there may be a continuous flow at any one time and over a very large domain, this does not mean that a water parcel entering at one end of the domain necessarily moves continuously to the other end. The motion pattern per se can be "moving" through the coastal circulation system, and we need to know more of how such "motion translation" can occur. Cross-flow processes need further study to understand fully the along-shore continuity topic.

- Based on our discussions, we know little about the sources and sinks of water masses involved in the poleward transport process. Some mention was made of a possible tie-in to the equatorial undercurrents, and, while this "connection" is a most interesting one to explore, its existence is not yet well documented, let alone understood. More work is needed because the source-sink question spills over into several of the topics previously mentioned; e.g.,continuity, transports, core depth behavior, and more.

- Of course, a long-standing issue in kinematic descriptions of poleward flows is that of undercurrents versus counter-currents. There is also an intermediate point of view, i.e., that countercurrents are due to "surfacing" of an undercurrent for short times or long times, over short or long distances. Here we encounter problems of communication, largely due to a lack of adequate data from which to derive a more precise description.

- Concomitant with continuity and transport topics is that of water mass anomalies associated with poleward flows. In many upwelling regimes the poleward currents transport anomalously warm, salty water, high in nutrients and low in dissolved oxygen. This is not a universal rule, for as Des Barton pointed out for the Northwest African case, the water mass carried in the poleward flow is relatively fresh, producing a salinity minimum in vertical profiles. In either case, we use the salinity and nutrient characteristics as tracers for the poleward flow.

- Then there seem to be some real puzzles associated with the along-shore descriptions, in the sense that the poleward flow must traverse transition zones which are themselves described in terms of water-mass transitions. Examples are (1) the fronts off the Gulf of California [where the California Current converges with the mainstream of the Equatorial Countercurrent (the latter diverted around the Costa Rican Dome to become itself a "poleward" stream) and both veer westward into the North Equatorial Current] and (2) the zone of extrusion of Mediterranean Intermediate Water westward from the Strait of Gibraltar.

Dynamics. There should be no surprises in the following list of controlling factors:

- The forcing functions include heat and mass fluxes, as discussed above. Equally important is the atmospheric wind stress and curl. A good deal of information has been developed on the space and time structure of the coastal wind stress fields. We are accumulating information on the along-shore position and seasonal variation in wind stress at size scales from local and regional to basin, and on time scales from the mesoscale and synoptic scale to the seasonal and interannual. But it is not yet clear that we have successfully incorporated these observations into our models. For one thing, we do not know how accurate the wind curl charts are or need to be, i.e., from the point of view of the sensitivity of the ocean's response to such forcing.

  River runoff has to be included for its key role in controlling local buoyancy flow. Several papers demonstrated the importance of the strong internal tides, particularly due to their breaking in the vicinity of the shelf break, which could lead to setup of radiation stresses to play a role in driving poleward flows. The subject provokes much discussion; one can immediately talk of these effects in terms of Sverdrup transports and poleward flows to the (unknown) extent that those dynamics are involved.

- We have heard much talk on various aspects of topographic control of poleward flows, particularly of capes and canyons. In the case of the Benguela regime, there is the influence of the Walvis Ridge, a submarine ridge that runs seaward from the slope. There seems to be other, less well appreciated, transverse ridges extending outward from continental margins, e.g., the Mendocino Escarpment. We ought to remain alert to the possible influence such submarine ridges may have on flows higher in the water column.

- Greg Holloway, Ken Brink, and others have made us well aware of the influence of topographic drag on poleward flows.

- Repeatedly, we have heard of the importance of the along-shore pressure gradient in the dynamics of poleward flow, both in observations and in models. The impression is "... if we just knew THE along-shore pressure gradient ...!" everything would come out fine; what we soon appreciate, however, is that there is a frequency dependence, and a horizontal wavenumber spectrum, to this gradient, so that, until we know much more about its behavior, many of our questions will remain unsettled. This subject is not well-treated in our theoretical models either.

- Coastal trapped waves offer a mechanism for transferring momentum and other properties along coastal boundaries. These occur together with along-shore pressure gradients, but one can encounter the latter without associated trapped waves. Similarly, coastal trapped waves can be associated with topographic drag, but can exist without the latter in the case of a smooth bottom.

- There is of course the dynamic factor of dissipation of turbulent energy. As yet there is not much of an empirical or theoretical basis for understanding the turbulent mixing as Julian McCreary parameterizes it in his models. (It is unfortunate that Rolf Lueck and Hidekatsu Yamazaki could not be here because they have actually made measurements of turbulent dissipation in poleward undercurrents, and would undoubtedly have added to the discussion of this aspect; however, they provided a manuscript to this volume in absentia.)

- The strong, meso-scale eddies that we see in eastern boundary current regimes entrain coastal waters. They may also entrain undercurrent waters. We do not know this for fact, but the possibility remains and this process ought to be kept in mind.

- As usual, issues of linear versus non-linear dynamics and of viscid versus inviscid fluid dynamics crop up. Even though the world is non-linear and viscid, there is always the hope and belief that the processes associated with the undercurrent may be somewhat linear and inviscid. On the other hand, we may just be victims of seduction to that which we want to see, as Greg Holloway, Ken Brink, and others have argued forcefully that non-linear processes are fundamental to the existence of poleward flows on eastern boundaries.

Energetics in Poleward Flows. In the event that the undercurrent is unstable, in the sense of baroclinic and/or barotropic instability, it probably plays a role in the genesis of the cross-shore mesoscale jets and eddies that are rather ubiquitous on eastern boundaries. If this transfer mechanism is substantial, we then face the greater task of evaluating the impact of poleward flows on the general ocean circulation of eastern boundary regions, with all their associated cyclic biological impacts or its role in offshore-directed fluxes of materials introduced at coastal margins. We saw a glimpse of this process in the paper by Batteen (Model Simulations of a Coastal Jet and Undercurrent in the Presence of Eddies and Jets in the California Current System), in watching the evolution of contorted streamfunctions at the level of the undercurrent during the genesis of an eddy. This was probably a glimpse into the real world.

Roles of Poleward Undercurrents in Ocean Science. I will try to identify some of the known, or probable, important roles and effects of eastern boundary poleward undercurrents in ocean behavior:

- In many instances, the poleward undercurrents provide "feedwater" for coastal upwelling regimes. Thus, we ought to advance research into the sources from which these undercurrents draw their water masses. It is in this sense also, that our studies gain strong inter-disciplinary harmonics and overtones. Our chemical and biological colleagues are dependent on us for even more detailed descriptions of the physics of poleward undercurrents than we have provided to date.

- In the same vein of interdependency, we ought to be aware that other transport processes are relevant; for example, biogeochemical transports associated with nutrient, oxygen, and other budgets. Clearly, the upcoming GOFS Program that plans to define global ocean fluxes of all major chemical constituents must depend heavily on the description of sources and sinks of poleward undercurrents, and how these feed into and extract from zones of rich production-like coastal upwelling centers. As Antoine Badan notes, the entry of warm waters into the Gulf of California [via poleward-flowing coastal waters from the south] have a curious effect of greatly increasing productivity.

- Of paramount importance is that we focus on learning how the observed "upwelling centers," fed by poleward undercurrents, are coupled into the mesoscale circulation features of eastern boundary regions; e.g., the squirts, jets, and eddies, both cyclonic and anticyclonic, that are ubiquitous in eastern oceans.

- The oceanographic community is gaining appreciation of the ocean's complex response to atmospheric forcing, as in the case of the El Niño-Southern Oscillation system. This response involves poleward-propagating disturbances from the equatorial region and sub-tropical, zonal, surface Ekman mass transports, as well as responses to anomalous atmospheric forcing in the mid-latitude and sub-polar zones. In other words, there are regional scales of motion through which the poleward undercurrent may pass.

- At least part of the mid-latitude El Niño response of poleward undercurrents seems to be connected with poleward-propagating disturbances. As another part, does the poleward undercurrent transport of warm water make a significant contribution to basin-scale heat budgets, for example?

- There are seasonal "transitions" that occur on eastern boundaries that are interconnected with poleward flows, but about which we have not developed substantive explanations. An example is the occurrence of the Davidson Current off California and Oregon-Washington in the fall-early winter months. Perhaps these phenomena will be better understood once we have answered a related puzzle aired many times in this workshop ... "does the poleward undercurrent regime consist of two distinct streams—the break-slope submerged flow stream and the shelf-shore surface flow, or a single stream that exhibits seasonal variation in depth and position?"

- There remains the uncertain role that the high primary productivity in eastern boundary zones may have on total ocean production. Lou Codispoti (Do Nitrogen Transformations in the Poleward Undercurrent off Peru and Chile Have a Globally Significant Influence?) has posed for us the problem that, whereas high production in undercurrent-fed upwelling centers is well-known, the denitrification that accompanies this phenomenon can lead eventually, over time scales of 100,000 years, to a depression in available nitrogen the ocean over, and therefore limit total oceanic production.

OBSERVATIONAL IMPLICATIONS IN PUC STUDIES

- A clear message was developed in this workshop: we absolutely require long time series [many years in length] of current measurements at several along-shore locations, so as to acquire a regional [ca. 1,000 km] view of flow regimes on the outer shelf and upper slope.

- Further, there is much value gained in short-term, areal "case studies," of the type outlined by Alan Bratkowich for the Central California regime, in which a variety of observing systems are deployed in a specific area for a specific time and experiment; these sensor packages produce shorter time series coupled with spatial images, current meter and drifter data, and so on.

- Perhaps as important as any other aspect of our future studies is the need to focus more attention on the "evolution" of vertical and horizontal structure in poleward undercurrents in response to well-documented forcing. We are still dealing with many fragments of information; we are limited by under-resolved descriptions of the structure, both in the vertical and horizontal. In the few cases where a good oceanic array has been deployed and the *in situ* measurements are adequate, the forcing fields are not adequately and correspondingly defined. It is the familiar story of questions asked after the fact.

- We ought to devise ways, and include them in our research programs, to test better the utility of so-called "easy indicators" as "proxy" monitors of the undercurrent behavior. I include among these such data series as (1) the upwelling index, (2) coastal sea level, and (3) coastal sea surface temperature.

- Lastly, we ought to explore the utility of "biotracers" in undercurrent research. Howard Freeland (Observations of the Low-Frequency Correlation off the West Coast of British Columbia, Canada) gave us one example in his use of a "zooplankton stream" in evaluating observations of physical variables off British Columbia.

THEORETICAL IMPLICATIONS IN PUC STUDIES

- Clearly, as McCreary has shown us, three-dimensional linear models are needed to produce poleward undercurrents. These lead to interesting implications for the influence of the wind stress curl, particularly in accounting for the surface poleward flow.

- We are aware that mixing parameterization plays a crucial role in these models, as does also thermohaline forcing [river outflow, rainfall, evaporation, etc.]. We are also aware of the need to incorporate coastal trapped wave phenomena into more modeling studies.

- Numerical simulation provides guidance in other ways, too, in that it helps both in planning and interpretation of real observations.

- Not well-developed at this workshop, but of equal value in guidance of future work, is the use of diagnostic analysis of analytical and numerical model output. Much more needs to be developed here. Such diagnostic analyses could help pinpoint dynamical balances, mixing zones, and more; hence, they would foster more significant interaction between the theoretical and observational efforts.

- A suite of topics that enter into modeling was dicussed: vortex-stretching, vorticity waves, bottom friction, and coastal topography. Greg Holloway and others (Topographic Stress in Coastal Circulation Dynamics) led us in a discussion of how coastal topography yields canyon upwelling and cape downwelling, and how poleward flow results from the interaction among these factors. Topographic drag issues were raised. Bob Haney described his exploration of the impact on slope circulation of such features as the Mendocino Escarpment.

OVERALL IMPLICATIONS

- Clear from our discussions is that both observational and theoretical modeling studies of poleward flows have reached useful, interesting levels of competence. It is now time to work even harder than before to bring the models and observations together. While the complexity of nature forces us to rely heavily on numerical simulations, this ought not to delay further theoretical studies on how to incorporate basic processes into these models. This consideration only underscores a need on the part of observationalists to enhance our ability to describe the pressure fields which form such important elements of models. Jack Whitehead reminded us of the potential that laboratory modeling has for studies of deep, poleward flow along the Mid-Atlantic Ridge—can we also look toward laboratory models for application to specific problems in coastal flows?

- The need for more diagnostic analyses, to obtain the structure and evolution of various fields, and dominant balances in the eastern boundary regimes, is re-emphasized.

- None of us at this workshop has addressed the task of "data assimilation" into models. We work with somewhat inadequate models and incomplete observations--this is the real world. What we must do next is to meld these techniques, piece-by-piece if necessary, to achieve the highest efficiency of the capacity of both employed together against the limitations of using each alone. Data assimilation is a process of inserting observations into fields computed by models, to serve as "constraints" upon the modeled processes themselves. In the process of assimilation, one faces the weaknesses that exist in each technique separately; subgrid seale processes are not simulated in the models, yet much of our observed data are point values influenced by subgrid scale processes--the problems of data assimilation are indeed not trivial ones.

- We might give serious thought to activating more fully the communications among our group, particularly insofar as long-term planning and effort are concerned. The rationale is that the resources required to advance the topic significantly, at least on the observational side, are formidable. We face great distances along the several coastlines where observations must be made, and these cross national boundaries. We face the problem that the time scales over which observations ought to be sustained can exceed the productive life span of an investigator. To progress boldly, we ought to collectivize bravely.

- We have heard a great deal of geographical explorations, with reports from every eastern boundary regime of the world ocean, including the Arctic Ocean. The excitement we find here is that poleward flows occur in all these situations. But we are coming to the end of pure geographical exploration. We ought now to change to some frame of comparison other than geographical. We require a system of comparative anatomy, so to speak. We need to devise a set of transforms through which the separate geographical cases might be folded into a "mother model space." This new parameter space will have to conform to physical laws of course, but it also ought to accommodate the diversity of cases that characterize our individual "study" areas. There are many of these individual characterizations: narrow shelf, wide shelf, shallow shelf with shallow break, prevailing wind directions, shoreline orientation and length, coastal topography broken with canyons and capes, and so forth.

POLEWARD FLOWS ALONG EASTERN OCEAN BOUNDARIES: An Introduction and Historical Review

Robert L. Smith
College of Oceanography
Oregon State University
Corvallis. OR 97331

As the first speaker at the workshop on "Poleward Flows Along Eastern Ocean Boundaries," I was asked to "get things started" by giving an historically based overview. I felt that exposing my naivete was not inconsistent with the task and stated that two questions hovered in my mind: What is meant by "poleward flows" in the context of this workshop?; and why does the topic warrant a workshop? My geographical focus is the eastern ocean boundaries at low- and mid-latitudes; other authors reddress this parochialism.

One reason for singling out poleward flow along the ocean's eastern boundaries at low- and mid-latitudes is that it is, in some sense, unexpected, i.e., not easily or fully explained by our present knowledge. We were brought up in the wind-driven school of ocean circulation in which the received view was that the global-scale wind system drove the large scale subtropical gyres and the local winds drove the coastal processes. Tomczak (1981) nicely summarizes it all: "The subtropical eastern boundary currents extend from the eastern boundaries of the oceans into their central areas. They are driven by the wind stress curl and modified by frictional boundary layers; compared to the basic circulation, these modifications are of only little influence on the overall dynamics. Along the eastern boundaries where the direction of the wind stress is meridional – while on the oceanic scale it is mainly zonal – coastal upwelling currents are found superimposed on the eastern boundary currents, driven by the meridional wind stress. These currents are elements of an independent dynamic process and not just modifications of the eastern boundary currents." Is poleward flow along the eastern boundary part of the large scale dynamics?, or part of the boundary effects (perhaps related to the coastal upwelling process)?, or due to yet another independent process?

All five of the ocean's eastern boundaries at the latitudes of the subtropical gyres, or the semi-permanent anti-cyclones of the atmospheric circulation, (Iberian Peninsula/Northwest Africa, Southwest Africa, South and North America, and the west coast of Australia) show evidence of poleward flow manifested in water properties, in the ecology, or in the movement of passive drifters. These occur in the region where the

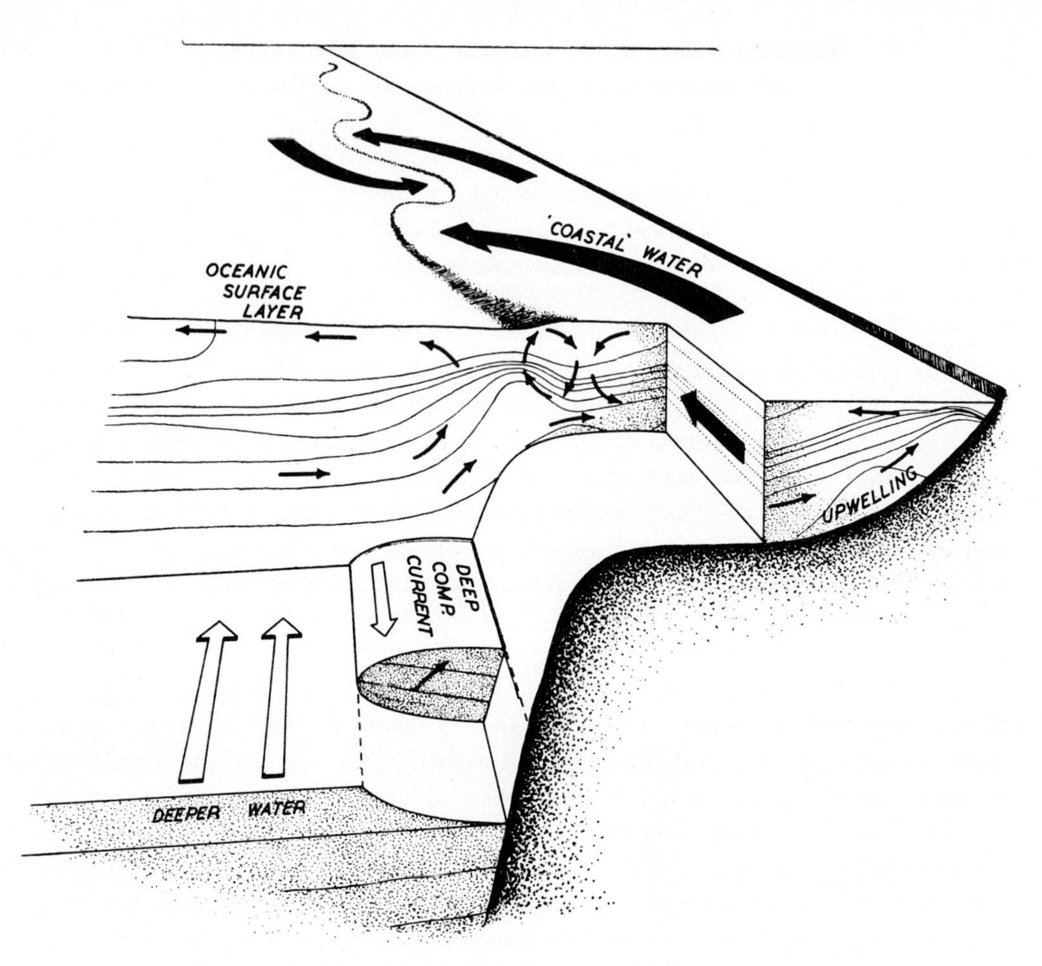

Figure 1: Schematic of circulation during upwelling in Benguela Current (southwest Africa) region; features shown appear to be common to all major coastal upwelling regions (mirror images in Northern Hemisphere). Note the sinuous frontal boundary indicative of eddies between upwelled coastal water and oceanic surface water, the circulation in vertical section, and subsurface counter-current or 'compensation' current. From T.J. Hart and R.I. Currie. The Benguela Current. Discovery Reports 31, 123-298, 1960. (Cambridge Univ. Press) Figure 97.

winds are equatorward and thus the poleward flow raises a dynamical question. But the poleward flow is more than simply a "dynamical curiosity" since it affects the distribution of heat, salt, chemical properties, the sediment transport and the life cycles of things that live in the sea. Indeed, until recently, the evidence for the existence of poleward flow along the ocean's eastern subtropical boundaries was just these effects. Perhaps, the existence of discernable effects is the warrant for having the workshop, and the criterion for considering a particular poleward flow feature in the workshop. The poleward flow does not have to be a continuous ever-flowing river, but clearly must have some temporal and/or spatial persistence to cause a discernible effect...or to even interest modelers. The poleward flow may be a counter-current (counter to the wind and expected surface flow) or a subsurface undercurrent at modest depths; the abyssal circulation is beyond our focus.

During the workshop several theoretical papers were given, revealing that several possible answers to the dynamical questions exist and that modelers are, finally, giving the topic serious consideration. Field experiments during the past decade or so, while rarely setting out to study poleward flow per se, have added to our descriptive knowledge...and added new questions for both observationalists and theoreticians. Present field studies over the continental margins off northern California (the Coastal Transition Zone and MMS studies) and the west coast of Australia (LUCIE: Leeuwin Current Interdisciplinary Experiment) are indirectly or directly, respectively, confronting poleward flow. Since these recent studies will be discussed in the papers of this volume, I shall return to the historical background.

The California Current region can perhaps be taken as paradigmatic. It gives evidence of both a countercurrent and an undercurrent, and has been widely, even if not adequately, observed. Jones (1918) wrote a monograph, "The Neglected Waters of the Pacific Coast," in which he recounts that George Davidson spent many years studying "the subject of coastwise currents. As a result of such study, Professor Davidson concluded that there existed, from 50 to 100 miles offshore, a south setting current of unknown width and velocity, and that inside of this, closely following the general trend of the coast, was a north setting current which he named the 'Davidson Inshore Eddy Current.' The data...were necessarily meager. He had observed such north setting currents at various anchorages while engaged on survey work along the coast, and had also ascertained that logs of the redwood (which do not grow north of California) were frequently found on the shores of Washington, British Columbia, and even Alaska, the wood being well know to the natives of these regions...Davidson was too good a scientist to attempt to promulgate a theory more definite than was justified by the available data..." Both Jones and Davidson were aware of the insufficiency of observations and "had in mind nothing more than a general tendency to a current flowing in a north direction, which tendency might at any time disappear or be reversed by the prevailing weather conditions." Tibby (1941), using water mass analysis, traced 'Equatorial Pacific' water up the coast to at least 45°N, and stated: "The transport of warm, saline Equatorial water to the north appears to be largely accounted for by the inshore northward flow of the California Coastal Counter Current. Present data indicate that this current exists throughout the year at about 200 meters and below, and that above 200 meters it is encountered only during the winter months when it appears at the surface as a narrow northward coastal current. It is within the region immediately adjacent to the coast and at intermediate depths that the highest percentages (of equatorial water) are found."

As hydrographic observations increased, it became clear that there was poleward geostrophic flow off central and southern California even when the winds were strongly southward. The CalCOFI Atlas No. 4 (Wyllie, 1966) shows northward flow at 200 m (relative to 500 m) throughout the year, except for April, from the tip of Baja, California to Cape Mendocino. Munk (1950) pointed out that the curl of the wind stress was positive (cyclonic) within 200 km of the central California coast and thus "the average wind distribution off Monterey is consistent with the existence of a south-flowing current offshore and a countercurrent inshore...a consequence of the local stress curl."

Beginning in the 1960s, physical oceanographical studies in the California Current region, from British Columbia to Baja, California, increased in number and employed the modern methods available to oceanography. The work from then to the present is reviewed by Huyer et al. (this volume). None of the results of recent studies contradict the above general picture. A strong tendency for poleward flow all along the coast is observed (Strub et al., 1987a). The strongest poleward flow seems to be at about 100 to 200 m over the slope (Huyer et al., this volume) with the flow further seaward being highly variable without a significant mean (Stabeno and Smith, 1987). The cyclonic wind stress curl is manifested in the long-term mean and seems to be a robust feature of the eastern boundary regions (Bakun, 1988). It is in the observations of the variability and the details of the structure, and in the details of the theories, that the richness and importance of modern work appears.

The Leeuwin Current region off western Australia is geographically and topographically the analogue of the other eastern boundary current regions and, as in those regions, the winds off western Australia are predominantly equatorward. However, the behavior of the ocean off western Australia seems to be quite unlike the other eastern boundary regions. There is no evidence of a regular continuous equatorward flow within 1000 km of the coast (Andrews, 1977) nor of coastal upwelling, despite the strong equatorward wind stress (Wooster and Reid, 1963). However, for us the region may be an exception that proves a rule. The observations are consistent in showing a poleward (southward) flow against the equatorward wind, over the continental slope, near the shelf edge, from at least 22°S to the southwest tip of Australia. An experiment (LUCIE: Leeuwin Current Interdisciplinary Experiment) had just begun at the time of the workshop and will provide an order of magnitude more observations (Church et al., this volume).

The earliest suggestion of southward flow of tropical water past western Australia came from the observations of tropical marine fauna (including coral) at subtropical latitudes (Saville-Kent, 1897). More recently, satellite tracked buoys (e.g., Cresswell and Golding, 1980) and the evidence for poleward flow in satellite VHRIR photos (Legeckis and Cresswell, 1981) emboldened Cresswell to name the poleward surface flow "The Leeuwin Current." Prior to LUCIE, the only detailed sections across the Leeuwin Current and direct subsurface current measurements were made by Thompson (1984). The axis of the poleward current (speeds up to about 50 cm/sec) seemed to coincide with a maximum in the surface mixed layer depth, which was near the shelf edge and sometimes exceeded 90 m. Below about 150 m, Thompson observed an equatorward (northward) undercurrent. Thompson (1984) argued that the poleward (southward) surface current was driven by the large along-shore pressure gradient, which is larger than in any other eastern boundary region, probably because the eastern tropical Indian Ocean has an open connection with the western equatorial Pacific (Godfrey and Ridgway, 1985). As Godfrey and Ridgway (1985; their Figure 12) show, the along-shore steric sea level slope is negative poleward in all eastern boundary regions, but greatest by far in the Leeuwin Current. It seems clear that it is the along-shore pressure gradient, and not the wind stress curl, that is the driving mechanism for the Leeuwin Current. The recent theories of McCreary et al. (1986) and Thompson (1987) model the Leeuwin Current and consider the effect of the along-shore pressure gradient. Much will be heard about the Leeuwin Current in the immediate future as a result of LUCIE.

The Peru Current seems the closest analogue to the California Current, with the historical literature referring to coastal currents, countercurrents, and undercurrents. The first "modern" study of the Peru Current was made by Gunther (1936) in his monograph entitled "A Report on the Oceanographical Investigations in the Peru Coastal Current," based on the cruise of the WILLIAM SCORESBY in 1931. The report is rich in history and scientific observations. Gunther observed a high salinity warm-water wedge which "had a breadth of some 50 miles and seemed to extend along the greater part of the Peruvian coast...(with) south flow...recorded off San Juan (15°S) and Northern Peru. Upon these facts, the conception of a counter-current is based, though its acceptance as a continuous counter-current leaves much to be explained." The text and the figures indicate that Gunther wasn't sure whether the counter-current(s) might not be part of 'swirls' (eddies). Perhaps Davidson had the same thoughts in mind when he named his current an 'Eddy Current.' Gunther also describes an undercurrent off Chile, based on the observations of high temperature and salinity and low oxygen.

The STEP-I Expedition on the HORIZON in 1960 was a latitudinally extensive study with necessarily wide station spacing. An analysis of the hydrography by Wyrtki (1963) and the drogue measurements by Wooster and Gilmartin (1961) provided evidence for both a Peru Undercurrent and a Peru Countercurrent. Wooster and Gilmartin state that an undercurrent, underlying the Peru Current at depths of several hundred meters, extends from northern Peru (about 6°S to 41°S) with its coastal edge paralleling the shelf edge. They establish its presence by direct measurements with parachute drogues, geostropic computations and analysis of water properties. Like Gunther, they find high salinity and temperature and low oxygen south of 15°S to be evidence of the undercurrent. Wyrtki, using the ship's wind data to obtain the Ekman transport, interprets the dynamic height field to obtain an absolute geostrophic current. He states that the undercurrent exists, but is weak compared to the Peru Countercurrent, which is distinct from, and extends farther seaward than the undercurrent. Wyrtki seems to assign no important role to the undercurrent stating that "south of 15°S upwelling is supplied by Subantarctic water flowing north with the Peru Coastal Current, and north of 15°S by Equatorial subsurface water flowing south with the Peru Countercurrent. It is shown that the Peru Countercurrent and the Peru Undercurrent, although both flow south, are different currents." A recent thorough analysis of a large collection of hydrographic data by Lukas (1986) suggests that water from the Equatorial Undercurrent reaches the coastal upwelling area off Peru by two paths: "One is along a line from the Galapagos Islands to the coast at 5°S (the Peru Countercurrent), and the other is along the equator to the coast, and then along the coast as the Peru-Chile Undercurrent. The Peru-Chile Undercurrent begins near 5°S; north of this latitude, the maximum southward current is at the surface." Silva and Neshyba (1979), using hydrographic data from STEP-I and other cruises, concluded that the "Peru-Chile Undercurrent can now be traced from its origin near 10°S off Peru to 48°S off Chile. The criterion for continuity is the association of a subsurface maximum in the southward geostrophic flow profile with the salinity maximum and oxygen minimum properties..." Thus, all the authors who have considered hydrographic data conclude that poleward flow is a major feature of the circulation (and perhaps a major factor in the ecology) of the Peru-Chile Current region.

The time-series current measurements from moorings off the Peru coast during the past decade (e.g., Brockmann et al., 1981; Brink et al., 1983) suggest that poleward flow is a nearly ubiquitous feature at depths of about 50 to 400 m on the shelf and over the continental slope from 5°S to 15°S. It is this poleward flow that supplies the upwelling water along the Peru coast (Brink et al., 1980). During El Niño the poleward flow initially intensifies and transports anomalously warm water poleward (Smith, 1983). The normal tendency for poleward flow over the continental shelf off Peru, in spite of remarkably steady and relatively strong equatorward winds, appears similar to the tendency for poleward flow on the central California shelf (Strub et al., 1987).

The Canary Current and the Benguela Current regions are reviewed in papers by Barton (this volume) and Shannon (1985), respectively. As in the other eastern boundary regions, early evidence for subsurface poleward flow in the Canary Current region came from the distribution of properties (see Wooster and Reid, 1962). There is now considerable evidence, including direct current measurements, for a poleward undercurrent off Northwest Africa and the Iberian Peninsula. A complete review of the recent work is found in Barton (this volume). Whether the subsurface poleward flow is continuous over this extent is an open question posed by Mittelstaedt (this volume).

Hart and Currie (1960) presented the first evidence for poleward flow along the coast of Southwest Africa, based on a cruise by the recommissioned WILLIAM SCORESBY in 1950. They calculated the geostrophic flow at 200 m relative to 600 m between about 22°S and 28°S and noted poleward flow along the edge of the continental shelf. From water mass analyses they inferred that the upwelling water on the shelf came from depths of 200 to 300 m, and referred to the poleward flow as a "compensation current" which provided the replacement source for the water that was upwelled. This is analogous to what is observed, at shallower depths, off Peru. While a number of papers present evidence of poleward flow off the west coast of South Africa (see Shannon, 1985), the region has not yet had the benefit of a large-scale coordinated field experiment that would help synthesize the previous observations. One can only hope that the political and social problems will be resolved in that troubled part of the world, and that Southwest African physical oceanography can again become an international field.

Now the reader should consider the papers in this volume. They show the advances made since the earlier work, but in general do not contradict the necessarily "broad-brush" and speculative results of the earlier investigators. Indeed, the wonderful schematic that Hart and Currie (1960, their Figure 97 and used as Figure 1 in this paper) present includes almost every mesoscale feature of research interest today. Today, as the papers in this volume demonstrate, we have a better knowledge of the spatial and temporal location and extent of the features, but we are still asking to what extent the features are dynamically inter-related. For example, the development of the 'squirts and jets' (cold filaments) seen off many eastern boundaries may depend on the existence of a poleward undercurrent (Ikeda and Emery, 1984). What is also demonstrated in this volume is that poleward flow is a common and important feature of the ocean's eastern boundary and that this topic is now ready for, and deserving of, a concentrated research effort. It is of a scale, geographically, and in both oceanographic and social-economic importance, that justifies the study of the poleward flow along the ocean's eastern boundaries being included in WOCE (see the What Next? summary in this volume).

From: Joseph L. Reid, Scripps Institution of Oceanography

On: Review and commentary to paper **POLEWARD FLOWS ALONG EASTERN OCEAN BOUNDARIES** by Robert L. Smith

"I wish to point out that the author in his opening paragraphs discusses only those poleward flows that lie in mid-latitude, alongside but opposed to the great anticyclonic gyres. The remarks made therein are meant to apply to those latitudes, but not necessarily to the eastern boundary currents of the higher latitudes, the Alaska Current, which is a part of a cyclonic gyre, and the poleward flow south of 40°S along the coast of Chile. These, of course, are not always opposed to an equatorward wind, and although they are currents at the eastern boundary, they seem to be part of the great west-wind drifts."

THEORETICAL UNDERSTANDING OF EASTERN OCEAN BOUNDARY POLEWARD UNDERCURRENTS

Allan J. Clarke
Department of Oceanography and
Geophysical Fluid Dynamics Institute
Florida State University
Tallahassee, Florida 32306-3048

INTRODUCTION

Mean flows along eastern ocean boundaries may be generated by several mechanisms, including tidal rectification [e.g., Georges Bank, Loder (1980)]; melting ice and rivers emptying into the sea [e.g., the northern Gulf of Alaska, Royer (1981)]; forcing by offshore mean density fields [e.g., the Leeuwin Current, Huthnance (1984), and McCreary et al. (1986)]; the nonlinear interaction of forced barotropic shelf waves over a continental margin [Denbo and Allen (1983)]; oscillatory flow over small scale bumps in the shelf bottom topography [Brink (1986), Haidvogel and Brink (1986), and Holloway (1987)]; and mean wind-stress and wind-stress curl [McCreary (1981), Wang (1982), Suginohara (1982), Philander and Yoon (1982), Suginohara and Kitamura (1984), McCreary and Chao (1985), and McCreary et al. (1986)]. We shall concentrate our discussion primarily on the dynamics of poleward-flowing undercurrents on eastern ocean boundaries. Consequently, since mean flows produced by melting ice and rivers are generally surface flows near the coast, this mechanism is not relevant. Tidally rectified flows only exist in special areas where tidal currents and bottom topography slope are large (e.g., Georges Bank), so this mechanism is also not important in the generation of the major poleward-flowing undercurrents. Generation of poleward flow by the nonlinear interaction of forced barotropic shelf waves would also appear not to be a relevant mechanism as the mean flows generated have been estimated to be very small, only ~0.1 cm s^{-1}. Mean flows produced by offshore mean density fields do produce flows of the right strength, but on eastern boundaries these offshore fields drive equatorward rather than poleward undercurrents (McCreary et al., 1986).

Of the two remaining mechanisms, let us first consider mean flow generation by oscillatory flow over small amplitude, small scale bumps in the shelf bottom topography. Coastally trapped waves always propagate poleward on eastern ocean boundaries, so when the oscillatory current is flowing equatorward, a coastally trapped lee wave disturbance can form behind a bump. Energy propagates off to infinity and, thus, creates a drag on the flow. When the oscillatory current is flowing poleward, no coastally trapped waves can remain stationary with respect to the obstacle, so there is no drag on the current. The asymmetry in the wave drag results in a poleward mean flow. Holloway (1987) discusses this generation in a more general context. For realistic parameters, barotropic theory gives currents of about 2.5 cm s^{-1} (Haidvogel and Brink, 1986), considerably smaller than observed undercurrent speeds (~10 cm s^{-1}). Transports are on the order of 10 percent or less than observed coastal undercurrent transports. Calculations still need to be performed for the more realistic stratified ocean case.

Undercurrents can also be driven by the wind. More work has been done on this mechanism than any other, and it will be discussed in the next section.

WIND-FORCED EASTERN BOUNDARY POLEWARD UNDERCURRENTS

McCreary (1981), Wang (1982), Suginohara (1982), Philander and Yoon (1982), Suginohara and Kitamura (1984), and McCreary and Chao (1985) have all considered wind-driven eastern boundary flows. Many fundamental aspects of the dynamics are illustrated by McCreary (1981), and this model is discussed below.

MCCREARY'S CONSTANT DEPTH, LINEAR MODEL

The equations of motion, linearized about a state of rest and horizontally uniform stratification, are

$$u_t - fv = -p_x + (\nu u_z)_z, \tag{2.1}$$

$$v_t + fu = -p_y + (\nu v_z)_z, \tag{2.2}$$

$$u_x + v_y + w_z = 0, \tag{2.3}$$

$$\rho'_t + w\bar{\rho}_z = (\kappa\rho')_{zz}, \tag{2.4}$$

$$p_z = -g\rho', \tag{2.5}$$

where the vertical eddy viscosity and diffusivity coefficients are given by

$$\nu = A/N^2, \quad \kappa = \sigma^{-1}A/N^2. \tag{2.6}$$

In these equations x, y, and z refer to distances eastward, northward, and vertically upward from the ocean surface at $z = 0$; u, v, and w are velocities in the x, y, and z directions; and t, f, p, $\bar{\rho}$, ρ', g, N, and σ refer, respectively, to time, Coriolis parameter, pressure divided by the mean (constant) water density, the water density in the absence of motion, the perturbation density due to the motion, the acceleration due to gravity, the buoyancy (Brunt-Väisälä) frequency and the Prandtl number. Note that in Eq. (2.4) McCreary uses an unconventional form for the diffusion of mass so that the problem can be separated into vertical modes.

The surface boundary conditions are

$$\nu u_z = \tau^x, \quad \nu v_z = \tau^y, \quad w = \kappa\rho' = 0 \quad \text{at } z = 0, \tag{2.7}$$

where τ^x and τ^y are the eastward and northward components of the wind stress at the ocean surface. The ocean is of constant depth H and at the bottom $z = -H$

$$\nu u_z = 0 = \nu v_z = w = \kappa\rho'. \tag{2.8}$$

Solutions of (2.1)-(2.6) can be written as sums of vertical modes $F_n(z)$ as

$$u = \sum_{n=1}^{\infty} u_n F_n, \quad v = \sum_{n=1}^{\infty} v_n F_n, \quad p = \sum_{n=1}^{\infty} p_n F_n,$$

$$w = \sum_{n=1}^{\infty} w_n \int_{-H}^{z} F_n \, dz, \quad \rho' = \sum_{n=1}^{\infty} \rho_n F_{nz}, \tag{2.9}$$

where the expansion coefficient functions u_n, v_n, p_n, w_n, and ρ_n are functions only of x, y, and t. The eigenfunctions $F_n(z)$ satisfy

$$(F_z/N^2)_z + F/c^2 = 0, \tag{2.10a}$$

$$F_z = 0 \quad \text{on } z = 0, -H, \tag{2.10b}$$

and are normalized so that

$$F_n(0) = 1. \tag{2.11}$$

The barotropic (n = 0) mode is missing from the sums in (2.9) as it typically makes a negligible contribution to the solution. Substitution of (2.9) into (2.1)-(2.5) and use of the orthogonality relation for the eigenfunctions leads to the following equations for the expansion coefficient functions:

$$(\partial/\partial t + \lambda_n)\, u_n - f v_n = -p_{nx} + X_n, \tag{2.12a}$$

$$(\partial/\partial t + \lambda_n)\, v_n + f u_n = -p_{ny} + Y_n, \tag{2.12b}$$

$$(\partial/\partial t + \lambda_n/\sigma)\, p_n/c_n^2 + u_{nx} + v_{ny} = 0, \tag{2.12c}$$

$$w_n = (\partial/\partial t + \lambda_n/\sigma)\, p_n/c_n^2, \quad \rho_n = -p_n/g, \tag{2.12d}$$

where

$$\lambda_n = A/c_n^2, \tag{2.13}$$

$$(X_n, Y_n) = (\tau^x, \tau^y)/D_n, \tag{2.14}$$

and

$$D_n = \int_{-H}^{0} F_n^2 \, dz. \tag{2.15}$$

We are interested in steady solutions $\partial/\partial t = 0$) driven by an along-shore windstress $\left(Y_n \neq 0, X_n = 0\right)$ Under these simplifications, u_n and v_n can be expressed in terms of p_n:

$$\left(\lambda_n^2 + f^2\right) u_n = -f p_{ny} - \lambda_n p_{nx} + f Y_n$$

$$\left(\lambda_n^2 + f^2\right) v_n = f p_{nx} - \lambda_n p_{ny} + \lambda_n Y_n \,.$$

Since the along-shore (y) scale is much greater than the offshore (x) scale and all modes contributing significantly to the response have

$$\lambda_n \ll f,$$

the above equations can be simplified to

$$u_n = -p_{ny}/f - \lambda_n p_{nx}/f^2 + Y_n/f, \tag{2.16a}$$

$$v_n = p_{nx}/f. \tag{2.16b}$$

Substitution of these expressions into (2.12c) gives the field equation for the pressure as

$$p_{nxx} + \beta\lambda_n^{-1} p_{nx} - \sigma^{-1} f^2 p_n/c_n^2 = 0 \tag{2.17}$$

where it has been assumed that the wind stress Y_n is independent of x, and we have taken $\beta = f_y$. This field equation is subject to the boundary conditions that the motion remains finite at large distances from the coast and that u_n vanish at the coast x = 0. In terms of p_n, these conditions are

$$p_n \text{ remains finite as } |x| \to \infty, \tag{2.18a}$$

$$\lambda_n p_{nx} + f p_{ny} = fY_n \text{ at } x = 0. \tag{2.18b}$$

The solution of (2.17) subject to (2.18) is

$$p_n = \exp\left(k_n x\right) \varphi_n(y), \tag{2.19}$$

where

$$k_n = \beta\left(2\lambda_n\right)^{-1}\left[\left(1 + 4\sigma^{-1} f^2 \lambda_n^2 c_n^{-2} \beta^{-2}\right)^{\frac{1}{2}} - 1\right], \tag{2.20}$$

and $\varphi_n(y)$, which satisfies

$$\varphi_{ny} + \lambda_n k_n \varphi_n/f = Y_n \quad , \tag{2.21}$$

is given by

$$\varphi_n = \exp\left(-\Lambda_n(y)\right)\int_{y_0} \exp\left(\Lambda_n(y_*)\right) Y_n(y_*)\, dy_* \quad , \tag{2.22}$$

with

$$\Lambda_n(y) = \int_{y_0}^{y} \lambda_n k_n / f \, dy_* \quad . \tag{2.23}$$

From (2.19) and (2.16)

$$v_n = k_n p_n / f \tag{2.24}$$

and

$$u_n = \left(Y_n/f\right)\left(1 - \exp\left(k_n x\right)\right) \quad . \tag{2.25}$$

The solutions for p, v, and u are now simply obtained by summing the modes as in (2.9).

DYNAMICS

As in the time dependent no dissipation case (Gill and Clarke, 1974), the velocity perpendicular to the coast is determined entirely by the local wind-stress (see (2.25)), while the along-shore velocity and pressure depend on wind-stress integrated along the coast (see (2.19), (2.22), and (2.24)). Thus, the along-shore velocity and pressure may be nonzero even in regions where the wind-stress is negligible.

To further understand the dynamics, first let us consider some idealized cases. For very strong dissipation (2.18b) reduces to

$$p_{nx} = fY_n/\lambda_n \quad ,$$

i.e.,

$$v_n = Y_n/\lambda_n \quad . \qquad (2.26)$$

In this case, the pressure and along-shore velocity are locally determined. At the other extreme, as $\lambda_n \to 0$ and dissipation is negligible, by (2.20), $k_n \to 0$,and, hence, in the limit of no dissipation

$$u_n = v_n = 0 \qquad (2.27)$$

and

$$p_{ny} = Y_n \ , \qquad (2.28)$$

i.e., there are no currents and the wind-stress is balanced by a pressure gradient. The initial value problem (Anderson and Gill, 1975) shows that this is the balance left after long Rossby waves have propagated westward from the initial wind-generated disturbance on the eastern boundary. When dissipation is included, the Rossby waves cannot completely escape from the eastern boundary, and nonzero currents are present.

Since $c_n \sim 1/n$ for large n, from (2.13)

$$\lambda_n \sim n^2 \quad . \qquad (2.29)$$

Consequently, we expect the dominant balance in (2.18b) to be (2.26) for the higher order modes. We would not expect these modes to contribute much to the alongshore velocity field because (2.26) and (2.29) show that for these modes

$$v_n \sim Y_n n^{-2} \sim n^{-2} \qquad (2.30)$$

and n is large. [The result $Y_n \sim n^0$ in (2.30) follows from (2.15) and results for the asymptotic eigenfunctions $F_n(z)$ in Clarke and Van Gorder

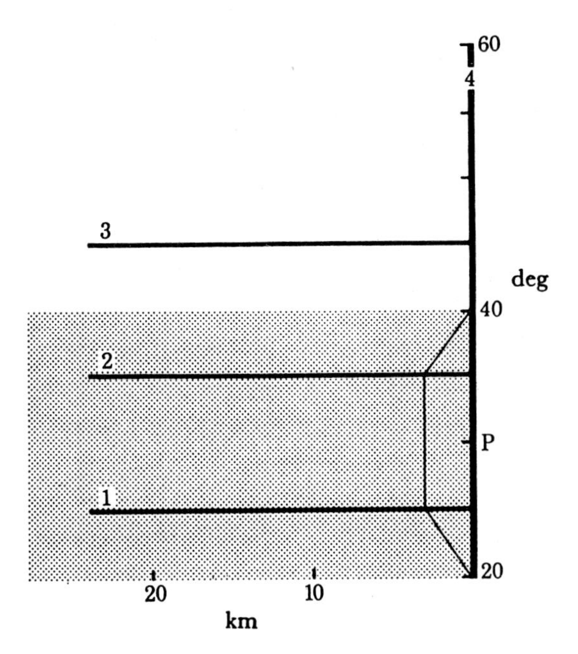

Figure 1: A schematic diagram showing the location of sections 1-3, the position of the coast, and the structure of the wind forcing. The shaded region indicates the latitudinal extent of the wind field, and the thin line its meridional profile. The wind field does not weaken offshore, and so it is without curl. Wind stress is entirely equatorward and reaches a maximum amplitude of 0.5 dyn/cm^2. (After McCreary, 1981).

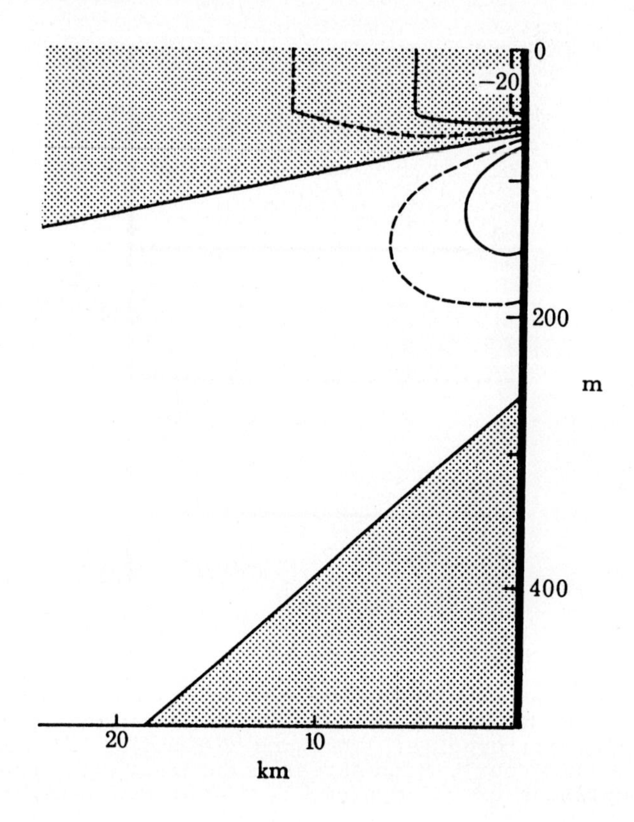

Figure 2a: Vertical sections of along-shore current along section 1 of Figure 1. The velocity contour interval is 10 cm s^{-1} and the dashed contours are ±5 cm s^{-1}. The shaded region indicates equatorward flow. (After McCreary, 1981).

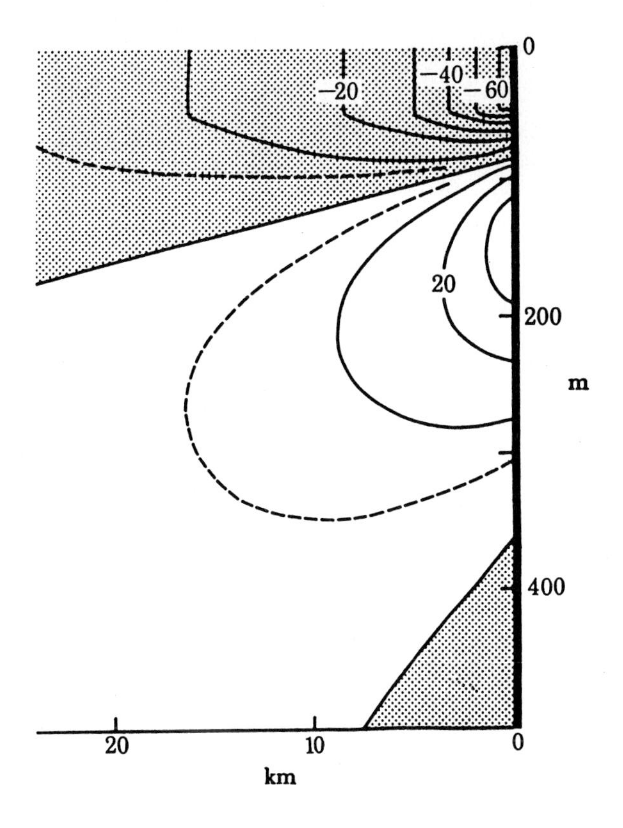

Figure 2b: As for Figure 2a, but along section 2.

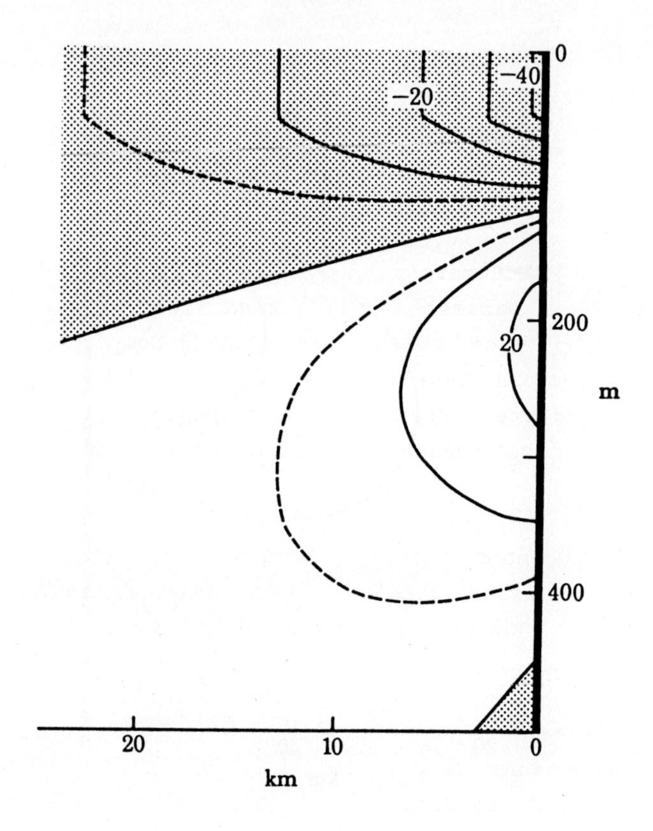

Figure 2c: As for Figure 2a, but along section 3.

(1986)]. For realistic parameters, the lower order modes satisfy (2.27) and (2.28), so again the contribution to the along-shore velocity is small. Therefore, the along-shore velocity field is principally due to the "intermediate modes" and these consequently set the horizontal and vertical scales for the motion.

Figure 2 shows McCreary's calculations of the along-shore velocity field for realistic parameters and a wind field resembling the steady component of the winds off the western coast of North America (see Figure 1). The undercurrent is clearly visible and has a strength and horizontal and vertical structure comparable with the observed undercurrent. The horizontal and vertical scales correspond to scales for the "intermediate modes" ($n \approx 6$).

The undercurrent's existence depends strongly on the existence of an alongshore pressure gradient; when calculations are carried out with $\partial/\partial y = 0$, no undercurrent forms. The undercurrent's existence is not, however, crucially dependent on ß-plane dynamics; the f-plane solution and the ß-plane solution are very similar.

Some observations (e.g., Mittelstaedt, 1978) suggest that the coastal undercurrent may deepen and weaken toward the poles. Figures 1 and 2 show that the model undercurrent deepened and weakened north of the wind patch. What is the physical reason for this? North of the wind patch the solution of (2.21) is

$$\varphi_n(y) = \varphi_n(y_0) \exp\left(-\Lambda_n(y)\right) . \tag{2.31}$$

Since $\Lambda_n \sim n^3$, the higher order modes are much more rapidly damped than the lower order ones, so north of the wind patch the motion is increasingly dominated by the lower order intermediate modes. These modes are less tightly trapped to the coast and have deeper z structures; so north of the wind patch, the undercurrent is deeper, less tightly trapped to the coast and, because there are fewer modes, it is weaker.

CRITIQUE

Suginohara and Kitamura (1984) and McCreary and Chao (1985) showed that, due to an equatorward barotropic shelf current induced by the wind, the continental shelf acts to strengthen the equatorward jet and to weaken or eliminate the undercurrent of the constant depth model. However, McCreary and Chao (1985) showed that the positive wind-stress curl which exists near the western coast of North America drives a coastal poleward shelf flow (as would be expected from Sverdrup balance). This poleward shelf flow counteracts the equatorward shelf barotropic flow and restores the undercurrent.

Other problems still exist, however. Figure 6 of McCreary and Chao (1985) suggests that when the bottom stress is realistic, the undercurrent is much weaker than observed. This is cause for concern because the wind-driven undercurrent must be strong enough to overcome likely equatorward undercurrents induced by offshore mean density fields (McCreary et al., 1986). In their study of the California Current System, McCreary et al. (1987) imposed a no-slip condition at the "coast" which strengthened the undercurrent, presumably because the equatorward shelf barotropic flow was lessened by the no-slip condition. But both the slip and no-slip conditions are not really justified, since they are not applied at the real coast, but rather at about 100 m depth near the edge of the true shelf topography. Another problem is that the models discussed are linear and nonlinear terms and are not negligible in parts of the flow. Finally, many of the important results depend on quite large values for the eddy viscosity and diffusivity, and we do not know whether it is reasonable to use such values.

RECOMMENDATIONS FOR FUTURE WORK

I believe that the theory of wind-driven mean eastern boundary flows is sufficiently developed, the theoretical results sufficiently encouraging, and observations sufficiently scarce that much would be gained from a large scale Eastern BOundary Circulation Experiment (EBOCE). Because of its proximity and the measurements we already have, the western coast of the United States should be the site of the experiment. Models already exist for a prior estimate of the flow to provide design criteria and focus for the experiment (e.g., McCreary et al., 1987). Some important unanswered questions about eastern boundary flows are listed below:

(i) The linear dissipation wind-driven models mainly discussed in this paper suggest that mean along-shore flows are not driven locally, but rather they are at least partially forced by an integral along the coast of the along-shore component of the wind-stress [see Eqs. (2.19) and (2.22)]. This suggests that eastern boundary currents are of large along-shore scale. What do observations say? Can present theory explain (c.f. the earlier discussion in section 2) observed changes in the width, depth, and strength of eastern boundary currents?

(ii) Current structures and strengths in the linear wind-driven models depend crucially on the use of large values for eddy viscosity and diffusivity coefficients. Would observations of turbulent fluctuations in the flow warrant the use of such large values?

(iii) Are the calculated mean eastern boundary current fields stable? The eddies, jets, and squirts observed along the U.S. west coast suggest that they are not. Perhaps the instabilities will result in the required large values for the eddy coefficients.

(iv) How important is the generation of mean flow by the mean large-scale offshore density field and by oscillatory flow over small bumps in the shelf and slope topography?

(v) Alongshore changes in the shelf and slope bottom topography will affect the structure and strength of mean or very low frequency flows. Can theory like that of Gill and Schumann (1979) predict such changes?

(vi) Pacific Eastern boundary flow is strongly affected by interannual variations associated with El Niño. At these frequencies, coastal sea level has been observed to propagate poleward at about 40 cm s^{-1} (Enfield and Allen (1980) and Chelton and Davis (1982)). Associated El Niño currents strongly affect eastern ocean boundary biology (e.g., Chelton et al., 1982). What is the horizontal and vertical structure of these El Niño currents? Does theory suggest that these currents can be estimated by using convenient and cheap sea level measurements at the coast?

ACKNOWLEDGEMENTS

This work was supported by NSF Grants OCE-8500669 and OCE-8515979. Paula Tamaddoni-Jahromi ably typed the manuscript.

The editors of this volume express appreciation to Dr. Ken Brink, Woods Hole Oceanographic Inst., for his helpful review and suggestions to the manuscript.

POLEWARD FLOW ALONG EASTERN BOUNDARIES: WHAT NEXT?

Adriana Huyer

TIME VARIABILITY OF POLEWARD UNDERCURRENTS

We need long-term (> 1 year) monitoring of poleward flows in each of the five subtropical eastern boundary regions, and very long records (> 5 years) in at least one region. Observations should include moored current measurements across the continental margin, repeated hydrographic and ADCP sections, and remote and in situ measurements of the wind and pressure fields. These observations might be situated at the eastern end of each of the WOCE principal zonal long-lines; i.e., at 28°S and 24°N in the Pacific, 30°S and 24°N in the Atlantic, and 30°S in the Indian Ocean.

ROLE OF POLEWARD UNDERCURRENTS IN GLOBAL AND GYRE-SCALE PROCESSES

We need experiments to determine the along-shore structure and continuity of an eastern boundary undercurrent, and to determine the associated transports of mass, heat, nutrients, and tracers. Observations should include current meter moorings at several latitudes, repeated hydrographic and ADCP surveys, and remote and in situ measurements of the wind and pressure fields. Designing the optimum experiment and interpreting results will require both theoretical and numerical modeling. These experiments will address the question of how the major gyres communicate with their eastern boundaries, and therefore they should be included in the WOCE Gyre Dynamics Experiment.

We also need to determine whether and how processes associated with poleward flow on eastern boundaries affect the global nutrient budgets: The high productivity of these regions gives them the potential for affecting chemical and biological cycles in the open ocean; e.g., the southeast tropical Pacific may play a major role in the global nitrogen cycle.

FORCING AND DYNAMICS OF POLEWARD UNDERCURRENTS

We need to continue to develop analytical, numerical, and laboratory models to increase our understanding of the roles of mixing, along-shore pressure gradients, wind-stress curl, form-drag, and wave-mean-flow interactions in driving undercurrents and countercurrents. We need to develop paradigms for choosing among different models, and to design field experiments that can discriminate among different forcing/response hypotheses.

A more detailed outline of the discussion under each heading follows:

I. Time Variability of Poleward Undercurrents

What To Do

(a) Long-term (> 1 year) monitoring of poleward flows in each of the five subtropical eastern boundary regions (viz., California, Canary, Benguela, Peru, and Leeuwin Current regions); very long-term (> 5 years) monitoring in one region.

(b) Comparative anatomy of poleward undercurrents in different "regions" with different physics; viz., latitude, climate (wind, runoff, and heat budget), and bathymetry. This might be accomplished by studies at several latitudes within one eastern boundary current region.

How

(a) Long-term current-meter moorings across the continental margin.

(b) Repeated hydrographic and ADCP sections across the margin and beyond.

(c) Measure wind and wind-stress curl fields (scatterometers and buoys).

(d) Measure along-shore pressure gradients (altimeters, hydrographic surveys, pressure, and tide gauges).

(e) Use support from local agencies and countries (including LDC's) where possible.

Where

(a) At the eastern end of each of the WOCE CORE-I zonal long-lines (i.e., at about 28°S and 24°N in the Pacific, 30°S and 24°N in the Atlantic, and 30°S in the Indian Ocean). At each of these locations, moored current meters should be in place for at least one year.

In at least one of these locations, current meters should be in place for five years or more.

(b) On the eastern margin of the gyre chosen as the object of the WOCE CORE III: Gyre Dynamics Experiment. (For the North Atlantic subtropic gyre, we would recommend an intensive study off both Portugal and North Africa.)

(c) Continue monitoring at sites presently (or recently) occupied;

(i) In the California Current:
- La Perousse Bank at 49°N (1983-1993?, by IOS, Canada).
- Off northern and central California (MMS studies, NPS).
- Off Oregon 43°N (OSU).

II a. Role of Poleward Undercurrents in Gyre Dynamics

What To Do

(a) Conduct process-oriented studies of poleward undercurrent kinematics and dynamics as part of the WOCE CORE-III Gyre Dynamics Experiment. (How do gyres communicate with their eastern boundaries?)

(b) Determine the along-shore structure and continuity of poleward undercurrent; explore evidence for flow instabilities.

(c) Determine transports of mass, heat, and tracers by poleward undercurrent/countercurrent.

Where

Off North Africa and Portugal assuming the North Atlantic subtropical gyre is chosen for WOCE CORE-III. (Ideally, similar studies would also be conducted off North America and South America).

How

- Subsurface drifters (RAFOS or Pop-Up).
- Moorings (conventional and ADCP).
- Altimeter and scatterometer (to measure surface pressure gradients and wind fields).
- Hydrography and tracers (dye?).
- Models.

II b. Role of Undercurrents in Global Chemical/Biological Budgets/ Transports

What To Do

Determine the transports and uptake/removal rates of nitrogen and carbon in the southeastern tropical Pacific.

Why

There is evidence that materials advected from the continental margin strongly affect global budgets of nitrogen, carbon, etc. (Chemical processes in the ocean interior may be influenced by what happens on continental margins.)

How

- Measure uptake/removal rates directly.
- Measure distributions and currents to estimate transports.
- Hydrographic and ADCP surveys.
- Surface and subsurface drifters.
- Conventional and ADCP moorings.
- Modeling.
- Take advantage of WOCE and GOFS plans.

III. Forcing and Dynamics of Poleward Undercurrents

What To Do

(a) Continue to develop the different theoretical and numerical models to understand undercurrent dynamics (e.g., McCreary, Brink, Haidvogel, Clarke, and Holloway).

(b) Continue to work on understanding the important physics contained in the results of numerical models.

(c) Develop diagnostics from theoretical and numerical models which can serve to focus observational efforts.

(d) Design laboratory experiments to test particular hypotheses.

(e) Develop paradigms for choosing among different models. (What feature(s) of each model can we expect to observe?)

(f) Design field experiments that can discriminate between different models. (Different models might give similar results for entirely different reasons.)

PART II:

GEOGRAPHICAL REPORTS:

THE ATLANTIC OCEAN
THE PACIFIC OCEAN
THE INDIAN OCEAN

THE ATLANTIC OCEAN

THE BAROCLINIC CIRCULATION OF THE WEST SPITSBERGEN CURRENT

Robert H. Bourke and Alan M. Weigel
Naval Postgraduate School
Monterey, CA 93943

ABSTRACT

The baroclinic circulation of the West Spitsbergen Current (WSC) is described based on a dense network of CTD stations acquired in September 1985. Near 78°N the baroclinic flow extends to 200 m with a transport of 1.3 Sv northward. Westward turning of the WSC to join the southward flowing East Greenland Current occurs from 76° to about 81° N, being interrupted between 79° and 80°N by the Molloy Fracture Zone. Near 80°N the WSC bifurcates to form an eastern branch which transports about 20% of its initial flow into the Arctic basin, observed as the warm, salty Atlantic Layer. No northward flow of WSC water is observed along the western flank of the Yermak Plateau. A cyclonic recirculation of the eastern branch of the WSC is noted east of the Yermak Plateau.

INTRODUCTION

During September 1985 the ice breaker NORTHWIND conducted an oceanographic measurement program in the northern Greenland Sea. A relatively dense network of CTD stations (Figure 1), approximately 10 to 20 km spacing, provided extensive information on the circulation and water mass structure of the polar waters of this region. It is the purpose of this paper to describe the characteristics of the relatively warm, saline West Spitsbergen Current, a poleward flowing eastern boundary current. An investigation of the baroclinic nature of the flow and a detailed assessment of the bifurcation of the current into a westward branch connecting with the East Greenland Current and a northeastward branch flowing into the Arctic basin were the two principal objectives of the cruise, and these features are described herein.

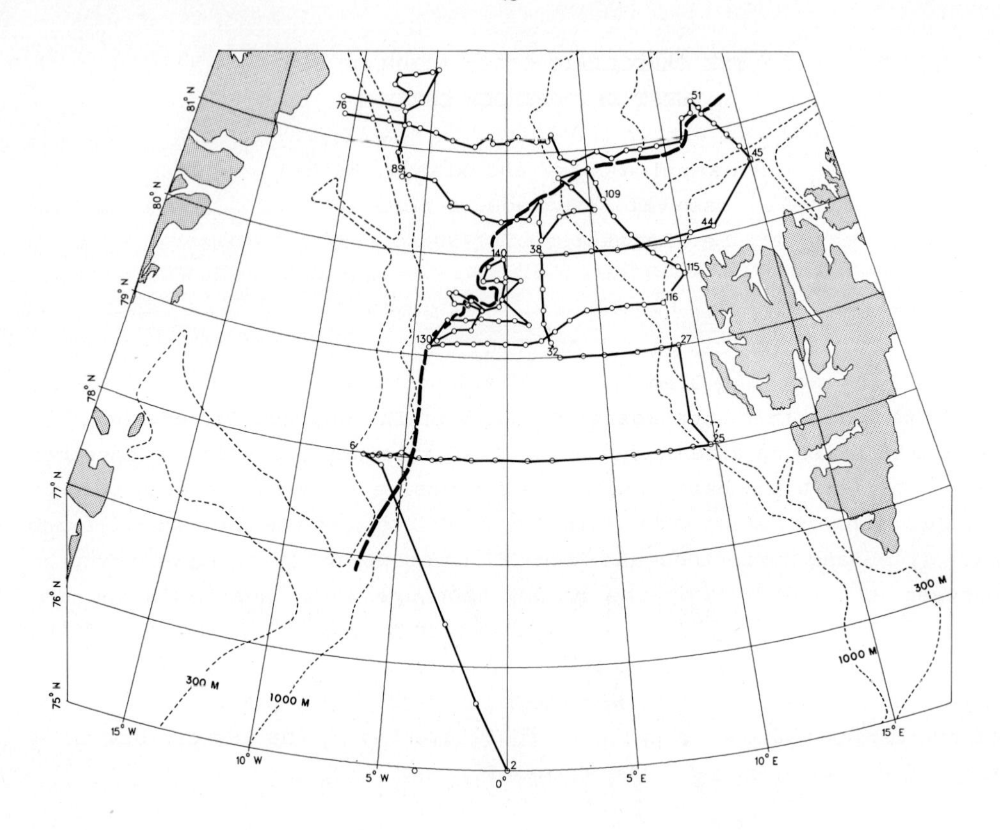

Figure 1: The NORTHWIND September 1985 cruise track and location of CTD stations. The heavy dashed line shows the position of the ice edge. The Yermak Plateau is the broad, shallow area to the northwest of the Svalbard archipelago near 80°N.

HISTORIC OVERVIEW

The circulation within the Greenland Sea is dominated by a cyclonic gyre centered on the prime meridian between 74°-76°N (Figure 2). The northward flowing West Spitsbergen Current (WSC) forms the eastern boundary of this gyre. Aagaard and Griesman (1975) have shown that the WSC provides the major inflow of mass, salt, and thermal energy into the Arctic basin. This flow enters the Arctic Ocean through Fram Strait, the narrow, moderately deep passage between Greenland and Svalbard Archipelago to the east. Along the Greenland coast the southward flowing East Greenland Current (EGC) forms the western boundary. Its characteristic features, e.g., temperature, salinity, speed, baroclinicity, depth distribution,

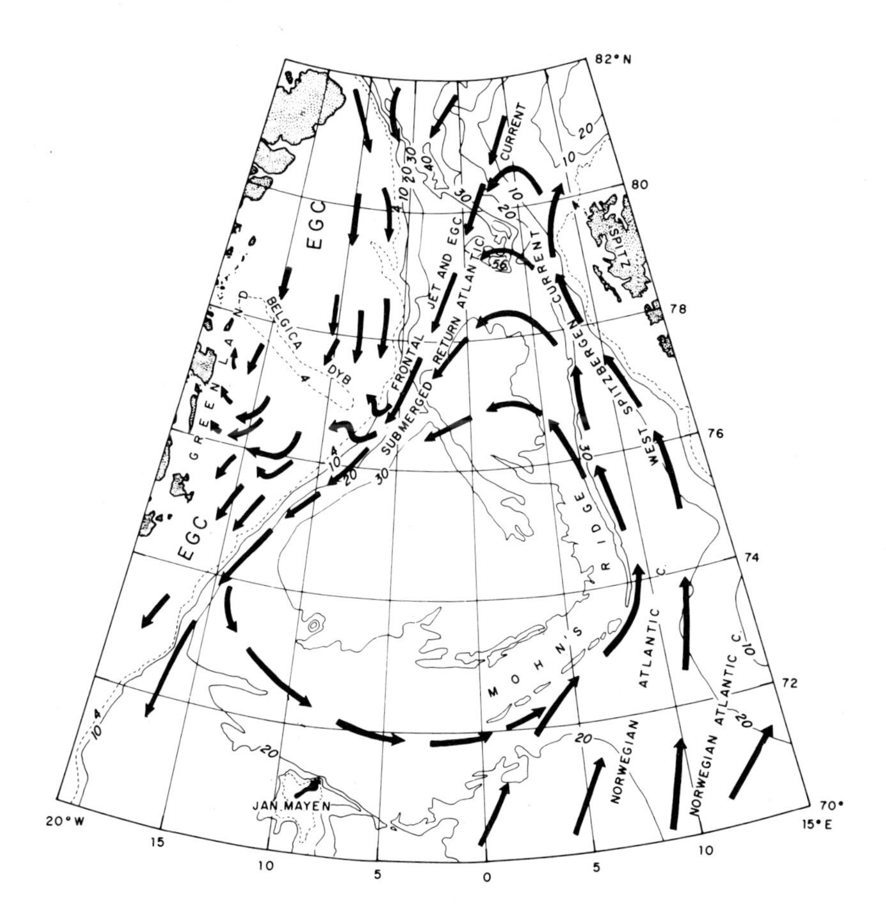

Figure 2: The general circulation and bathymetry of the Greenland Sea (from Paquette et al., 1985). The complex bathymetry of the Molloy Fracture Zone is seen near 79°N, 3°E.

etc., are almost completely in opposition to those of the WSC, a contrast which dramatically illustrates the differences between eastern and western boundary currents. The eastward zonal flow known as the Jan Mayen Polar Current closes the circulation to the south while the westward turning of the WSC into the EGC closes the circulation to the north in a flow known as the Return Atlantic Current (RAC) (Gladfelter, 1964; Paquette et al., 1985).

The WSC is formed when warm, saline water from the North Atlantic enters the Norwegian Sea and flows northward along the continental slope as the Norwegian Atlantic Current. Off the southern tip of Spitsbergen near 74°N it mixes with colder, fresher water from the Jan Mayen Polar Current and the East Spitsbergen Current and continues flowing north along the west coast of Spitsbergen. Over the continental slope the flow is relatively deep, extending to about 800 m (Hanzlick, 1983). North of about 75°N it loses heat and is diluted as it spreads out in a thin layer over the top of the Greenland Gyre. Between 79°N and 80°N the WSC bifurcates. An inshore branch turns northeastward crossing the Yermak Plateau where it eventually sinks to about 200 m depth as it passes under the ice edge and enters the Arctic basin (Perkin and Lewis, 1984; Johannessen et al., 1983). This branch of the current is the principal supply for the warm Atlantic layer observed throughout the Arctic Ocean (Aagaard et al., 1987). The second branch of the current turns westward initially following the western flank of the Yermak Plateau, probably as a result of conserving potential vorticity in this highly barotropic current (Quadfasel et al., 1987). This branch then mixes with the southward flowing EGC as the RAC. The RAC forms the eastern boundary of the East Greenland Polar Front (EGPF), the strong front separating cold, dilute Polar Water (PW) from warmer, saltier water of Atlantic origin (Bourke et al., 1987).

Hanzlick (1983) summarized the transport calculations of earlier investigators. Based on hydrographic data from 1933 to 1960, as gleaned from various sources (Hill and Lee, 1957; Kislyakov, 1960; and Timofeyev, 1962), the mean baroclinic transport across the WSC was estimated to be 3 to 4 Sv northward. Greisman (1976) estimated the mean total transport, using long-term direct current measurements obtained during 1971-72, as about 7 Sv northward, with a 10.7 Sv maximum in September and a 3 Sv minimum in February. Based upon hydrographic data obtained during this period, Greisman estimated the mean baroclinic transport to be 3.5 Sv, or about one-half of the total. During 1976-77 Hanzlick (1983) estimated a mean transport of 5.6 Sv from current meter data, with a maximum in the fall and a minimum in the spring. The mean baroclinic transport for the same period was only 1 Sv, significantly less than in 1971-72.

The WSC is rich in temporal and spatial variability, perhaps its principal characteristic feature. Greisman (1976) and Hanzlick (1983) have shown that both volume flow and core temperatures of the WSC are variable over time scales that range from days to years. For example, Hanzlick showed evidence, based largely on hydrographic data, that the interannual variability in the volume and heat transport was of the order of 50%. Shorter-term variability is noted in the direct current measurements made in Fram Strait during 1984-85 (Aagaard et al., 1985a) which showed a northward flow in the core of the WSC in excess of 0.4 m/s in the latter half of August 1984. By the first week in September 1984, the flow had reversed, with southward speeds of approximately 0.05 m/s being observed. An important conclusion which arises from these various data sets is that the changes between successive yearly surveys may in fact be due to short term variability and are not true interannual changes. Various investigators have also shown evidence of spatial variation in the WSC flow and temperature. Dickson (1972) and Hanzlick (1983) indicate this variation takes the form of banded structures or northward flowing filaments with the scale of the lateral structures being about 15 km.

The mean wind stress over the Greenland-Norwegian sea region is apparently the major factor affecting the interannual and shorter time scale variability of the WSC (Greisman, 1976; Hanzlick, 1983). Greisman and Aagaard (1979) suggested that the WSC is driven by large scale variations in the atmospheric forcing over the Greenland and Norwegian Seas on time scales of less than one month and that there was considerable month-to-month variation in the forcing. Hanzlick suggested that baroclinic instability and coastally trapped waves were also possible contributors to shorter time scale variability.

Water Mass Characteristics

Four water masses are found in the WSC. Their basic properties have been defined by Coachman and Aagaard (1974), but modifications to the temperature and salinity limits have frequently been imposed. Atlantic Water (AW) extends from the surface to about 800 m and is warmer than 0°C. Surface temperatures in summer may be higher than 6°C, while during the winter the surface temperatures are typically warmer than 2°C keeping the west coast of Spitsbergen ice free. The salinity of the AW increases in the upper 100 m so that in the core of the WSC the salinity may be greater than 35.2. Below 100 m the salinity decreases to about 34.95. Warm water, properly called AW, can also be found with salinities as low as 34.7. This low salinity fraction of AW, termed Atlantic Surface Water by Swift (1986), is primarily a surface phenomenon and is the result of dilution by ice melt. The area of low salinity is generally found near the ice edge or in more southerly areas where ice melt water has been advected.

Beneath the AW is found Greenland Sea Deep Water (GSDW). This water mass forms a large dome at the center of the Greenland Gyre with temperatures colder than 0°C and salinity between 34.87 and 34.95. Carmack (1972) proposed that GSDW was formed by convective subsurface modification of AW. Carmack and Aagaard (1973) and Aagaard et al. (1985b) suggest that chimney formation and double-diffusive mixing between PW and AW may also be involved with the production of GSDW.

From the CTD data acquired between 6 and 26 September 1985, a series of vertical cross sections, or transects, have been constructed to characterize the properties and flow of the WSC. The AW core of the WSC will be traced with the 3°C isotherm in order to distinguish AW from colder (< 3°C) Atlantic Intermediate Water (AIW), a cooled and slightly diluted variant of AW associated principally with the EGC (Paquette et al., 1985; Bourke et al., 1987). The lower salinity limit for AW was earlier defined as 34.9, which is also the upper limit for AIW. To further distinguish it from AIW in the following transects, AW will be traced using water more saline than 35.0.

From a transect along 78°N, the structure of the northward flow of the WSC along the west coast of Spitsbergen can be observed (Figure 3). In this figure AW can be seen as a wedge-shaped core along the coast, extending seaward to a distance of 60 km (near Station 22) and to a depth of almost 275 m near the coast. Associated with the warm wedge is a core of saline (> 35.1) water between 50 and 100 m. Seaward of the wedge the warm water spreads out, for almost 200 km, into a surface layer 50-75 m deep. The saline water defined by the 35 isohaline is also found as a tongue spreading out almost halfway across Fram Strait, but generally deeper than the 3°C isotherm, at 100 to 400 m. At the western end of the wedge, the surface salinity decreases to less than 34.9 as the AW mixes with PW of the EGC and becomes diluted. The temperature-salinity features of the WSC observed in Figure 3 are quite similar to those observed by Bourke et al. (1987) in August 1984, suggesting minimal interannual variability was experienced during these two summers. As the WSC flows farther northward,the warm wedge remains closer to the Spitsbergen coast becoming thicker, extending in depth to 300 m. The doming of the deep and intermediate waters in the center of the Greenland Gyre is seen in the downward sloping of the isotherms towards the coast. Figure 3 and other transects off the west Spitsbergen coast illustrate that in August the warm core of the WSC is at the surface. However, observations made later in the year show that surface cooling forces the warm core subsurface between a 100 and 400 m depth (Johannessen et al., 1983).

The distribution of AW in transects which cross the WSC between 78° and 80°N appears to be controlled, at least in part, by bathymetry. In Figure 3 the isotherms and isohalines, although slightly offset to the right, appear to be pushed up by the 2000 m isobath defining the Knipovich Ridge. The core of the WSC is located to the west of the shelf break, while the largest part of the 35.0 isohaline tongue is located over the western slope of the ridge. In this and other transects, the western slope of the ridge appears to control the western limit of AW becoming more predominant with increasing latitude.

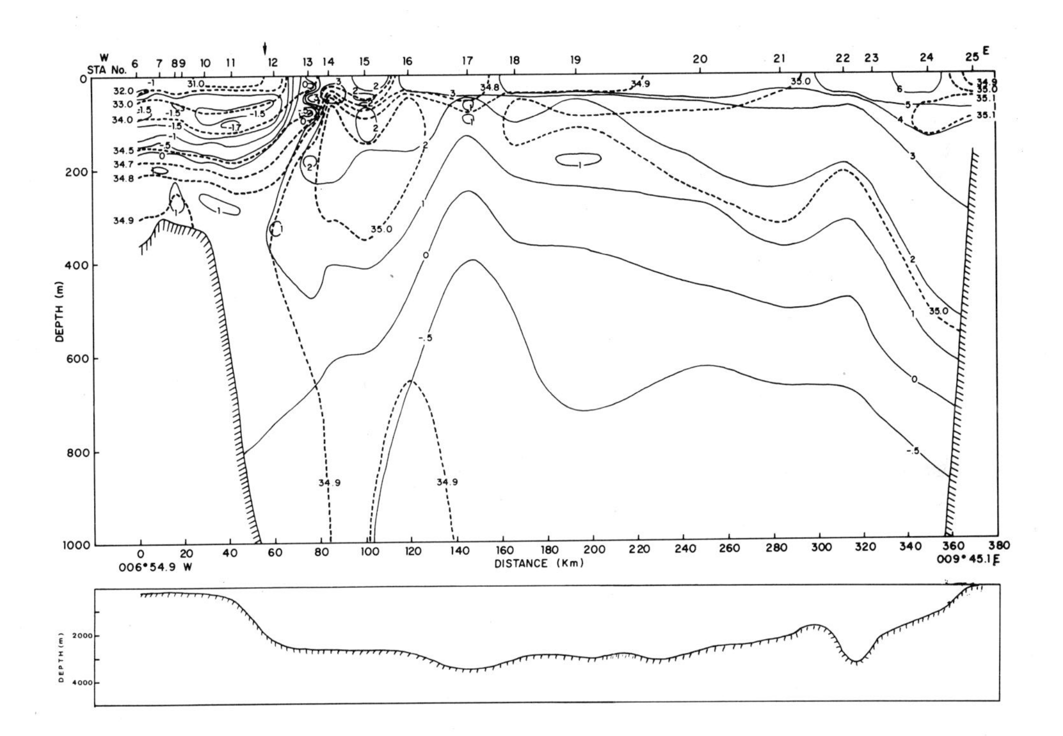

Figure 3: A temperature (solid line) and salinity (dashed line) transect along 78°N. The 2000 m feature between Stations 21 and 22 is the Knipovich Ridge. The arrow at the top indicates the location of the ice edge.

Branching of the WSC

Just south of 80°N, the WSC branches into two streams. The left branch turns westward and appears to sink below the cold surface water. The right branch turns eastward crossing the Yermak Plateau and flows along the north coast of Spitsbergen and into the Arctic basin. The branching appears to be controlled in part by bathymetry. A transect (Figure 4) which extends northwestward along the ridge of the Yermak Plateau shows that the warm saline core, defined by the 3°C isotherm and the 35.1 isohaline, is positioned along the northwest coast of Spitsbergen with the 35.0 isohaline extending beyond the 1000 m isobath into deep water. The properties of the eastern branch of the WSC can be seen in a transect which extends northwestward from the continental shelf north of Spitsbergen into the Arctic basin (Figure 5). AW is found here as a narrow stream, within 40 km of the coast and in shallow water. The 35.0 isohaline, however, is found to extend well away from the coast and is evidence of the broad thrust of AW, now cooled to form AIW, propagating north and eastward into the Arctic basin to form the Atlantic layer.

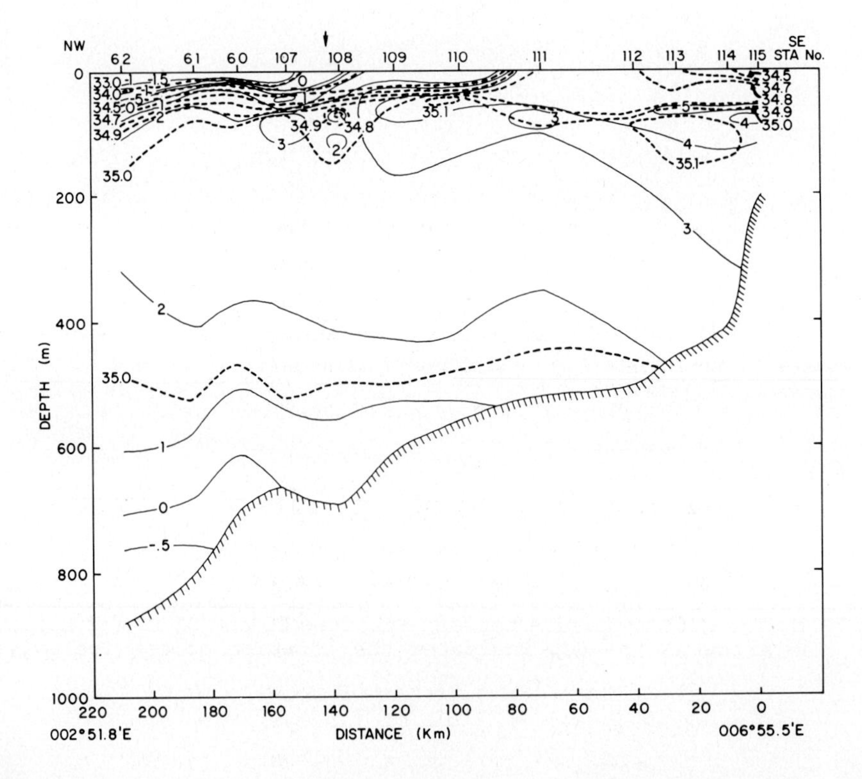

Figure 4: A temperature-salinity transect along the top of the Yermak Plateau.

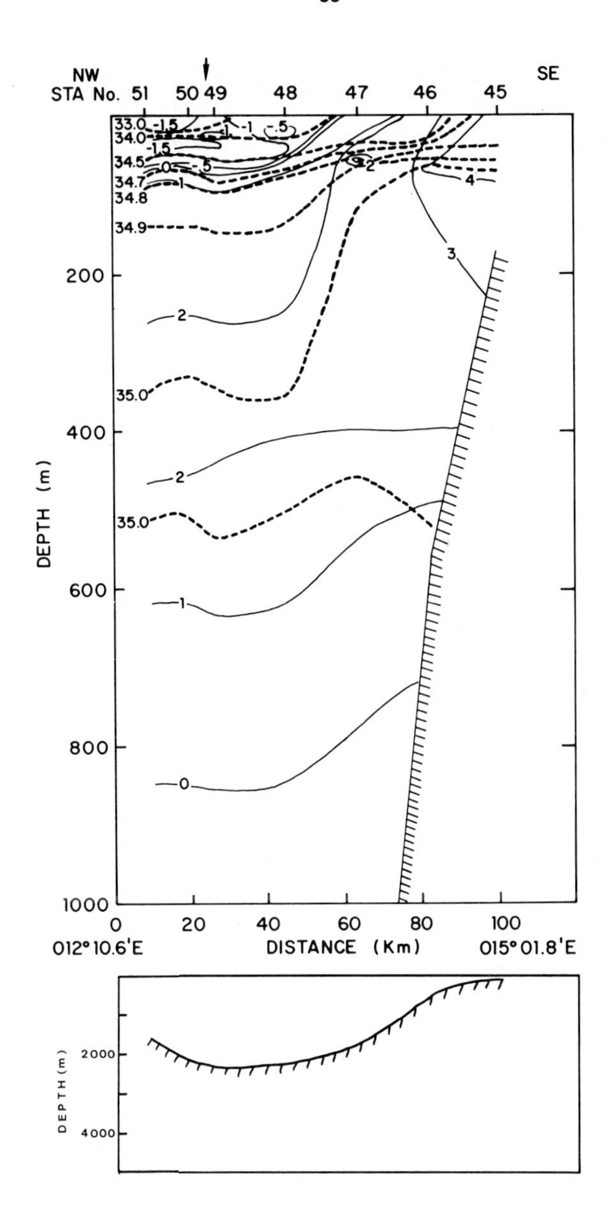

Figure 5: A temperature-salinity transect north of Spitsbergen. The WSC is seen to hug the north coast and shelf of Svalbard.

As previously discussed, south of 80°N, AW is found along the west Spitsbergen coast close to the 1000 m isobath. North of this latitude the left branch of the WSC leaves the shallow water along the west slope of the Yermak Plateau, turns westward into deep water, and joins the southward flowing EGC. The westward turning of the WSC can be traced by following the AW where most of the baroclinicity occurs. The northward extent of the westward turning is found just north of 81°N where only a small trace of AW at 100 m near the prime meridian is seen. This is illustrated in a transect which crosses through the center of the westward turning along the 2°E meridian (Figure 6). In this transect the core of the warm westward-turning WSC is seen centered at 100 m between Stations 101 and 99, just at the ice edge, and well to the north of the seamounts between Stations 34 and 36. This is the farthest north that warm AW is found. Westward turning is also indicated south of the complex bathymetry of the Molloy Fracture Zone south of Station 33.

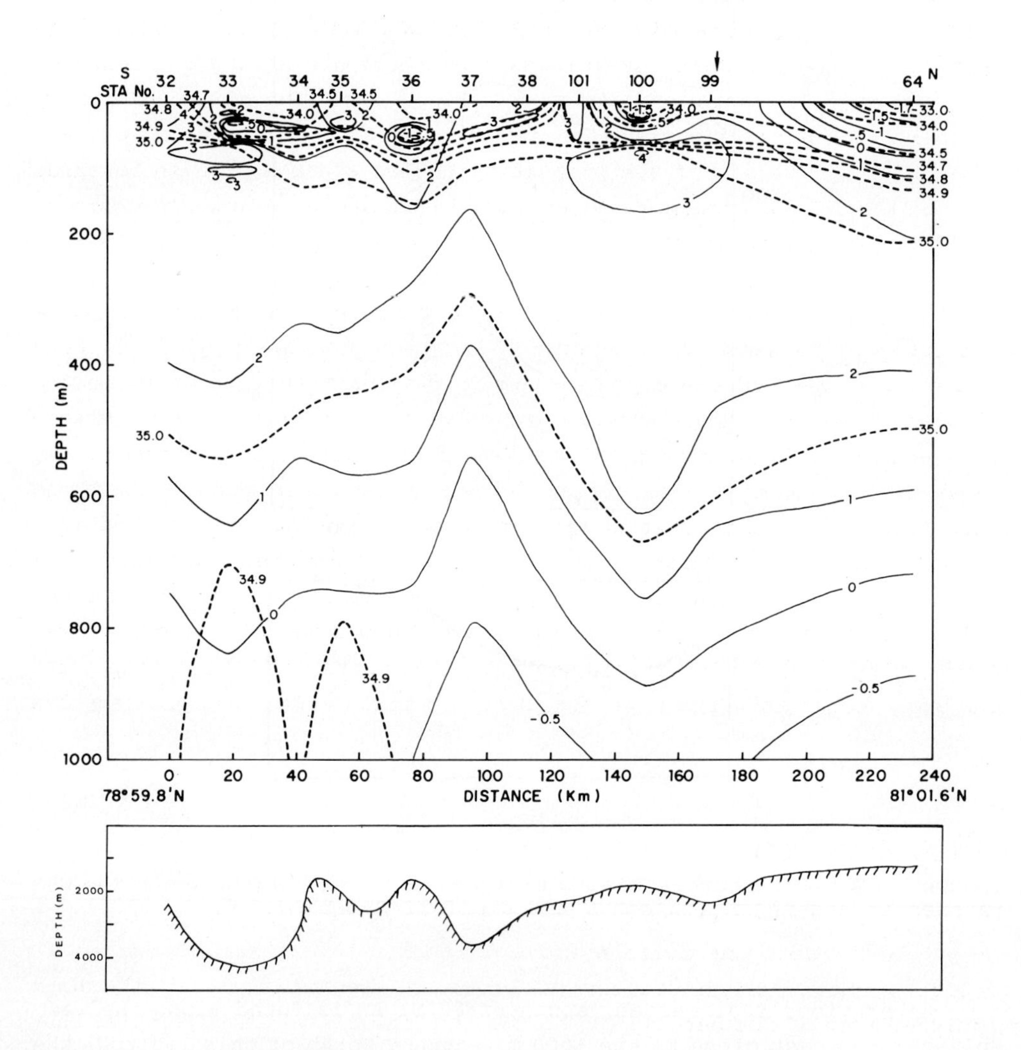

Figure 6: A temperature-salinity transect along 2°E. Westward flow of the WSC is indicated by the lens of 3°C water between Stations 37 and 99 and also south of Station 33.

The core of the warm westward turning is found below the surface and can be accounted for in two possible ways. The warm saline AW may have submerged under the cold, fresh PW. This could occur if the salinity of the AW were high enough to offset the density of the cold PW. The AW would then sink as it turned westward into the EGC. Another possibility is that the AW is cooled and freshened by the presence of the ice edge as it turns westward. This would then result in finding a temperature maximum below the cold surface layer.

Horizontal plots of water mass properties may also be used to describe the branching and westward turning of warm AW. Although plots of surface properties are of limited usefulness because of the strong influence of the ice edge, Aagaard and Coachman (1968) have shown that plotting isotherms on a sigma-t surface can be effective to illustrate the westward turning. An analysis of the property profiles at each station revealed that in the WSC the maximum temperature in the water column (the core of AW) was nearly coincident with the maximum salinity. The density at the depth of maximum temperature, expressed as sigma-t, has a mean between 27.9 and 28.0. For this reason, and to compare with historical data presented by Aagaard and Coachman (1968), the temperature of the 28.0 sigma-t surface and the depth of the 28.0 sigma-t surface were plotted (Figures 7 and 8). The westward spread of the 2°C isotherm can be seen between 80° and 81°N, with 81°N as the maximum extent of the westward flow. Aagaard and Coachman demonstrated virtually identical features based on data from 1965, although the maximum temperatures were warmer in 1965. The depth of the 28.0 sigma-t surface perhaps shows the best evidence of westward turning based on horizontal distributions of temperature and salinity. This figure indicates that a zone of westward turning has taken place north of 78°N with little westward flow north of 81°N. To the north and west, behind the ice edge, the depth of this surface was greater than 500 m.

Dynamic Topography

Dynamic topographies derived from the September 1985 cruise were computed for both the surface and 150 decibar levels referenced to 500 decibars, the average deepest depth of the sensor and a depth which included most of the baroclinicity. The surface dynamic topographies show the weak baroclinicity of the WSC in comparison to the EGC (Figure 9). The northern limb of the cyclonic Greenland Sea Gyre can be seen in the southern part of Fram Strait near 79°N. The dynamic height gradient across the EGPF would suggest a baroclinic flow of between 0.20 and 0.30 m/s, while the sluggish WSC baroclinic flow is 0.06 to 0.08 m/s. This figure

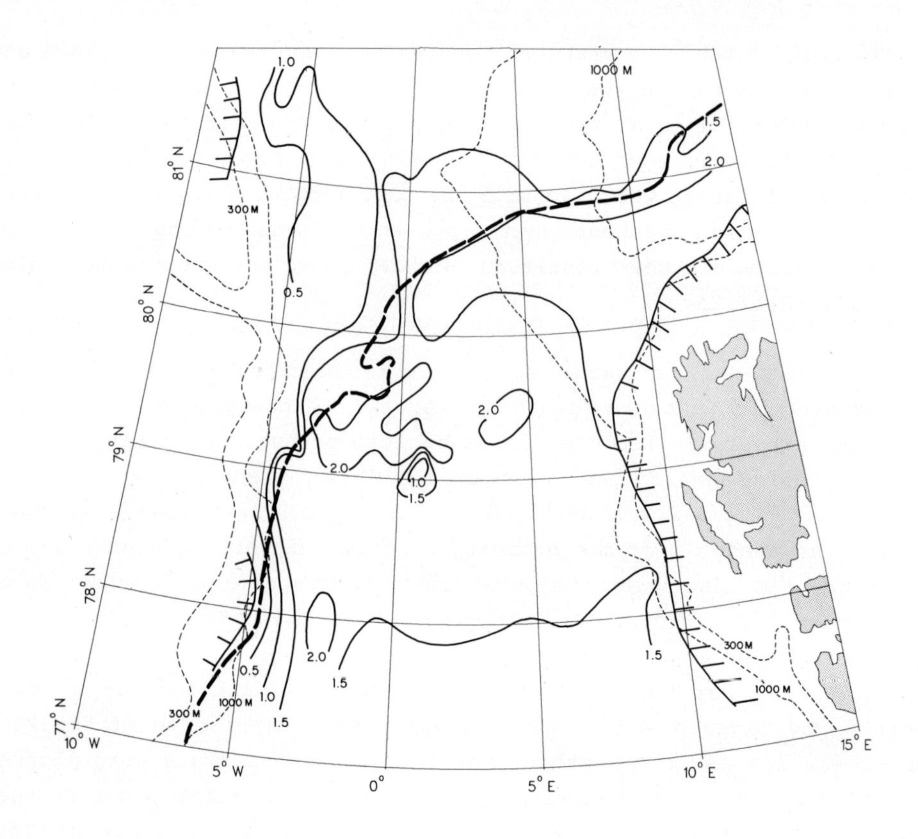

Figure 7: The temperature of the 28.0 sigma-t surface.

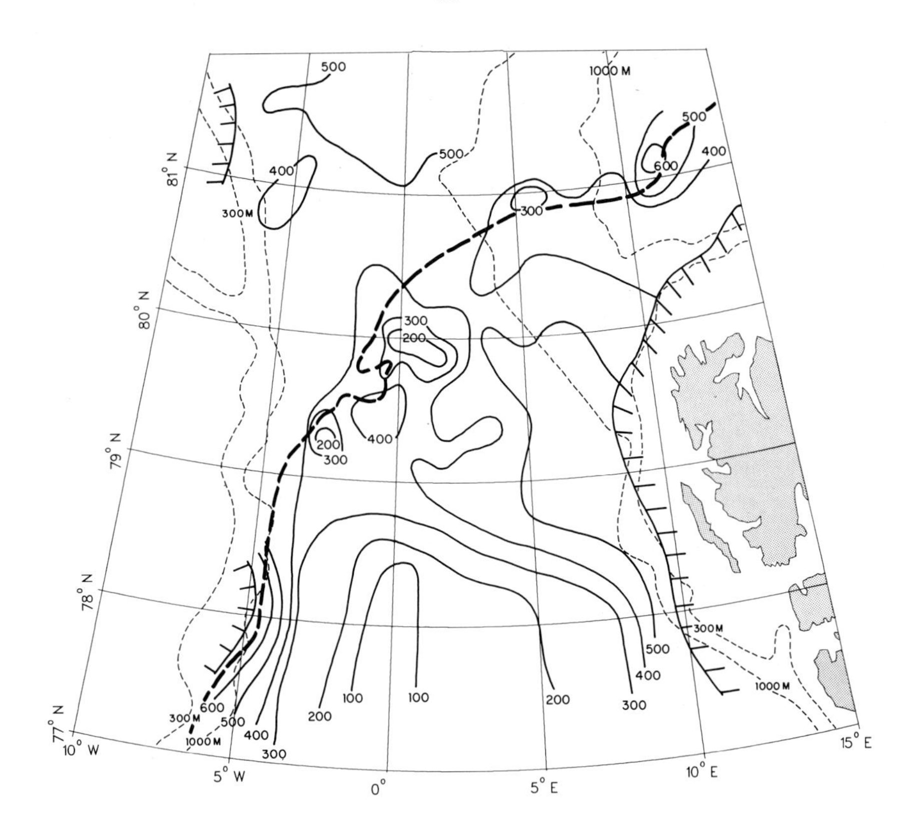

Figure 8: The depth of the 28.0 sigma-t surface illustrating the westward turning of the WSC.

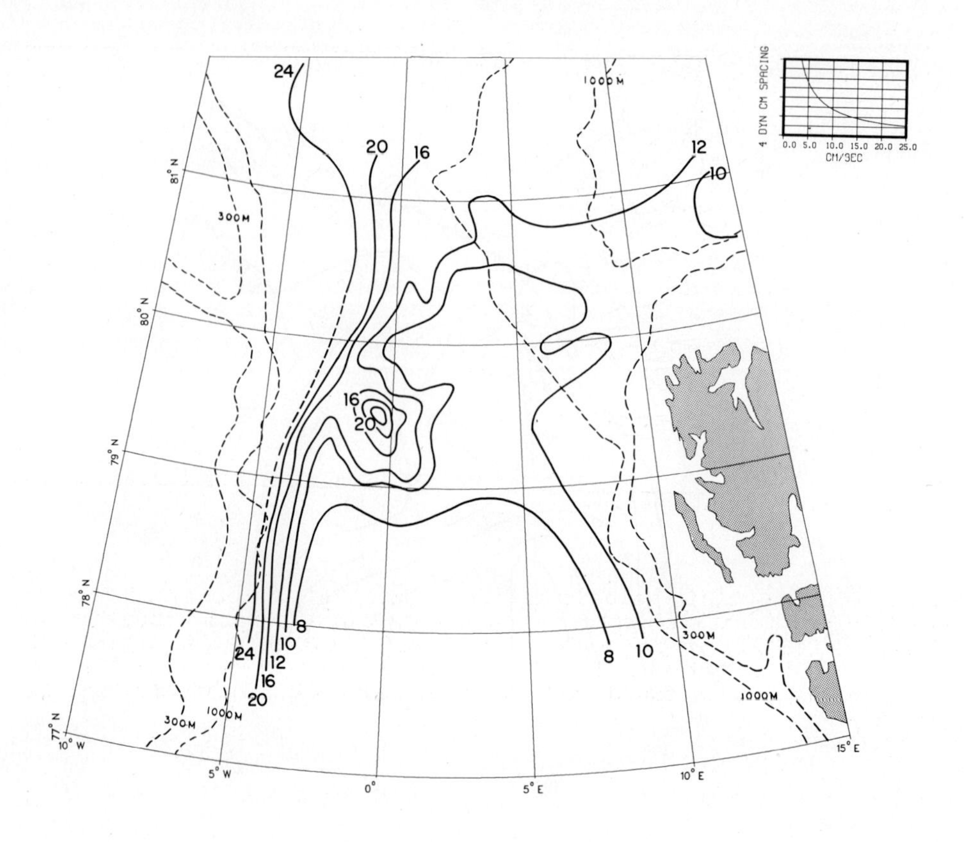

Figure 9: The surface dynamic topography referenced to 500 decibars (in dynamic centimeters).

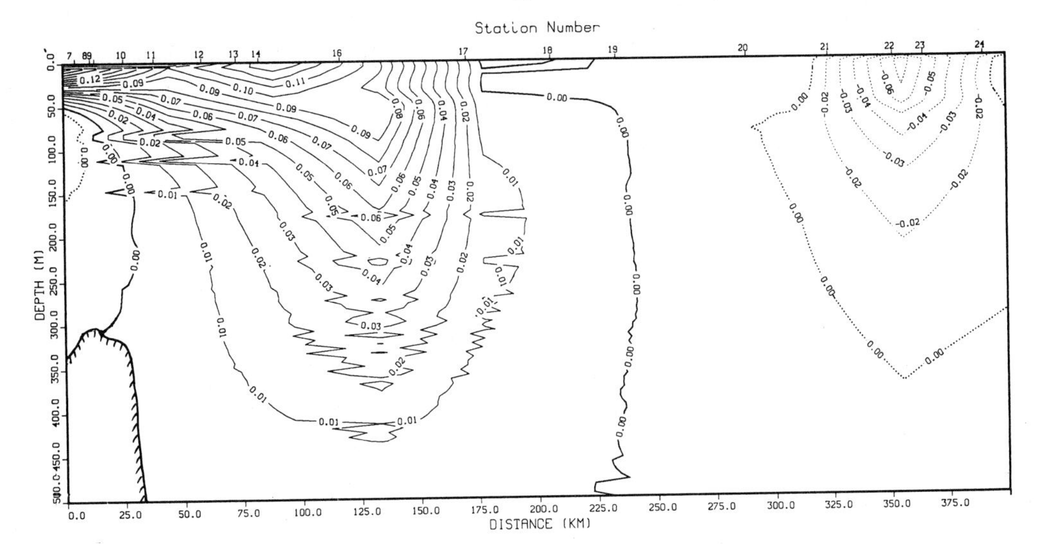

Figure 10: Baroclinic velocity section along 78°N. Southward flow is indicated by positive isotachs spaced at 0.01 m/s intervals, but at 0.02 m/s intervals in the high speed jets. The northward flowing WSC is observed near Stations 21 and 24. The RAC is centered near Station 14, while the EGC is found over the Greenland Shelf between Stations 7 and 10.

illustrates the branching of the WSC in the vicinity of the Yermak Plateau just north of 80°N. Westward flow is indicated between 79° and 81°N with the core of the westward turning crossing 5°E just north of 80°30′N and ultimately merging with the southward flowing EGC. This figure and dynamic topographies at 150 decibars show that the surface and subsurface components of the WSC turn westward at or south of 81°N.

To further investigate the geostrophic velocity and volume transport in Fram Strait, several vertical baroclinic velocity sections were constructed. A section along 78°N, based on the properties of Figure 3, shows the weakly baroclinic WSC with a core speed of 0.07 m/s located over the center of the Knipovich Ridge (Figure 10). Little baroclinic flow is noted at depths below 200 m. The northward transport resulting from this flow is 1.3 Sv. In contrast, the EGC in the extreme left part of the figure shows strong baroclinic flow over the Greenland continental shelf with a speed of almost 0.2 m/s. The core of the RAC is located at a depth of 50 m at Station 15 with a maximum speed of 0.13 m/s. The combined transport for this section for the EGC and RAC is 3.1 Sv southward. Figure 10 also shows that the circulation about the center of the Greenland Gyre, as observed between Stations 18 and 21, has virtually no north-south component of flow at this latitude. This lack of flow is supported by year-long current meter measurements, made across Fram Strait near this latitude (Aagaard et al., 1985a), which indicate that the flow is weak but predominantly westward.

Farther northward, between 79° and 80°N, the baroclinic flow of the WSC becomes even weaker exhibiting northward speeds of 0.02 to 0.03 m/s, based on geostrophic calculations. The northward transport is 0.5 Sv, indicating that some of the flow has turned westward between 78° and 79°N. A section based on the temperature-salinity distribution of Figure 4, which extends along the ridge of the Yermak Plateau, illustrates the branching of the WSC into two streams. One branch flows north and east along the Spitsbergen continental shelf, and the other flows northwest following the western slope of the Yermak Plateau. The northward speed in both branches is 0.02 to 0.03 m/s, while the net transport for this section is 0.5 Sv northward.

Currents computed from the section in Figure 5, which extends northwestward from the north coast of Spitsbergen, show the WSC positioned over the continental slope with a speed of 0.02 m/s. The transport through this section is 0.1 Sv toward the northeast, indicating that approximately 20% of the baroclinic flow of the WSC has turned into the Arctic around the northern coast of Spitsbergen. Southerly flow is observed farther offshore between Stations 46 and 49 suggesting the existence of a cyclonic filament of this branch. Such a recirculation is plausible due to vortex stretching as the current moves off the shelf into deeper water.

Current speeds calculated from a section along 81°N (Figure 11), all of which is behind the ice edge, indicate the maximum speed of the WSC over the Yermak Plateau is less than 0.01 m/s, i.e., the northward baroclinic component of the WSC has virtually disappeared. The existence of a northwestward component of flow over the plateau, as inferred to maintain continuity with the flow crossing 80°N, cannot be demonstrated due to the orientation of the hydrographic stations. However, Aagaard et al. (1987) report a similar absence of baroclinic flow over the plateau based on late November 1977 data.

The above section also illustrates the rapid (0.16 m/s) southward flow of the EGC, centered between Stations 67 and 69 just to the west of the prime meridian. The southward flow observed just to the east of the EGPF, between Stations 60 and 63, is of some interest as this flow incorporates a small amount of AW and AIW and may be considered the first trace of the RAC. This section and other nearby stations indicate that there is no substantial baroclinic component of the western branch of the

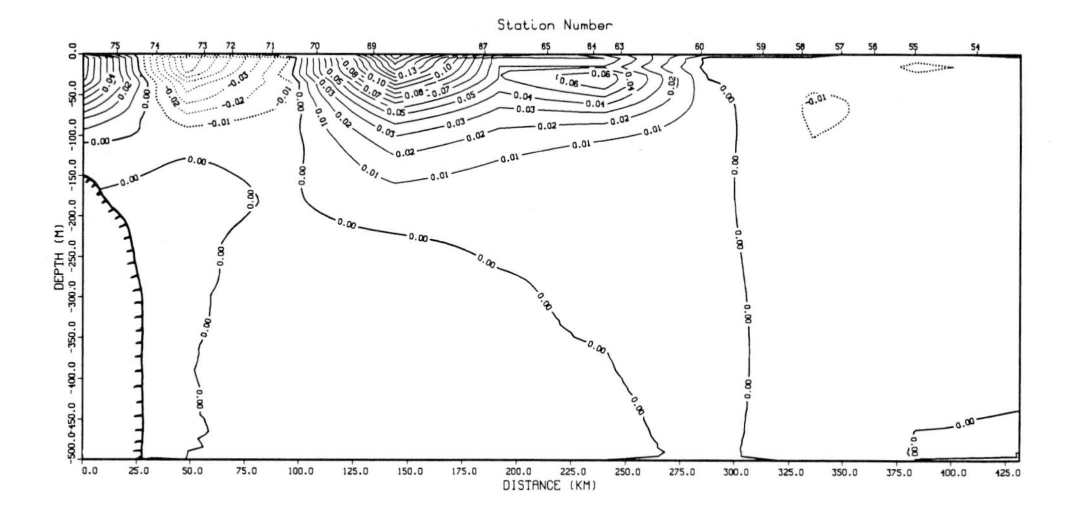

Figure 11: Baroclinic velocity section along 81°N. The southward flowing EGC is centered near Station 69. The southward flow near Station 64 is the first trace of the RAC.

WSC which continues northward past 81°N following the contours of the western flank of the Yermak Plateau to continue into the Arctic basin as suggested by Perkin and Lewis (1984) and Johannessen et al. (1983). The existence of such a northward component is questionable in light of reinterpretation of the Perkin and Lewis data which indicate a northward component cannot be substantiated.

A meridional section along 2°E between 79° and 81°N (Figure 12) shows that the core of the westward turning is predominantly north of 80°N, between Stations 38 and 99, with a speed of 0.03 m/s. Westward turning is also observed to take place south of 79°N. The transport for the northern core of westward turning is 0.4 Sv, while that south of Station 34 is about 0.1 Sv westward. The separation of the westward turning into two filaments may also be seen in the dynamic topography of Figure 9 and in the depth of the 28.0 sigma-t surface (Figure 8). The separation is attributed to the presence of the Molloy Fracture Zone (Perry, 1986), a region of complex bathymetry which includes the Molloy Deep, with depths exceeding 5000 m, and several seamounts rising to within 1500 m of the sea surface. The waters in between the two westward excursions of the WSC are cooler and less saline and may be associated with a cyclonic vortex rotating in the vicinity of the Molloy topographic structure (Quadfasel et al., 1987; Johannessen et al., 1987).

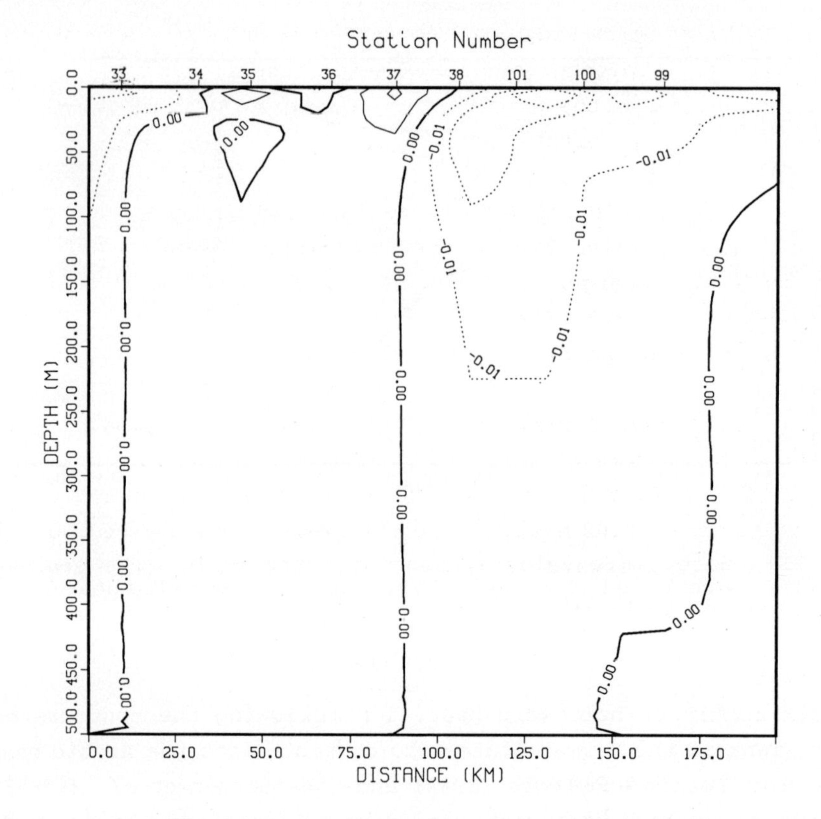

Figure 12: Velocity cross-section along 2°E showing westward turning of the WSC north of 80°N (Station 38) and south of 79°N (Station 33). The flow is weak and variable in between possibly due to the presence of the Molloy Fracture Zone.

CONCLUSIONS

The water properties and circulation of the West Spitsbergen Current (WSC) have been examined based on a dense network of CTD stations occupied by NORTHWIND in September 1985. Our principal findings are summarized below.

1. The WSC is found as a weakly baroclinic flow in several streams along the Spitsbergen continental slope. Northward geostrophic velocity at the surface along 78°N is found to be 0.07 m/s. Little baroclinic flow is observed at depths below 200 m.

2. Bathymetric steering appears to play an important role in the location of the separate streams of the WSC.

The Knipovich Ridge, a 2000 m deep feature to the west of the Spitsbergen continental shelf, exerts an influence on the distribution of water mass properties, especially north of 78°N.

3. North of 80°N the WSC splits into two branches, with approximately 20% of the baroclinic flow following the Spitsbergen continental shelf into the Arctic Basin and 80% of the flow turning westward over the western slope of the Yermak Plateau.

4. Some westward turning of the WSC takes place south of 79°N. A core of westward turning was located between 80° and 81°N along the prime meridian with baroclinic speeds of 0.02 m/s. Several filaments of AW were found turning westward between 79° and 80°N, although these contributed little to the net westward transport.

ACKNOWLEDGEMENTS

The authors are pleased to acknowledge the support of the Arctic Submarine Laboratory, Naval Ocean Systems Center, San Diego, CA, for their continued support of our marginal sea-ice zone studies, the latest under contract number N66001-86-WR-00131. We are grateful for the help provided by the officers and men, especially the marine science technicians, of the USCGC NORTHWIND.

From: Mary L. Batteen, Naval Postgraduate School, Monterey, CA

On: Review and Commentary to paper **THE BAROCLINIC CIRCULATION OF THE WEST SPITSBERGEN CURRENT**, by R.H. Bourke and A.M. Weigel

This is an interesting study and review of observations of the West Spitsbergen Current (WSC), which forms the eastern boundary of the cyclonic (subpolar) gyre circulation of the Greenland Sea. Unlike typical eastern boundary currents, e.g., the California or Peru Currents, which flow along the eastern boundaries of anticyclonic (subtropical) gyres, this surface current is relatively warm and saline and flows poleward (rather than being cold and fresh and flowing equatorward). Because of these characteristics, it would be of scientific interest to compare and contrast the characteristics of the WSC with those of typical eastern boundary coastal undercurrents, since these flows are also typically warm and saline and flow poleward. One could also make comparisons with other poleward surface currents, such as the Alaska, Davidson, or Leeuwin Currents. Insofar as the currents are wind-driven, as, for example, the Alaska Current, one would expect that the WSC and these currents would be dynamically similar. In the case of the Leeuwin Current, however, since recent results (McCreary et al., 1986) indicate that the Leeuwin Current is driven by thermohaline rather than by wind forcing, the WSC and the Leeuwin Current would not be expected to be dynamically similar.

Bourke and Weigel describe the water properties and baroclinic circulation of the WSC, based on a "relatively dense" network of CTD stations. Even though the station spacing is about 20 km, it is not dense enough to resolve eddies, which have horizontal space scales as small as 5 to 10 km in the Arctic region. However, the station spacing is dense enough to resolve the larger scale features such as the WSC.

In Bourke and Weigel's comprehensive historical overview of the WSC, they mention that the mean baroclinic transport can be about one-half of the total, but about 20% of the total is more typical. This suggests that the barotropic transport is quite significant. Bourke and Weigel also call the WSC a "highly barotropic" current, when reviewing the results of Quadfasel et al. (1987) which pertain to a branch of the WSC that follows the western flank of the Yermak Plateau. These results indicate that the barotropic component of the WSC can be quite substantial. Since Bourke and Weigel's findings also show that bathymetric steering and topographic effects can influence the WSC, and it is well known that barotropic effects are well felt by bottom topography, it is essential that future observations of the WSC include both the baroclinic and barotropic components of the velocity field.

Overall, as summarized in the conclusions, this paper brought out some important points, which may also be applicable to other poleward flows on eastern boundaries. I thought the point that changes in the WSC flow field can be construed as interannual rather than as short-term variability was especially worthy of note.

A BRIEF SKETCH OF POLEWARD FLOWS AT THE EASTERN BOUNDARY OF THE NORTH ATLANTIC

T.A. McClimans
Norwegian Hydrotechnical Laboratory
and
Division of Port and Ocean Engineering
Norwegian Institute of Technology
Trondheim, Norway

The circulation of the surface waters of the North Atlantic is shown in Figure 1 (Sverdrup, et al., 1942). The eastward drift current at mid-latitudes shows a secondary flow as shown in Figure 2, with a poleward component at depth (Dietrich, et al., 1980). Impingement of the drift current upon the European Continental Shelf results in both an equatorward flow toward the Mediterranean and a poleward flow past the British Isles. Three branches of this current system join to form the Norwegian Current (Figure 3). The early work of Helland-Hansen and Nansen (1909) implied a transport of 3 Sv, but newer results (Gould et al., 1985) imply a flow of ~7-1/2 Sv with maxima in winter. The large poleward flow of warm Atlantic Water provides a favorable climate for Norwegians. A branch of the Norwegian Current flows to the east in the Barents Sea while the major part drives north through the Fram Strait past Svalbard, diving under the brackish, ice-covered surface layer in the Arctic.

Most of the fresh water runoff from northern Europe flows to the Baltic Sea. This is the major source of the poleward flow of the Norwegian Coastal Current of Figure 4 (Braathen and Saetre, 1973). This current is colder in the spring and warmer in the fall. Satellite images reveal how large mesoscale eddies cause an interaction with the Norwegian Current, especially north of Stad (Figure 5) (NOAA-6). Some of this flow is engulfed in the Norwegian Current while much of it follows the coast eastward to the Siberian Shelf.

The third poleward flow of interest and perhaps a major driving force for the poleward circulation is the dense bottom flow generated by salt rejection during ice formation in the shelf seas of the Arctic Basin (Figure 6) (Midtun, 1985). This process extracts fresh water (in the form of ice flows) which joins the equatorward flow mainly within the East Greenland Current. All of these flows follow the continental slopes to the right due to the earth's rotation.

The hydrographic sections in Figure 7 (Dietrich, et al., 1980) show the currents described in this brief sketch.

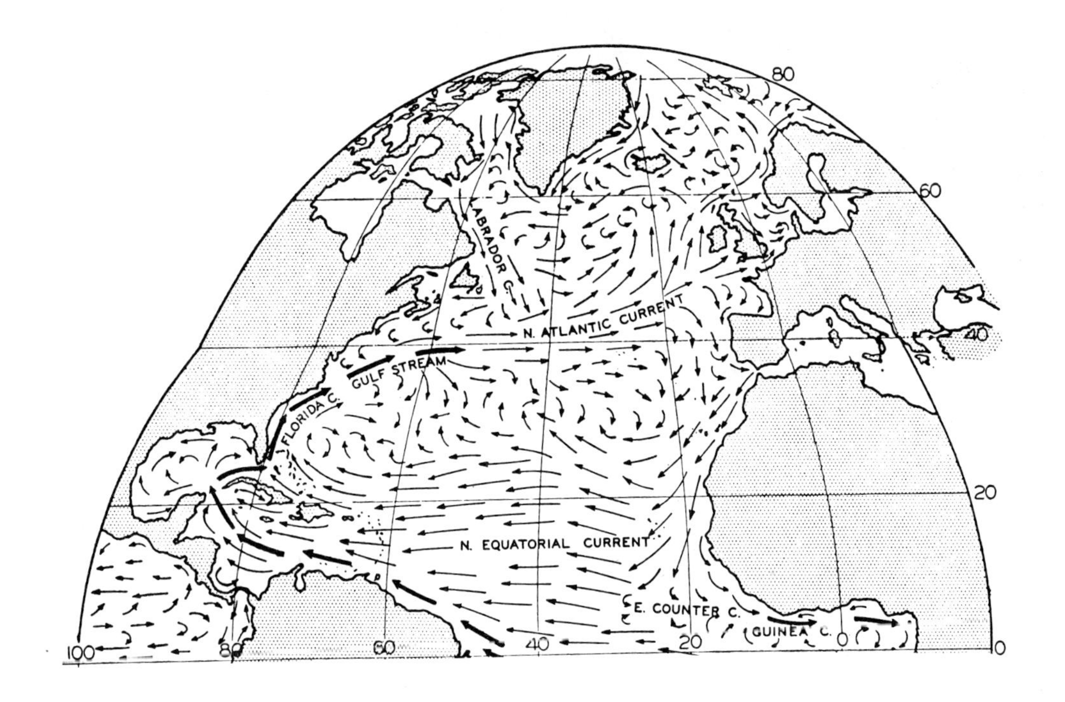

Figure 1: Surface currents of the North Atlantic in February-March. (Sverdrup, et al., 1942)

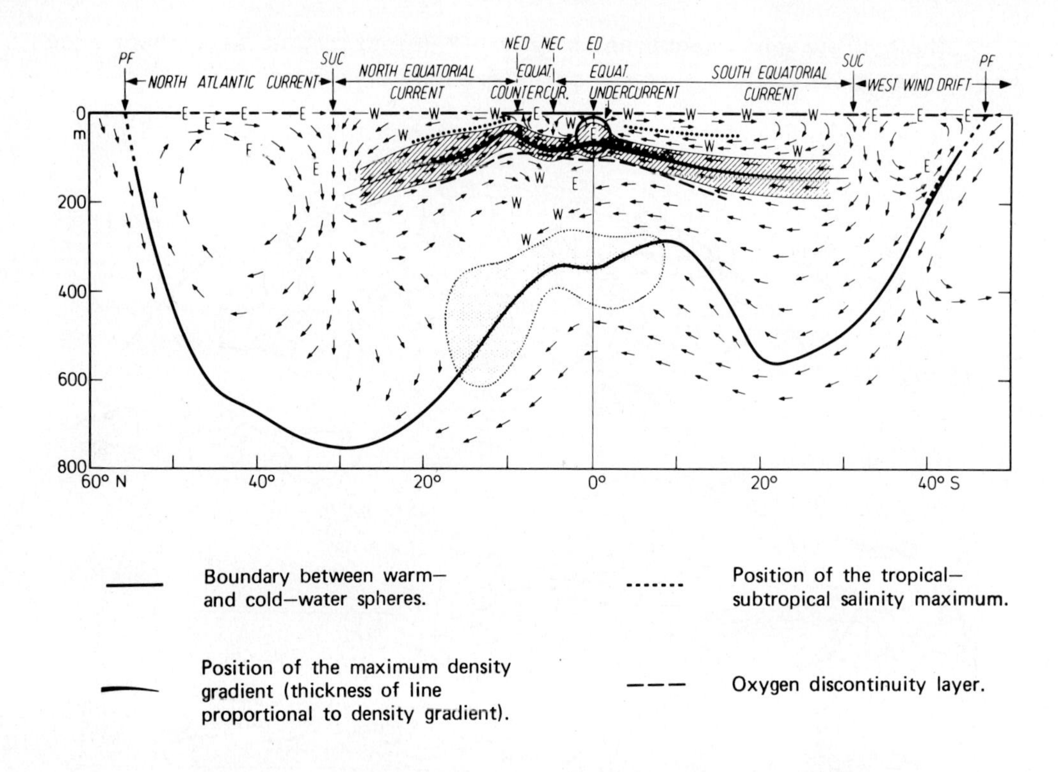

Figure 2: Longitudinal section of the poleward circulation of the Atlantic. (Dietrich, et al., 1980)

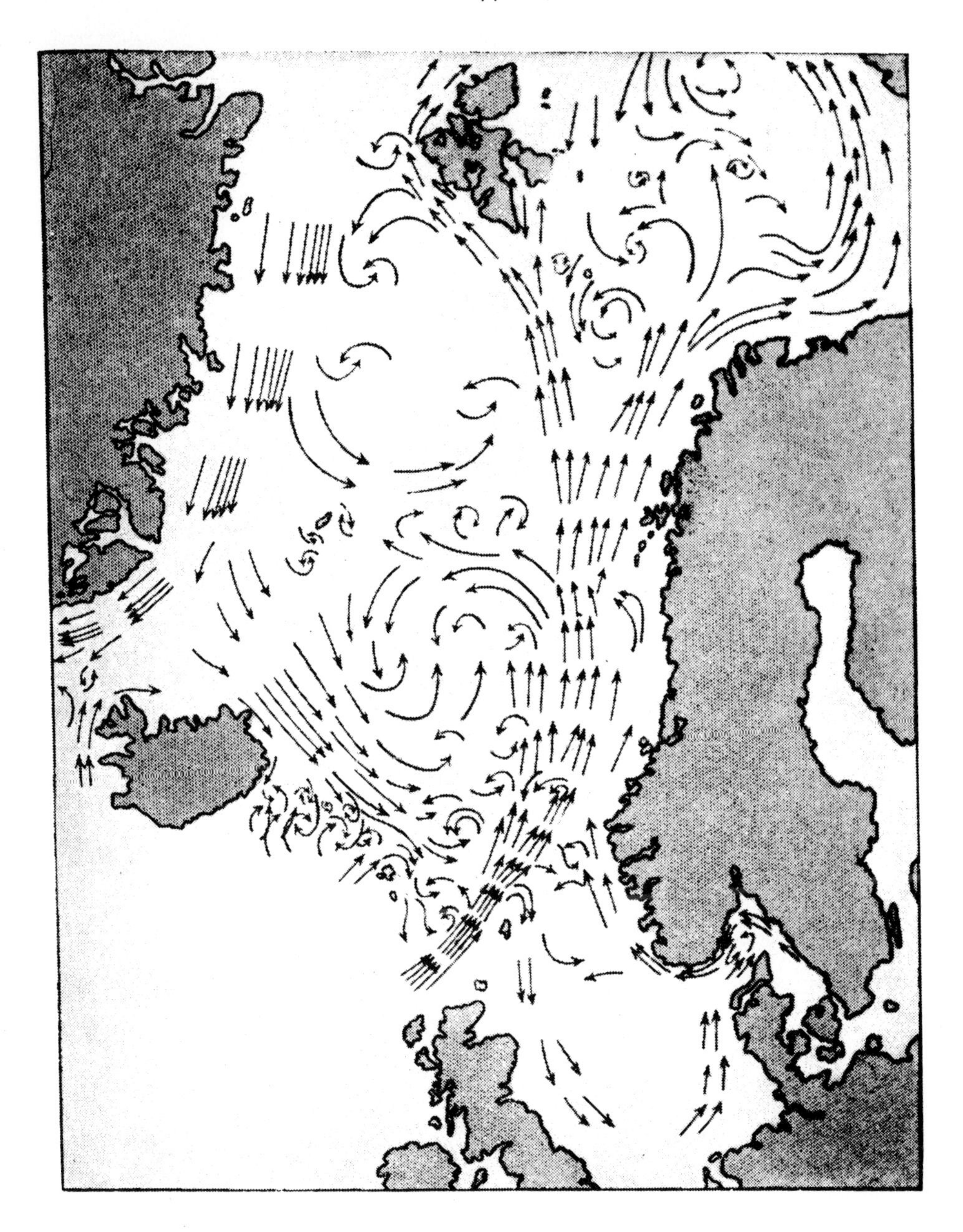

Figure 3: Surface currents of the Norwegian Sea. (Helland-Hansen & Nansen, 1909)

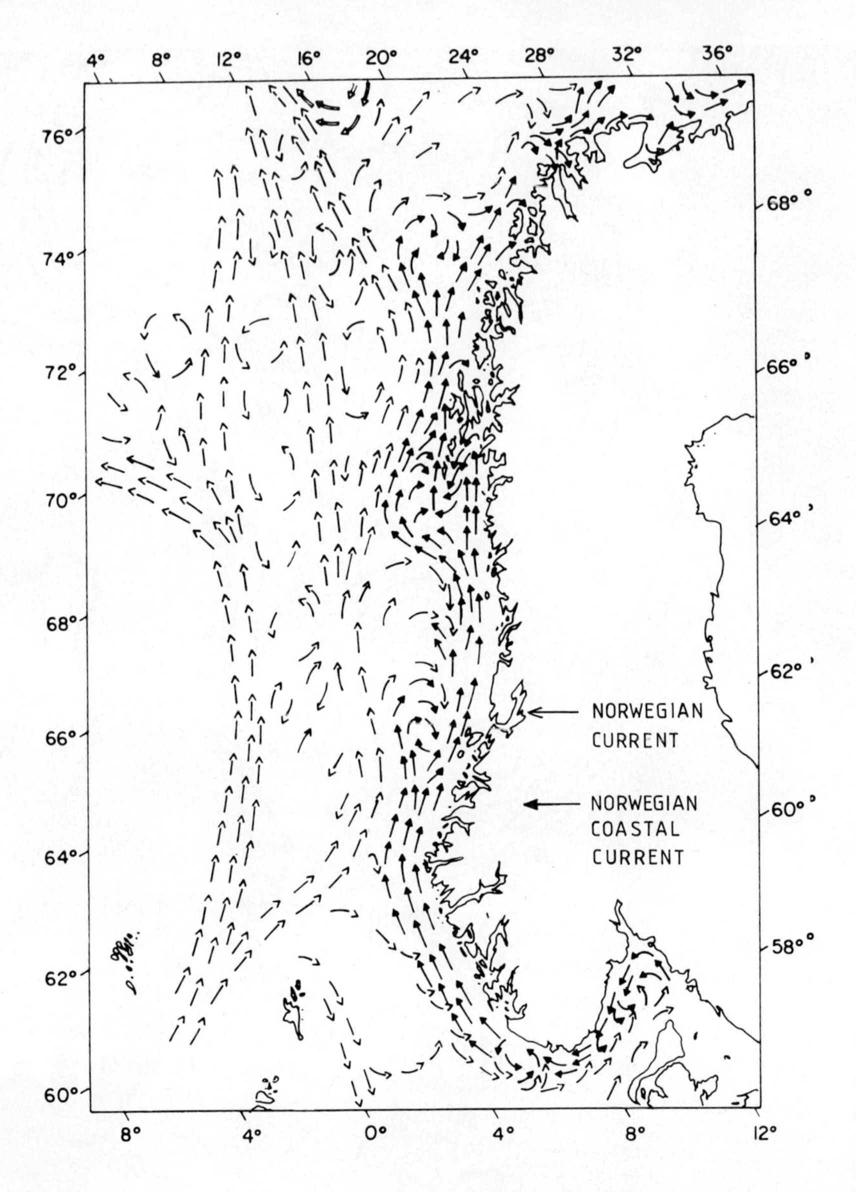

Figure 4: Surface currents west of Norway. (Braathen & Saetre, 1973)

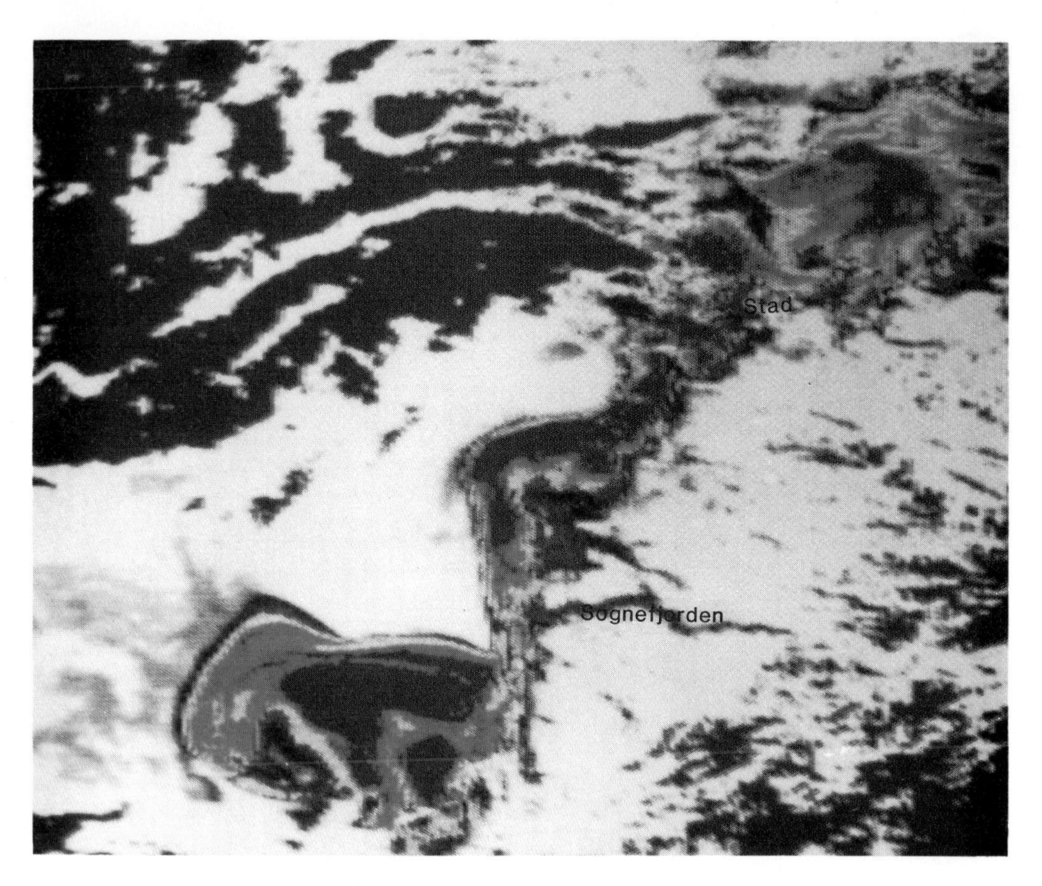

Figure 5: NOAA thermal image of the Norwegian Coastal Current off the west coast of Norway. (Data from Tromsø Telemetry Station)

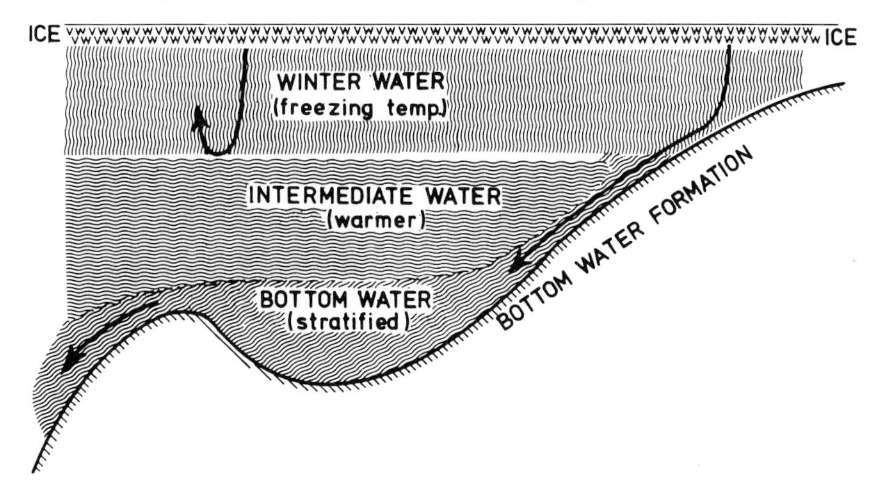

Figure 6: Illustration of dense water formation. (Midtun, 1985)

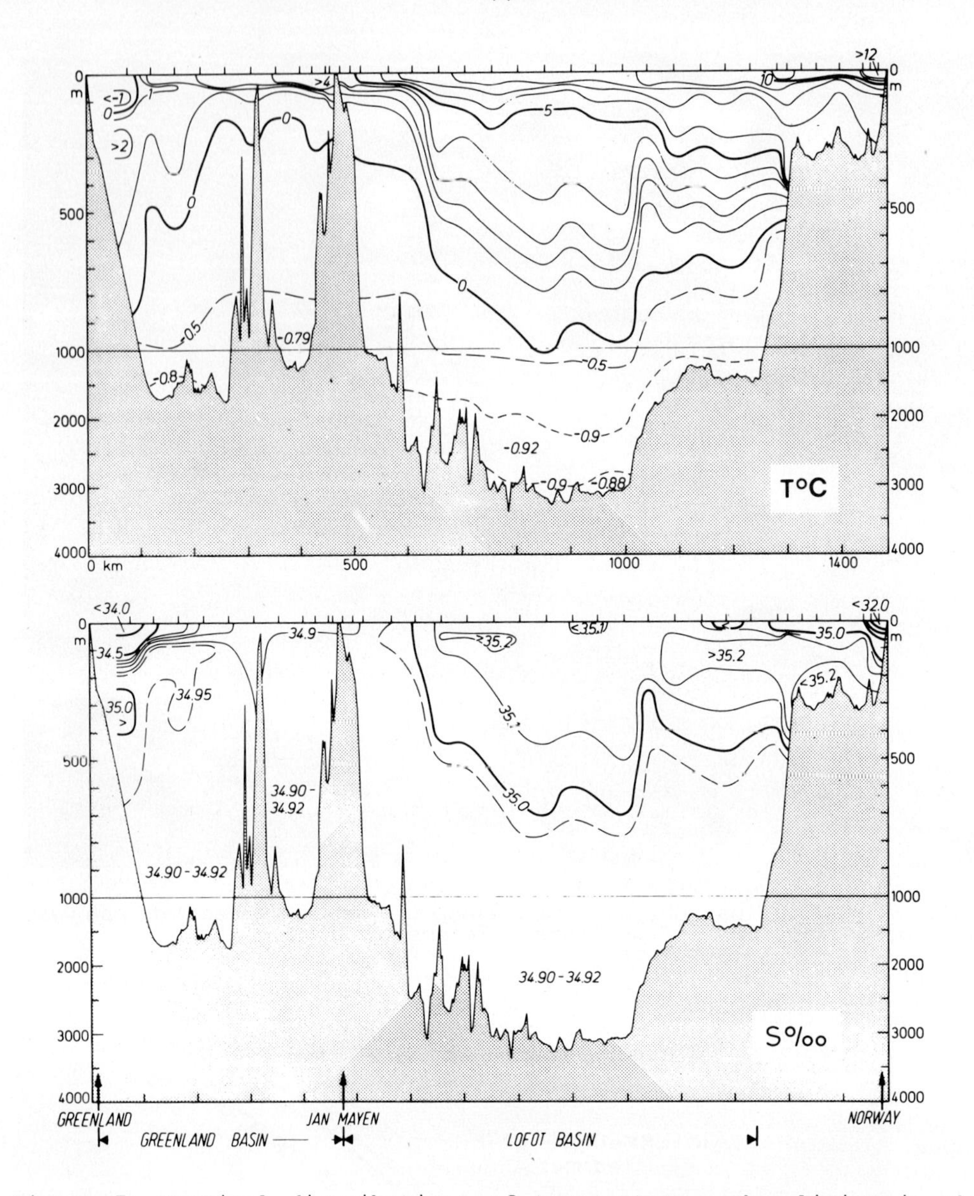

Figure 7: Vertical distributions of temperature and salinity in the Greenland and Norwegian seas in the winter of 1958. (Dietrich, et al., 1980)

From: J.M. Huthnance, Proudmand Oceanographic Laboratory, UK

On: Review and commentary to paper **A BRIEF SKETCH OF POLEWARD FLOWS AT THE EASTERN BOUNDARY OF THE NORTH ATLANTIC,** by T. A. McClimans

This "brief sketch" outlines the main contributions to poleward flow in the eastern North Atlantic and the Norwegian Sea. More detail regarding the poleward flow past the British Isles, and specifically along the continental slope, is given in Huthnance and Gould (this volume). Some of this 7 1/2 Sv (referred to) flows along the upper continental slope from west of Shetland north-eastwards and then south-eastwards into the Norwegian Trench along its western slope, before eventually turning across the Trench to join the poleward flow adjacent to the Norwegian Coast. Such flow appears to be considerably guided along the depth contours.

The fresh water runoff from northern Europe includes a contribution from the southern North Sea and German Bight (notably from the Rhine and Elbe outflows) joining the Baltic outflow at the Skagerrak.

It should perhaps be emphasized that none of the flows are steady; transports may vary seasonally by 50 percent or more and cease altogether on a few occasions during a year's current measurements; moreover, the large eddies alluded to may severely distort the path of flow from any long-term mean of simple form.

ON THE NORTHEAST ATLANTIC SLOPE CURRENT

J.M. Huthnance
Proudman Oceanographic Laboratory
Bidston Observatory
Birkenhead, Merseyside L43 7RA, United Kingdom
and
W.J. Gould
IOS Deacon Laboratory
Wormley, Godalming, Surrey, GU8 5UB, United Kingdom

ABSTRACT

A year-round current flows northwards along the steep continental slope of the NE Atlantic. Transport estimates average $1.5 \times 10^6\ m^3 s^{-1}$ west of the UK, but increase greatly in the Faeroe-Shetland Channel. There is evidence of larger transports in winter. Density and associated longshore pressure gradients are thought to force the flow.

INTRODUCTION

In the context of poleward boundary currents in the eastern Atlantic, we are here concerned with flows around the west and north of the British Isles. Currents further south are considered elsewhere, the nearest observations being off Portugal (Ambar and Fiuza, 1987). By considering a slope current specifically, a distinction is made excluding a broader northwards drift of the upper waters over the Rockall Trough and Plateau (e.g., Ellett et al, 1986; also see "Discussion" below). To the north, we also discuss a slope current in the Faeroe-Shetland Channel. However, additional flows and recirculation there, and the Channel's distinctive deep thermocline and division from the Rockall Trough by the Wyville-Thomson Ridge, warrant separate treatment.

SOUTH OF THE WYVILLE-THOMSON RIDGE

Observations from 47-1/2° to nearly 60°N at the Wyville-Thomson Ridge are listed and results plotted in Huthnance (1986, Table 1 and Figure 1). In addition, drogues have been tracked along the Rockall Slope and further north (Booth and Meldrum, 1987a and b) and southwest of Britain (R.D. Pingree, personal communication) where currents were again recorded in 1984 in 500 m and 1500 m depth near 47-1/2°N, 6-1/2°W.

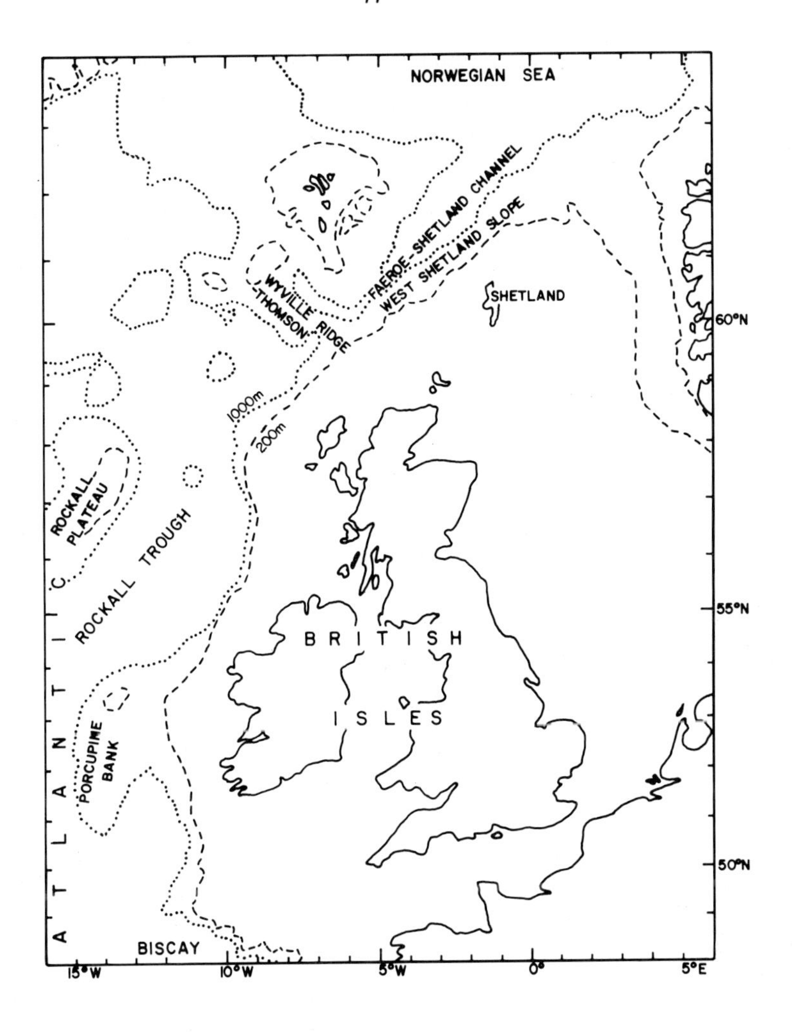

Figure 1: Locations referred to in text.

Mean currents are found closely following the depth contours from 47-1/2°N to 60°N along the continental slope. Speeds range from 3 to 30 cm s^{-1}(values exceeding 10 cm s^{-1} are mostly north of 57°N), but transport estimates are more consistent, between 1.2 and $2.2 \times 10^6 m^3 s^{-1}$ inshore of the 2000 m depth contour. The mean currents are essentially barotropic (albeit decreasing with depth) and are confined to the continental slope width 20 to 50 km, being strongest usually in 500 to 1000 m water depth. Below tidal frequencies, the currents over the slope are relatively steady in that they rarely reverse, unlike the local meteorological forcing and associated currents over the shelf. However, there is some evidence of a winter maximum in transports.

Transports appear to persist over 1000 km or more. However, true continuity (the movement of identifiable water) is not established to the same extent by the few drogue tracks available. In this sense, with current moorings also lacking, the flow off Porcupine Bank near 52°N might not be continuous with flows to the north or south. Indeed, salinity patterns north of Ireland show oceanic water penetrating onto the shelf and entrained in on-shelf flow, representing a shoreward loss from the slope currents, implying a gain from the oceanward side (if the slope current transport is maintained from south to north) and arguing against true continuity. The along-slope persistence of slope current fluctuations can be estimated—up to 500 km—from simultaneous records only in the northern Rockall Trough. This estimate is for fluctuations of period exceeding two days or actually on the shelf; coherences are poorer for shorter periods.

NORTHEAST OF THE WYVILLE-THOMSON RIDGE

The oceanographic regime north of the UK is markedly different from that further south. Water masses of Atlantic origin overlie the deep cold Norwegian Sea water (temperatures on the order of -0.5°C). Sandwiched between are intermediate waters of Norwegian Sea polar front origin (Meincke, 1978).

The strong thermocline between the Norwegian Sea water and the typically 8-9°C Atlantic water is domed towards the axis of the Faeroe-Shetland Channel and intersects the sea bed on the West Shetland Slope at around 500 m. The main core of the warm saline Atlantic water is confined between the thermocline and the continental slope. This picture is confirmed by classical hydrographic and CTD stations and by satellite infrared imagery of sea surface temperatures.

There have been many estimates of the volume transport of water masses in the Faeroe-Shetland Channel, but almost all suffer from the usual uncertainties of the assumption of a reference level of no motion. Where direct measurements are available, they are for the most part of short duration (< one month).

Estimates of the inflow of Atlantic water to the Norwegian Sea derived from geostrophy alone and with a variety of assumed reference levels give mean values on the order of $2 \times 10^6\ m^3 s^{-1}$ and from these figures Aas (1977) suggests a seasonality related to the strength of the wind stress having maximum transports in winter. The analysis of Aas suffers from the scarcity of winter observations (27 estimates from September to April compared with 126 for the remaining four summer months). Estimates by

Dooley and Meincke (1981) based on the combination of geostrophic calculations with brief current measurements give an estimate for the transport of Atlantic water in the slope current of 4.1×10^6 $m^3 s^{-1}$ in September 1973.

More recent year-long direct measurements of currents over the continental slope northwest of Shetland in 1983-4 showed an annual mean northeastward (along-slope) transport of 7.8×10^6 $m^3 s^{-1}$ with highest values in January/February ($\sim 9.5 \times 10^6$) and minima in June/July/August ($\sim 5.0 \times 10^6$) (Gould, Loynes, and Backhaus, 1985). These figures probably represent an (perhaps 10%) overestimate of the Atlantic water transport since they include some water which is clearly colder than 8°C, but they are a good indication of the transport of the slope current. The mean transports from these direct measurements are close to the values estimated for the Atlantic inflow to the Norwegian Sea calculated on the basis of heat budget by Worthington (1976) and McCartney and Talley (1984).

The large increase in volume transport from $1\text{-}2 \times 10^6$ $m^3 s^{-1}$ to 7×10^6 $m^3 s^{-1}$ on crossing the Wyville-Thomson Ridge suggests that there is an effective mechanism for entraining the additional Atlantic water from the west into the boundary flow. The mechanism is not clear, but there is circumstantial evidence of a persistent northward flux of Atlantic water across the entire length of the Wyville-Thomson Ridge in the permanence of warm water over the ridge axis. The added inflow implies a mean current on the order of 6 cm s^{-1} across the ridge.

Besides the seasonal signal in the slope current transport in the Faeroe-Shetland Channel, there are shorter-period fluctuations with periods in the range from 2 to 20 days but with the most pronounced spectral peaks in the current records being near 5 and 14 days.

DISCUSSION

Various possibilities for currents along the northeast Atlantic boundary include a diffusive boundary layer as described by McCreary (1981) and the results of forcing by rectified tides, winds, freshwater runoff, or longshore density gradients. Discussion by Huthnance (1984, 1986) suggests that any slope current forced by tidal rectification is concentrated close to the shelf break and is only a small fraction of the total observed; freshwater runoff is also thought to be weak in this context. McCreary's (1981) boundary layer width of several hundred kilometers (the distance offshore from the eastern boundary for baroclinic Rossby wave decay by thermal diffusion) corresponds to the breadth of the

Rockall Trough and perhaps to the overall northwards drift there (Ellett et al, 1986) rather than to the slope current specifically described here. Winter wind-forcing is suggested by mean northerly currents found from November to April over several years just east of the shelf edge at 57°N (Ellett and Booth, 1983). Otherwise, the current is confined to the slope rather than the shelf, and its relatively steady character (with only rare reversals) suggests density forcing. The mechanism (northwards driving by a pressure gradient imbalance associated with less-dense lower-latitude waters "standing high") is discussed by Huthnance (1984, 1986).

A near-geostrophic balance in cross-slope momentum is expected for the nearly steady flow; friction is relatively weak. Hence the vertical structure corresponds to cross-slope density gradients, which are set by a balance between northward advection and cross-slope diffusion. Northward advection of warmer water is apparent in hydrographic sections and satellite infrared images (Booth and Ellett, 1983; Huthnance, 1986); the resulting baroclinicity tends to reinforce the northward flow at upper levels offshore, without apparently altering the flow's essentially barotropic character. Near-barotropic geostrophic flow also corresponds with the observed close following of depth contours, and hence, with approximate conservation of along-slope transport.

Along-slope persistence of identifiable water is a separate question, however. For example, barotropic and almost frictionless flow would conserve transport well, but would be consistent with large cross-slope diffusion, with water properties advected only short distances along the slope (Huthnance, 1986). Thus, the observations of a poleward current off Portugal (Ambar and Fiuza, 1987) and west of Britain do not establish continuity of a northeast Atlantic slope current in this sense.

Along-slope persistence of fluctuations in the current is another question again, but in theory (Huthnance, 1987) corresponds to the frictional decay distance of coastal-trapped waves, being also the along-shelf evolution scale for establishing the slope current's transport and form after a barrier (say). The eventual transport, balancing forcing and friction, evolves after the decay of coastal-trapped waves accommodating the different (barrier) geometry. Observed scales of 500 km or less for along-slope coherence suggest that second- and higher-mode coastal-trapped waves are significant in the slope current's evolution; the first mode decay distance is 0(1000 km) (Huthnance, 1987) and is better related to coherence over the shelf (rather than the slope).

Stability of the slope current is suggested in Huthnance (1986).

The Wyville-Thomson Ridge apparently marks the beginning of a much larger slope current transport to the northeast. Evidence discussed by Huthnance (1986) suggests that the slope current from the south is partly deflected at depth along the Wyville-Thomson Ridge, but gains overall as a broad upper-level flow from Rockall Trough is deflected anticyclonically over the Ridge (conserving potential vorticity) to follow the West Shetland Slope. Recirculation in the Faeroe-Shetland Channel is also thought to contribute.

Plans for the next year or two include current moorings and drogue tracking, to investigate continuity of the slope current, from the UK south-west approaches (48°N) northwards to Porcupine Bank (52°N) and to the south in the Bay of Biscay. A more intensive cross-slope mooring array and acoustic doppler profiling is also planned, northwest of Scotland, to better define the slope current structure there for comparison with models.

ACKNOWLEDGEMENTS

The co-authors express appreciation to Dr. D.J. Ellett, Dunstaffnage Marine Research Laboratory, Scotland, for review and suggestions to this manuscript.

THE POLEWARD UNDERCURRENT ON THE EASTERN BOUNDARY OF THE SUBTROPICAL NORTH ATLANTIC

E.D. Barton
Marine Science Laboratories
University College of North Wales
Menai Bridge, Gwynedd LL59 SEY
United Kingdom

ABSTRACT

Evidence of a persistent poleward undercurrent has been reported frequently in studies off the African and Iberian coasts. Most direct observations originate from investigations into upwelling between 17 and 26°N. The undercurrent appears to be strongest at about 300 m depth, but extends down through the Central Water Mass layer to about 1000 m. Off the Iberian Peninsula, the Mediterranean Water layer is also moving poleward, while off Africa it appears to be spreading equatorward. The undercurrent has a width of less than 100 km and a mean speed of about 10 cm s^{-1} at the level of strongest flow. South of Cap Blanc (20°N) and off northern Portugal and Spain, poleward flow extends for at least part of the year to the sea surface to form a countercurrent. Many reports are from single moorings on the continental slope, and there have been few observations of the offshore and depth structure of the flow. Moreover, most observations have been over periods of only a few weeks, and so variability of the flow is poorly defined.

INTRODUCTION

A poleward undercurrent may be defined as a persistent, poleward flow of restricted width and thickness, which is bound to the continental slope, runs counter to the dominant regional circulation, and may at certain times and locations extend upwards to the sea surface. It is generally accepted that such a flow exists along the eastern boundary of the North Atlantic, but knowledge of the flow is sketchy. The arguments for its existence as a continuous entity are based upon relatively few direct current observations, some interpretations of temperature and salinity data, and a degree of speculation.

The locations of direct observations of an undercurrent (Figure 1) are concentrated in the area around 20°N off northwest Africa. Despite some twenty years of interest in the region, investigations have been limited, in general, to short term CTD surveys and current meter deployments of a few weeks duration. After the period of intensified observational effort in the northeast central Atlantic during the seventies, interest has waned partly as a result of the territorial disputes in the area.

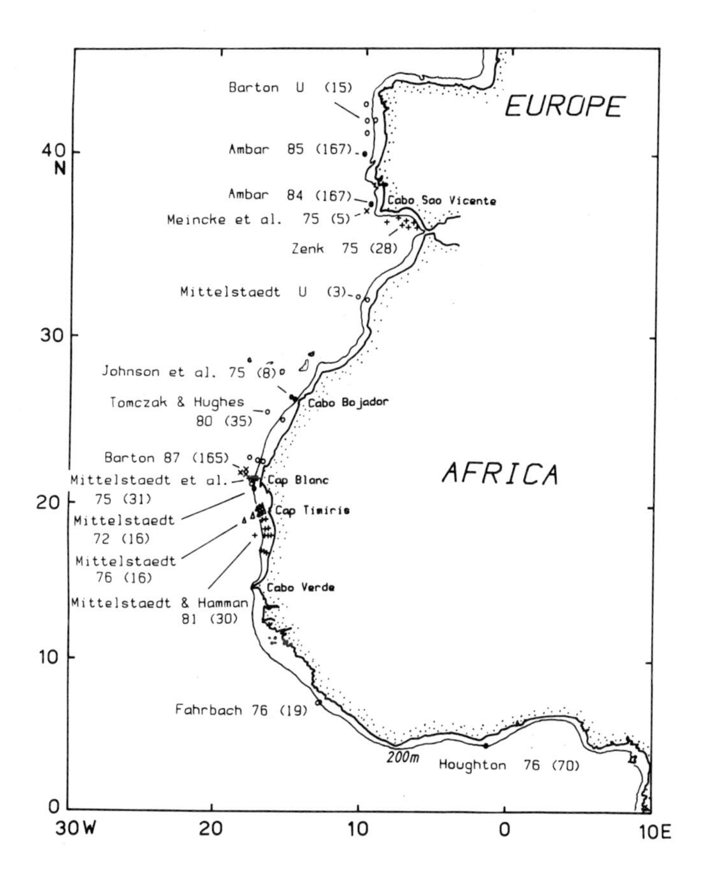

Figure 1: Locations of direct current observations referred to in the text. Author, year, and length of observation period in days are indicated 'U' signifies unpublished data.

The lack of long time series observations of the undercurrent is a major shortcoming because the eastern margin of the North Atlantic exhibits strong seasonality in response to the annual north-south shift of the Azores High. Coastal upwelling, driven by the Trade Winds, varies strongly with time of year throughout the region in question (Wooster et al., 1976). Estimates of mean monthly surface drift indicate northward flow at the surface during the absence of upwelling, both off Portugal and south of Cap Blanc (Figure 2). Moreover, significant interannual variability is apparent in observations of nearshore temperature off Cap Blanc between 1955 and 1982 (Arfi, 1987).

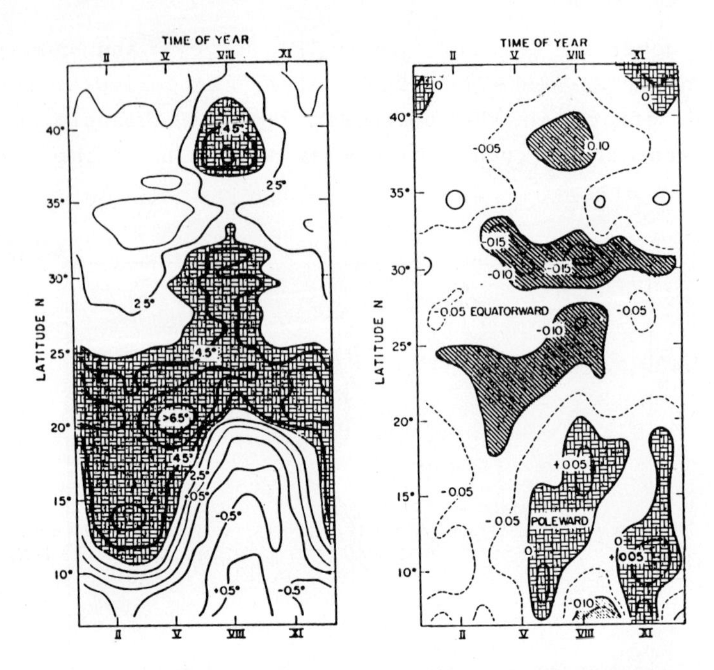

Figure 2: Monthly variation of surface zonal temperature anomaly and along-shore current calculated from wind stress on the eastern boundary of the subtropical North Atlantic (Wooster et al., 1976).

INDIRECT EVIDENCE

The first mention of an undercurrent per se off northwest Africa is by Wooster and Reid (1963), who stated that it appeared to be a common feature of eastern boundary regions. They noted that the frequent downward inclination of isopleths below 200 m depth towards the coast was indicative of a geostrophic northward current around that level. Montgomery (1938) produced charts of the salinity distribution on various sigma-t surfaces, which indicated northward flow between about 200 m and the sea surface along the slope south of Cap Blanc. Defant (1941) also deduced poleward flow at shallow levels along the African coast up to 20°N from geostrophic calculations. In neither case was there any indication, however, of a continuous flow along the eastern margin to the north of Cap Blanc.

As interest in upwelling off northwest Africa was burgeoning in the early seventies, a number of workers searched for evidence of an undercurrent in temperature-salinity data from extensive regional surveys. Tomczak (1973) and Hughes and Barton (1974) traced an anomaly of low salinity at about 300 m depth from Cap Vert, 15°N, to Cabo Bojador, 26°N (Figure 3). To the north of Cap Blanc, the dominant water mass in the layers above 700 m is the North Atlantic Central Water, but to the south is found the lower salinity and slightly cooler South Atlantic

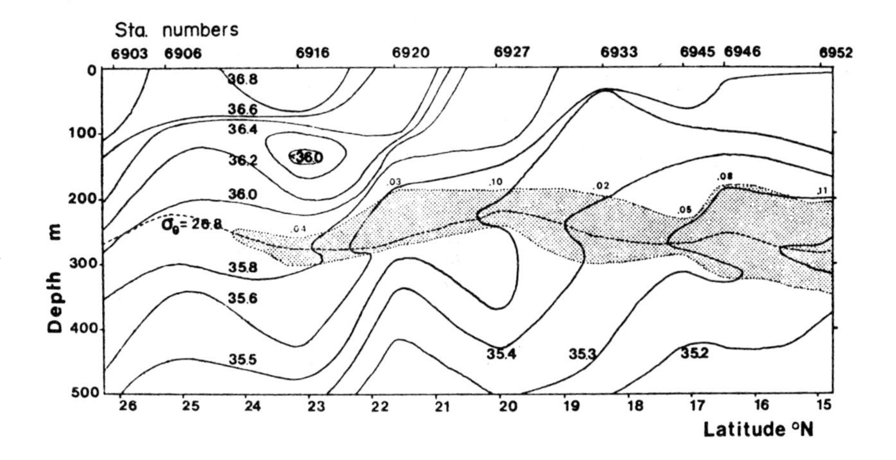

Figure 3: Along-slope section of salinity. The potential density surface $\sigma_\theta = 26.8$ is dashed, the salinity anomaly is shaded, and its magnitude is indicated above (Hughes and Barton, 1974).

Central Water. Lower salinity waters (and nutrient tracers, Gardner,1977) are apparently advected northward along the slope in the undercurrent. From the distribution of the salinity anomaly, it was concluded that the undercurrent was restricted to within less than 100 km of the coast and centered around 300 m depth. Beneath the core of the undercurrent, it seems that a slow northward drift exists to depths as great as 1000 m, where traces of Antarctic Intermediate Water have been identified (e.g., Fiuza and Halpern, 1982). Deeper still lies the presumed southward spreading Mediterranean Water layer centered at 1200 m.

Subsequent work by Tomczak and Hughes (1980) on mixing between the North and South Atlantic Central Water masses, clearly demonstrated a core of fresher water of southern origin being advected poleward along the slope at several latitudes. North of Cap Blanc, traces of southern water were found at depths down to at least 600 m (Figure 4). However, the same authors concluded that horizontal exchanges often resulted in the advection of isolated bodies of differing mixtures of the Central Waters by the undercurrent. In sections to the south of Cap Blanc, Hamann et al. (1981) showed that the southern water mass extended completely up to the sea surface.

This and other accumulated evidence, largely from water mass analysis and geostrophic calculations by Mittelstaedt (1972), Hughes and Barton (1974), Fraga (1974), Allain (1970), and other workers gave rise to an idealized circulation scheme shown in Figure 5 (Mittelstaedt, 1983). A permanent cyclonic eddy south of Cap Blanc produces a slope countercurrent

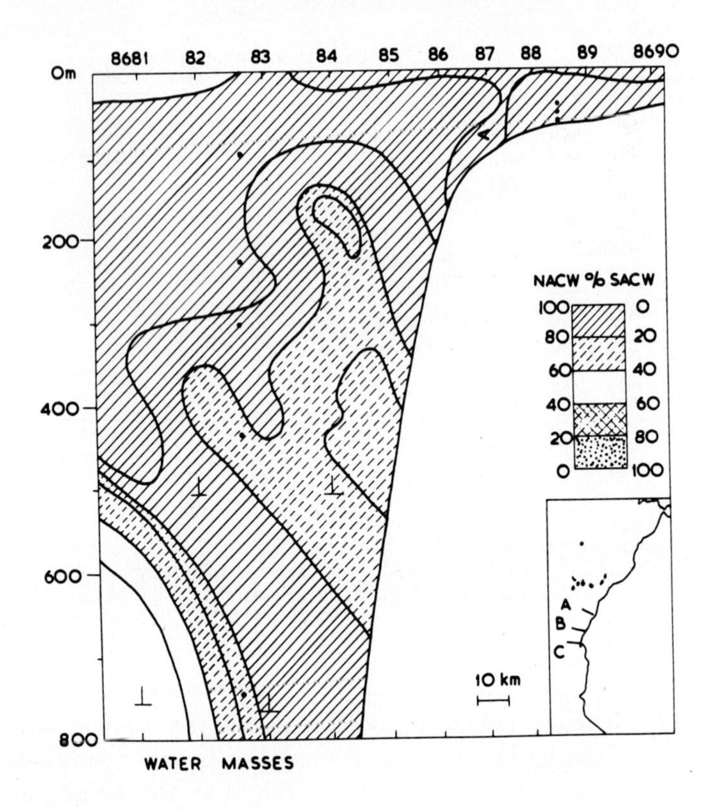

Figure 4: Percentage of North and South Atlantic Central Waters near 25°N (line A inset). The water masses are defined by conditions at 21°N and 25°N, respectively (Tomczak and Hughes, 1980).

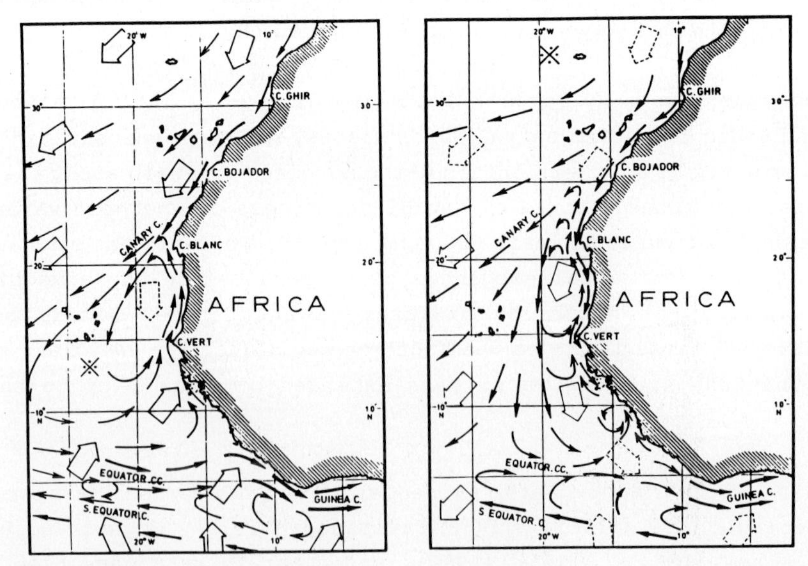

Figure 5: Summer (left) and winter (right) surface circulation. Dominant winds are shown as open arrows. Dashed arrows represent weaker, intermittent winds (Mittelstaedt, 1983).

throughout the upper layers. During summer, when the trade wind belt recedes northward and upwelling ceases, the counterflow extends inward to the coast; but during winter upwelling it is separated from shore by an equatorward coastal current. The main flow of the Canary Current separates from the coast near Cap Blanc. Evidence for this general scheme of surface flow can be found in Stramma's (1984) objective maps of annual mean surface geostrophic transport, although detail over the slope is not resolved by the analysis. At depths around 300 m, it is supposed that the slope undercurrent continues northward despite the recirculation southward of the surface waters.

DIRECT OBSERVATIONS

Near and south of Cap Blanc, the above idealization indicated that during the upwelling season (winter) three bands of meridional current could be differentiated. A section of 24 current meters on six moorings slightly south of Cap Blanc revealed the depth and offshore structure of the flow during three weeks in February-March 1972 (Mittelstaedt, 1976). Equatorward flow was evident both nearshore over the continental shelf and offshore in the southward drift of the Canary Current (Figure 6). The maximum of poleward flow appeared at about 200 m, but might have been shallower in the unsampled surface layers. Poleward flow was confined within 100 km of the shelf break and extended down through the Central Water layer to about 700 m. Similar conditions were seen in February 1975

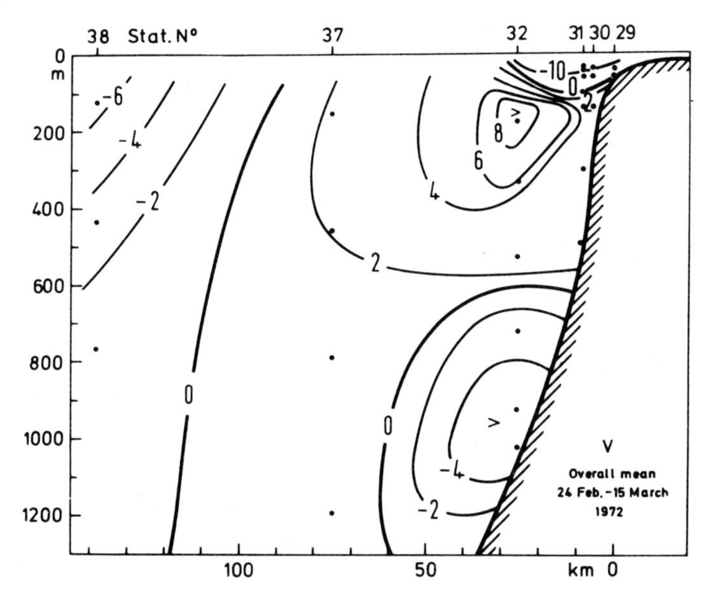

Figure 6: Mean along-shore flow at 19°30'N, February 24 to March 15, 1972. Solid dots mark current meters (Mittelstaedt, 1976).

(Barton and Hughes, 1982) within 100 km of the shelf in the area of 23°N just north of Cap Blanc.

Direct evidence of the along-shore confluence in the upper layers near Cap Blanc was reported by Barton (1987). Currents were observed at 22°N in the upper 350 m layer between November 1981 and April 1982 at three moorings less than 30 km apart and over 100 km offshore. The mean velocities at all sampling depths indicated an offshore diversion of weak flows converging from the north and south (Figure 7). The absence of observations at deeper levels and closer to the slope left unresolved the relationship between the supposed northward continuation of the slope undercurrent and the observed near-surface recirculation.

The JOINT-I current section, also near 22°N, was observed between February and April 1974 within 100 km of shore. A poleward flow restricted to the layers below 100 m depth within 30 km of the slope was observed (Barton et al., 1976). The maximum velocity reached a value of about 10 cm s^{-1} in the core of the undercurrent centered near 200-300 m (Figure 8). Further offshore, the mean along-shore flow was weak and actually poleward in a shallow eddy-like feature. On the continental shelf, strong equatorward flow representative of the coastal upwelling jet was seen. A similar picture was reported for eight days observations off Cabo Bojador, 26°N in August 1972 (Johnson et al., 1975); but strong

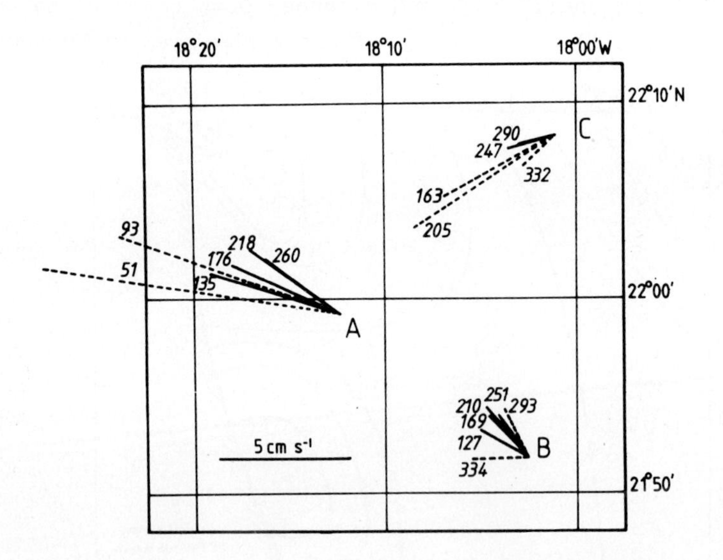

Figure 7: Mean current vectors November 1981-April 1982. Incomplete records are indicated by broken lines. Observation depths in meters are shown (Barton, 1987).

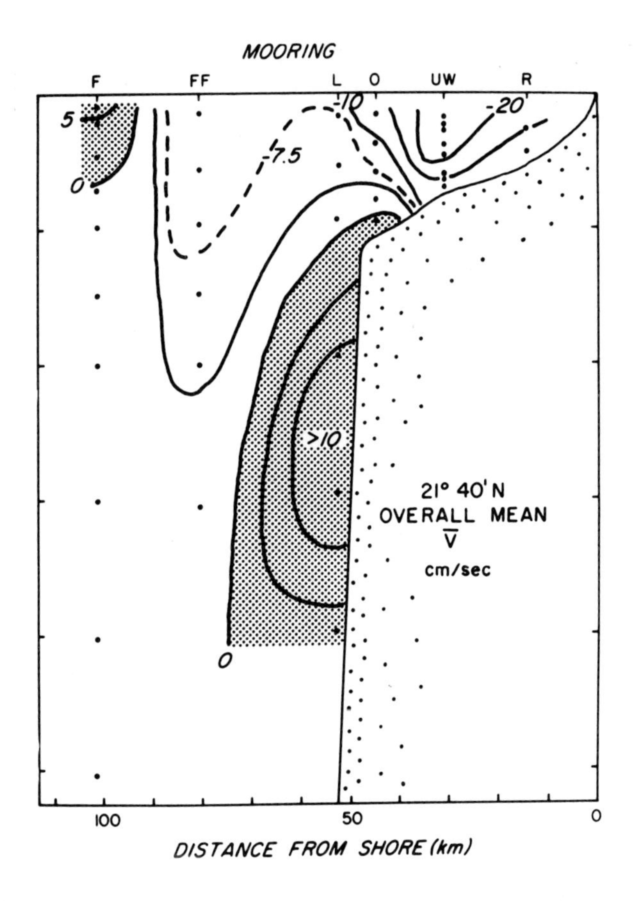

Figure 8: Mean along-shore flow at 21°40'N, February to April 1974. Solid dots mark current meter depths (Barton et al., 1976).

equatorward winds had enhanced the equatorward flow, resulting in a weak undercurrent centered at 400 m within 10 km of the slope.

In general, it has been found off Africa that poleward flow does not extend onto the continental shelf during the upwelling season, unlike off Oregon (Huyer, 1976). At times, however, this does occur when the equatorward winds weaken sufficiently. Mittelstaedt and Hamann (1981) reported that an array of moorings on the slope and shelf south of Cap Blanc between 17 and 19°N revealed an interlude of poleward flow at all but one of twenty-six current meters during a wind relaxation (but not reversal). This was also briefly evident during the JOINT-I experiment at times of weak equatorward winds (Mittelstaedt et al., 1975).

Outside the region of 17 to 26°N, there have been even fewer reports of the undercurrent off Africa. Fahrbach's (1976) roughly three weeks of observations revealed episodes of poleward flow at 7°N off Sierra Leone at depths between 80 and 900 m. In the Gulf of Guinea, almost three months of observations by Houghton (1976) revealed persistent subsurface flow towards the west, i.e., in the sense of the poleward undercurrent, beneath a shallow thermocline in 87 m of water on the shelf edge. More to the north, Mittelstaedt (1987) documented brief current observations during 60 hours off Morocco in April 1983. These indicated poleward flow at depths below about 500 m in 730 m of water, while temperature and salinity properties were consistent with a poleward tendency down to the depth of the Mediterranean Water layer.

Off the Iberian Peninsula, there have been a number of reports of poleward flow at various depths. Meincke et al. (1975) observed northward flow at levels between 234 m and 2046 m in 2450 m of water 100 km west of Cabo Sao Vicente, while Swallow et al. (1977) found a general northward flow in the upper 1500 m during January and February at a distance of

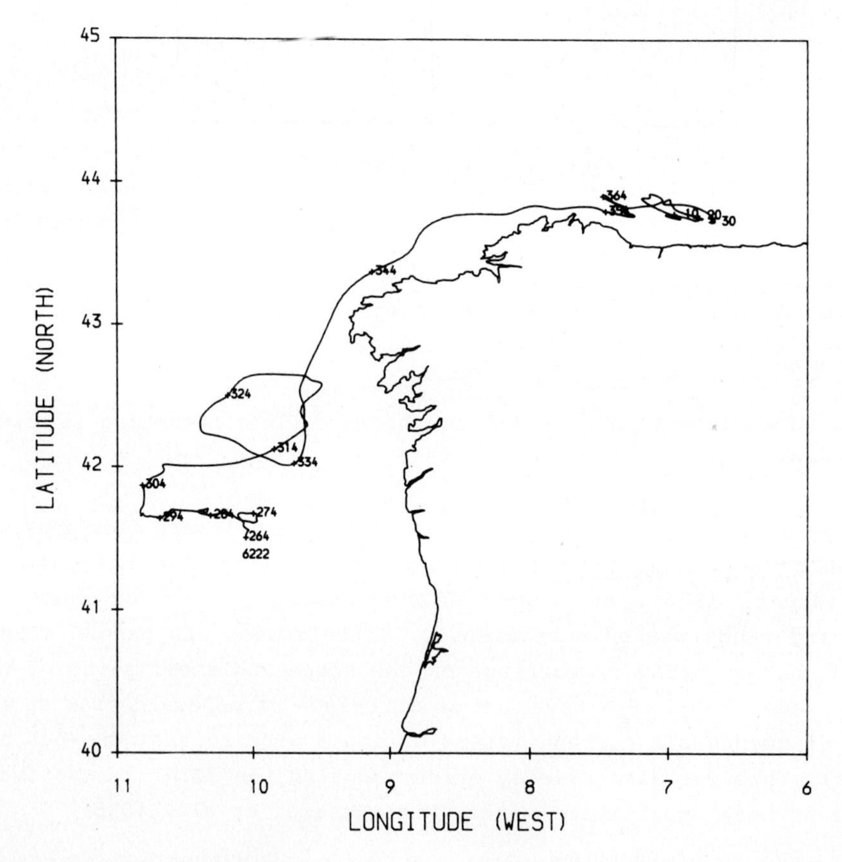

Figure 9: Trajectory of a drifter drogued at 5 m, September 1986 to February 1987. Positions every ten calendar days are marked.

95 km offshore. Closer to shore, two six-month-long series of observations were executed off Portugal in 1100 m of water by Ambar (1984 and 1985). At latitude 37°20'N, a mean flow of 10 cm s^{-1} towards the north was found over the period December to May at 750 m. The speed of the along-shore northward component decreased with decreasing depth to almost zero at 200 m. At 40°N, a more uniform mean northward flow in excess of 3 cm s^{-1} was found between 200 and 750 m between July and January. Recent current meter observations off northern Portugal and Spain showed weak poleward flow between 50 and 600 m over the slope in September 1986. Argos drifters, drogued at 5 m and released at the moorings, moved poleward along the slope around Cape Finisterre into the Bay of Biscay (Figure 9).

The eastern boundary of the North Atlantic is complicated by the physical discontinuity where the Mediterranean Outflow occurs. The warm, saline outflow from the Strait of Gibraltar itself constitutes an undercurrent (e.g., Thorpe, 1976), which is constrained to hug the northern margin of the Gulf of Cadiz. The path of the outflow through the Gulf has been investigated by several workers (e.g., Ambar and Howe (1979); Zenk (1975); Madelain (1970)). These workers have documented the splitting of the outflow into several cores which flow west at different depths to turn northward around Cabo Sao Vicente. Ambar (1983) has shown that another shallow vein of Mediterranean Water was identifiable at depths as shallow as 400 m off southern Portugal, while Fiuza (1982) has detected the same water mass on the outer shelf. The impact of the Mediterranean Outflow on the presumed northward continuation of the poleward undercurrent along the African coast has not been considered.

DISCUSSION

There are a number of unanswered questions about the nature of the undercurrent and its role. For example, it is unclear whether one can meaningfully distinguish between an undercurrent and a countercurrent. In certain locations, a counterflow adjacent to the coast may appear to be topographically generated, while in others it may appear as the temporary surfacing of a permanent undercurrent during a period of poleward wind stress. One might also ask whether the enhanced poleward flow observed bound to the continental slope further north in the subpolar gyre represents the same feature. It seems likely that a combination of different forcings is responsible for what appears as a single phenomenon.

The observations of an undercurrent reviewed here are spread quite widely along almost the entire eastern boundary of the subtropical gyre, but there are large gaps between sightings. Tentative conclusions are that the speed of the undercurrent core is about 10 cm s^{-1}, that it has a width of 30 to 100 km, and that it is concentrated at about 300 m, but extends downwards through the Central Water and Antarctic Intermediate Water layers. Off Africa, the lower limit of northward motion is the Mediterranean Water layer, but off Portugal northward flow may extend much deeper. However, many reports are based on single moorings located in various water depths; and the offshore and deep structure of the flow over the slope has been determined in only a few cases.

With few exceptions, observations have been limited to several weeks duration. Consequently, little is known of the temporal variability of the undercurrent. In particular, the seasonal and interannual fluctuations remain essentially unsampled, and even the several day time scale is poorly sampled. This could well result in misunderstanding of basic features of the flow regime. One example might be the apparent northward deepening of the undercurrent core, which is reported near the surface at Cap Blanc at 200 m near 22°N, at 400 m near 26°N, and below 500 m near 32°N. These observations were made in different years, in different months, and over short periods, and therefore may be unrepresentative.

Finally, the continuity of the flow is unknown. The effectiveness of the undercurrent as a transporter of, say nutrients or fish larvae, is undetermined. Although the tracing of water mass anomalies has been used as evidence for the existence of the undercurrent, it has been shown that these do not uniquely identify the flow at any particular time and place. Horizontal exchange with surrounding waters is evident from the presence of isolated bodies of anomalous water within and without the undercurrent. Moreover, wherever the poleward flow is shallow enough, vertical exchange by upwelling onto the shelf may occur.

While further study of the existing data base may well advance understanding of the undercurrent, there is a clear need for systematic, long term observations of the eastern boundary of the North Atlantic.

ACKNOWLEDGEMENTS

This research was supported by Natural Environment Research Council Grant GR3/5872. The drifters were provided courtesy of the United States Office of Naval Research.

From: M. Tomczak, The University of Sydney, Australia

On: Review and commentary to paper **THE POLEWARD UNDERCURRENT ON THE EASTERN BOUNDARY OF THE SUBTROPICAL NORTH ATLANTIC,** by E.D. Barton

This review of the poleward undercurrent along the continental slope of northwest Africa and southern Europe shows rather convincingly how little is known about this prominent feature of the Canary Current upwelling system, despite the high level of field activity in the area during the seventies and early eighties. I pointed out some years ago that most of the field projects were only loosely connected with the systematic investigation proposed by the Intergovernment Oceanographic Commission (Tomczak, 1979) and did not result in a comprehensive data set. The review by E.D. Barton does its best to extract what can be learned about the undercurrent from rather disparate data. Hopefully, it will stimulate oceanographers to go back to the area and produce some long time series of currents and temperature. The poleward undercurrent of the eastern North Atlantic Ocean remains one of the permanent but highly variable features of the oceanic circulation which have not yet been covered by long-term current measurements.

The R.V. "Alexander von Humboldt" was used between 1970 and 1984 by oceanographers from the Institute for Marine Research at Rostock-Warnemünde to collect data in the Canary Current upwelling region in a very systematic way. The early cruises, from 1970 to 1974, all followed a standard station pattern and formed the basis for an investigation into seasonal upwelling variability (Wolf and Kaiser, 1978). The data set consists exclusively of hydrographic casts, taken along seven sections roughly perpendicular to the coastline, with stations spaced about 35 km apart (Figure 1). This does not make it particularly suitable for the study of a feature believed to be between 30 km and 100 km wide, which may explain why Barton's review ignores the data set completely. However, following his suggestion that "further study of the existing data base may well advance understanding of the undercurrent" I extracted from the R.V. "Alexander von Humboldt" data the density difference between adjoining stations next to the continental shelf at 100 m, 200 m, and 300 m depth (at deeper levels differences were too small to give reliable results). The station pairs used are shown in Figure 1. Linear interpolation between the nearest samples was used to determine σ_t at the selected depth levels. The sign of the density difference was graphed as a function of latitude and month, with no regard to the year when the data were obtained. This should give a reasonable pattern if interannual variations are small

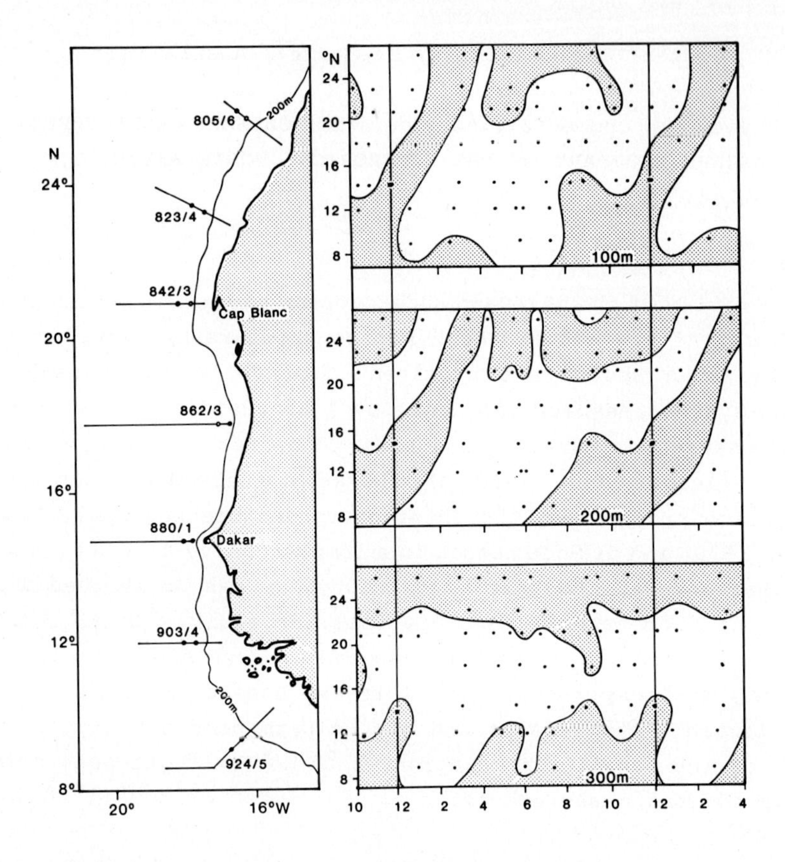

Figure 1: Standard sections of R.V. "Alexander von Humboldt" cruises 1970-1974 and station pairs used for the calculation of the zonal density gradient (left); sign of the zonal density gradient at three depth levels as a function of latitude and season (right). The area with positive gradients (isopycnals slope downwards towards the coast) is indicative of the presence of an undercurrent and is shaded. The areas outside the vertical lines are repeats of the data inside.

compared to seasonal variations. Support for this assumption was presented by Schemainda et al. (1975).

Several points should be kept in mind when looking at the graph. Interannual fluctuations are neglected. The distance of the station pairs from the continental slope is not uniform. The distance between stations in some pairs may not be sufficient to resolve the undercurrent. Consequently, while the shaded area in Figure 1 is strong evidence for the presence of the undercurrent, the current may also be present at times where the analysis does not show it. The points that can be made from the graph are:

(a) The undercurrent is generally deeper north of Cap Blanc (21°N) where it was found during all months at 300 m depth and in most cruises at 200 m depth. This is in agreement with Barton's conclusions.

(b) South of 14°S, the undercurrent is distributed more uniformly throughout the 100-300 m layer but appears to be varying in strength, being strongest in late summer when upwelling is confined to the region north of 18°N (Schemainda et al., 1975).

Note however that data in the southern part are sparse and other interpretations are possible.

(c) Coinciding with the southward progression of the upwelling regime during autumn, the strongest undercurrent signal is observed to move northward, being confined to the upper 200 meters. The same northward movement can be seen in the analysis of alongshore surface flow by Wooster and Reid (1976; figure 2 in Barton's review), but it appears there about four months earlier than in this analysis for the 100-200 m depth layer and is confined to the region south of 18°N. This may indicate a vertical migration of the poleward current in response to the southward progression of the upwelling.

These results are as tentative as Barton's conclusions, but the combined information will hopefully stimulate new oceanographic activity in the region and lead to a more cohesive data base.

I thank Eberhard Hagen for helpful input to these comments and Neil Trenaman for data processing and analysis.

THE SUBSURFACE CIRCULATION ALONG THE MOROCCAN SLOPE

Ekkehard Mittelstaedt
Deutsches Hydrographisches Institut
Postfach 220
2000 Hamburg 4 (FRG)

ABSTRACT

CTD data and short-term current data collected on METEOR Cruise No. 64 in January and April 1983 off Morocco are used to discuss the wind-induced upwelling situation at 32°N during the time of observation and the subsurface circulation off the Moroccan coast.

During the station work at 32°N, the vertical cross-sections of temperature and salinity clearly suggest upwelling during January and April. The short current records over about 60 hours in April indicate offshore flow in the surface layer. At depths around 250 m, the water moves onshore toward the shelf compensating the upwelling. The maximum depth from where the water ascends up the slope onto the shelf is deeper than 300 m.

Below 500 m depth the observed currents have a north component of a few centimeters per second extending down to the upper boundary of the Mediterranean water at about 1000 m depth. This poleward-going undercurrent is the deep part of the same subsurface flow which occurs along the continental slope off Mauritania. But in the south, the undercurrent extends from depths around 1000 m upwards to the lower boundary of the wind-driven surface layer or even to the surface. The deeper part of the undercurrent (approximately between 700 m and 1000 m) carries Antarctic Intermediate Water (AIW) toward the north. The upper part has a subsurface velocity maximum between 100 and 400 m depth and is characterized by substantial admixtures of South Atlantic Central Water (SACW), off Senegal and Mauritania from about 14°N to 21°N. This flow represents the major subsurface source for the waters upwelling along the shelfbreak and the coast.

Further north, between 21°N and 24°N, the upper undercurrent hits the Canary Current flowing toward the southwest at these latitudes. The convergence produces the frontal zone between NACW and SACW and forces the poleward-going under/counter-current within the upper 400 m to flow parallel to the Canary Current and the frontal zone toward the southwest.

The deeper part of the undercurrent merges below the Canary Current and continues its way toward the north. The transport of AIW along the continental slope can be identified at depths between 800 and 900 m as far north as the Canary Islands. At these latitudes the identity of AIW vanishes due to mixing, but the deep northward-going flow still occurs north of this latitude at 32°N.

INTRODUCTION

During her Cruise No. 64, from January to April 1983, RV METEOR investigated, among others, the waters off Morocco. The station work comprised oceanographical, chemical, and biological observations. The investigations concentrated along a section across the Moroccan continental slope at about 32°N (Figure 1). This section was surveyed twice: the first time in January, the second time in April 1983. STD-data and short-term current measurements at 32°N from this cruise, as well as other information, are used here:

- to describe the actual upwelling situation in January and April 1983, and
- to discuss the indications of a poleward-going undercurrent off Morocco.

Weikert (1986) pursues the same objective using certain mesopelagic species of zooplankton sampled on METEOR Cruise No. 64 as a tracer for the undercurrent.

OBSERVATIONS

Methods

Temperature and salinity distributions are measured by means of a CTD-probe (bathysonde). The data are adjusted to reversing thermometer-temperature readings and bottle data (for salinity) taken with the CTD-probe at every station. Two simultaneous moorings (5 to 8 April) provide short-term current information, recorded by means of Aanderaa instruments.

Hydrographic Conditions

The vertical sections of temperature and salinity at 32°N suggest upwelling over the upper slope and the shelf in January, as well as in

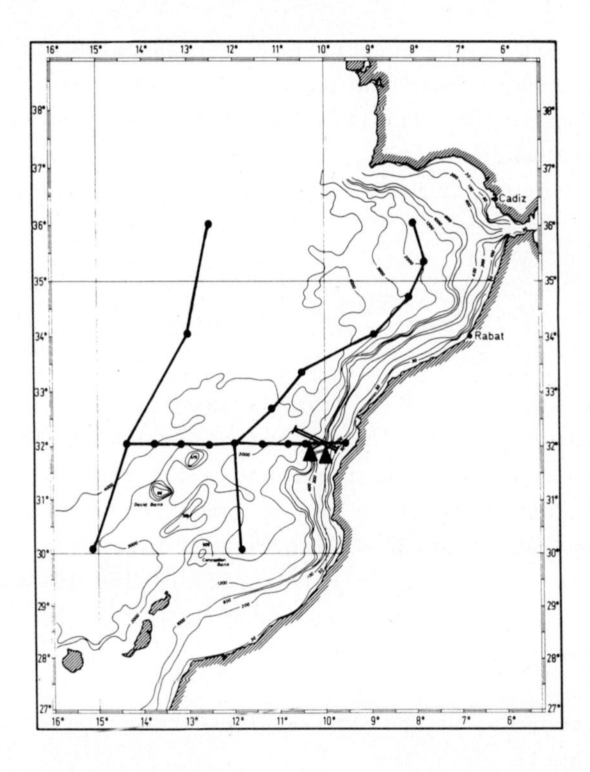

Figure 1: Stations and sections carried out by RV METEOR during winter 1983 and 1973. The double bar marks the section surveyed during January and April 1983. The two triangles give the positions of the short-term current meter moorings from 5 to 8 April 1983. The black dots and the thick lines represent a cross-section along 32°N and two meridional sections surveyed during February 1973 in the framework of CINECA.

April (Figure 2). The lower figure, showing April's near-bottom temperature at the shelf break, indicates that the upwelling has been stronger at this time because of stronger northeast winds before and during the observations. During the measurements in April, the winds blew from the NNE with a strength of 6 Beaufort (22 to 25 knots). The water temperature of the near-surface layer over the shelf corresponds to the temperatures at a depth of about 150 m, 20 to 30 nautical miles offshore of the shelf break. The near-bottom water temperature and salinity over the outer shelf compares with that one at a depth of about 200 m (in January) to 300 m (in April). Taking in account diffusion while the water ascends onto the shelf, the maximum upwelling depth is somewhat deeper than 300 m.

Due to horizontal advection, the salinity of the upper layer is higher over the shelf in April than in January. The temperature/salinity characteristics of the upwelled water agree fairly well with North Atlantic Central Water (NACW).

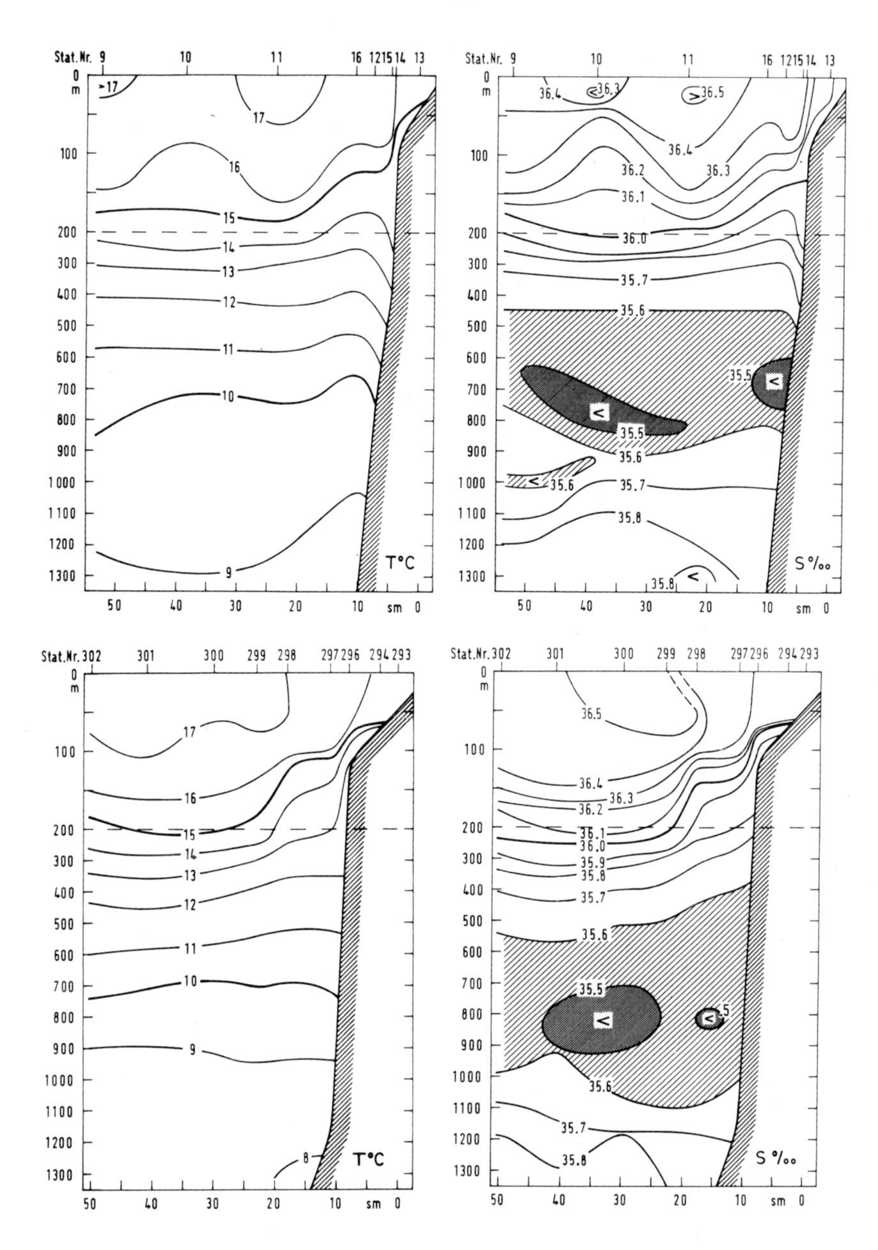

Figure 2: Sections of temperature and salinity at about 32°N from 15 to 16 January (above) and 5 to 6 April (below) 1983. The salinity minimum is hatched.

Below the shelf break the isotherms tend to incline downwards toward the slope instead of upwards, as in the layer above, due to upwelling. The down-warping within a narrow boundary zone along the slope is especially pronounced during January, when upwelling is weaker than it is in April. In January the salinity distribution also shows downtilted isohalines at the slope between 200 m and 600 m depth, but not in April when the signal is weak in temperature as well.

The downwarping of the isolines is the signature of a poleward-going undercurrent.

In general, salinity decreases with depth down to about 800 m. The salinity minimum (< 35.5 ^{o}/oo) occurs between 600 m and 900 m and forms the lower boundary of the NACW which extends up to the surface layer.

Underneath the NACW there is a distinct salinity maximum at depths between 1200 m and 1300 m, which signalizes the Mediterranean outflow. A hydrographic survey along 32°N ten years earlier, February 13, 1973, exhibits basically the same phenomena close to the slope (Figure 3):

- upwelling near the shelf break from a depth of about 200 m,
- a vertical "divergence" of the isolines close to the slope within the layer between 400 m and 600 m (in salinity down to 1000 m depth), and
- a core layer with maximum influence of Mediterranean Water between 1200 and 1300 m depth.

Currents

Two moorings, one on the shelf and the other at the slope, were supposed to record the currents at 32°N over three months, from January to April. Unfortunately, we discovered in April that the moorings had disappeared. To our knowledge, no current measurements are available from the continental slope off Morocco, so far. In order to get at least some data from around 32°N, we deployed another two moorings, one on the shelf (water depth 70 m) and one at the slope (water depth 730 m) for short-term measurements. The moorings stayed out from April 5 to 8, over 56 to 60 hours, while METEOR did station work. They consisted of an instrument leg with subsurface buoyancy, ground line, buoy line, and surface buoy. The instruments were Aanderaa current meters; the sampling interval was five minutes.

RESULTS

At the slope the records are dominated by the semi-diurnal tidal currents (Figure 4). Their amplitudes are highest near the bottom and in the near-surface layer. In both layers, maximum speeds in connection with

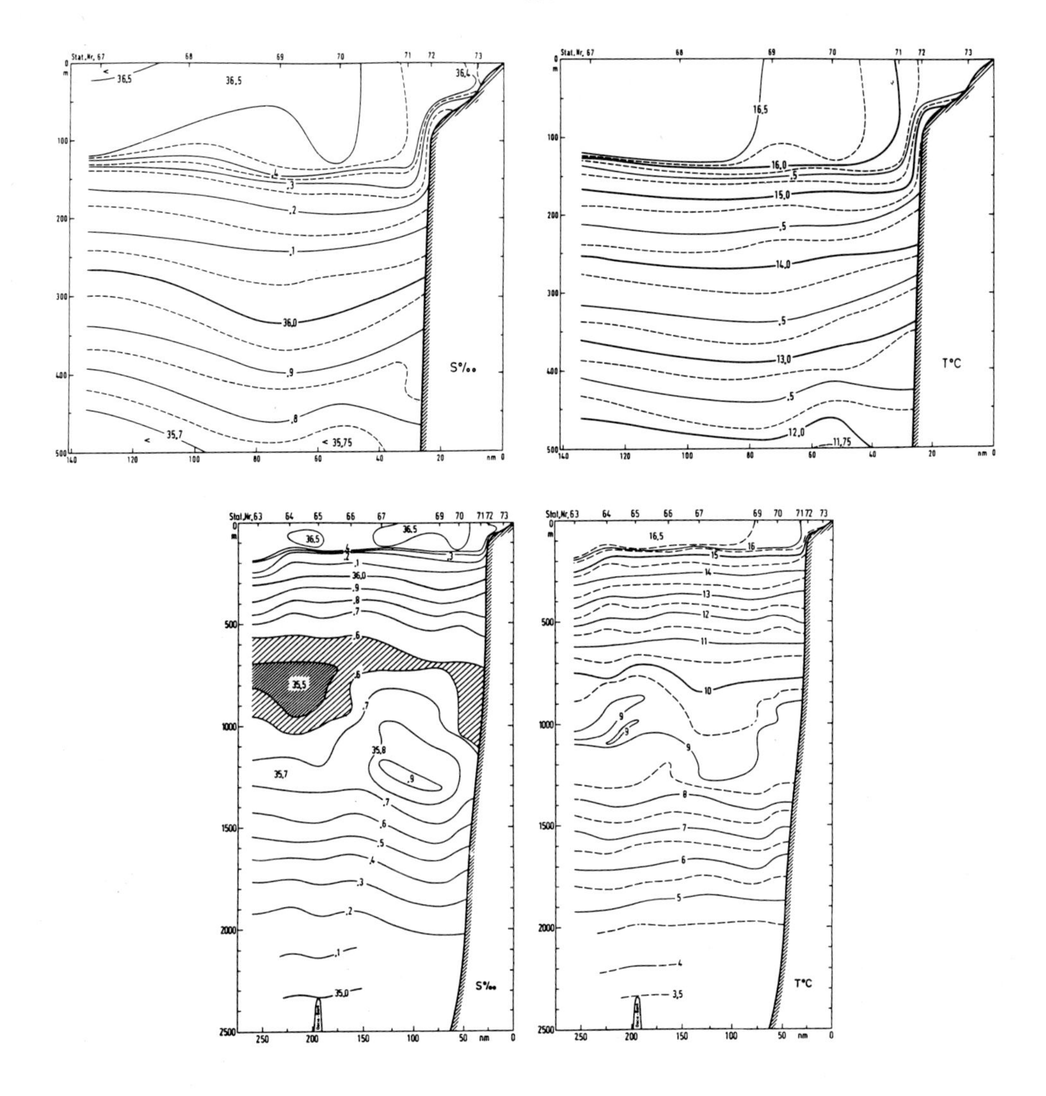

Figure 3: Sections of salinity and temperature at 32°N from 12 to 14 February 1973 (Huber, Mittelstaedt, Weichart, 1977). Distributions within the upper 500 m near the shelf break (above). Total sections (below). The salinity minimum is hatched.

the tidal currents exceed 40 cm s^{-1} (hourly mean). The tidal speed variations of the two layers are opposite to each other. While in the surface layer, the semi-diurnal variations of current directions are covered by a fairly strong mean flow, the records 20 m above the bottom exhibit strong periodical changes of current directions. At intermediate layers, the tidal current speeds are distinctly weaker than within both boundary layers near the surface and the bottom.

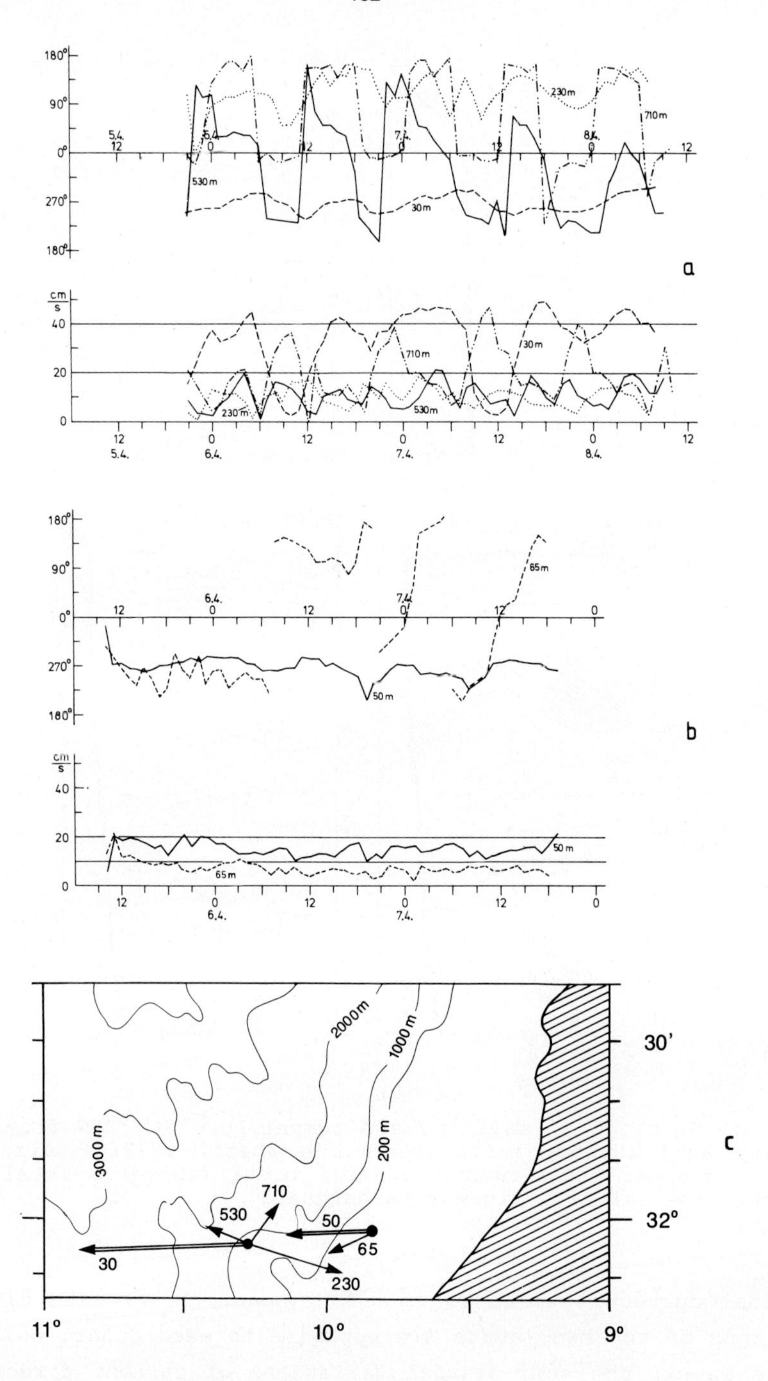

Figure 4: Current data from 5 to 8 April 1983: a.) Time series of direction and velocity (hourly means) over the slope at the observation depths of 30 m, 230 m, 530 m, and 710 m. b.) Time series of direction and velocity (hourly means) over the shelf at the observation depths of 50 m and 65 m. c.) Mean flow averaged over the observation period. Double arrows mark velocities above 10 cm s^{-1} (for numbers see Table 1).

Table 1
Mean Flow from April 5 to 8, 1983

	Observation Depth (m)	Flow Direction	Velocity (cm s^{-1})
Shelf Mooring	50	267°	15
(water depth: 70 m)	65	239°	4
Slope Mooring	30	268°	31
(water depth: 730 m)	230	115°	8
	530	290°	3
	710	34°	4

Within the lower layer over the shelf, tidal fluctuations are relatively small.

This qualitative description agrees with a number of tidal current analyses off Northwest Africa further south (Fahrbach, 1976; Horn et al., 1976; Gordon, 1978; Mittelstaedt et al., 1980).

In this report our interest concentrates on the mean water motions. The flow averaged over the observation period is given in Table 1 and Figure 4C.

The strong offshore flow near the surface indicates a pronounced upwelling situation during the period of observation, which is supported by the hydrographic section (Figure 2) from the same time.

The eastward-going and onshore current, as observed at the depth of 230 m, is most likely the subsurface compensation flow feeding the upwelling at the shelf break from below. At 530 m the flow is westward with a slight north component, and at 710 m the water movements are directed toward the NE with velocities around 4 cms^{-1}. We assume a northward deep flow within the layer, say, between 500 m and 1000 m along the continental slope off Morocco, to be a characteristic feature and not just a random event during our observations (see discussion below).

With this assumption, the vertical structure of the subsurface flow along the continental slope at 32°N can be described as follows:

1. A flow toward southerly directions between 1000 m and 1400 m depth (Mediterranean Water). These deep water movements are basin-wide and not restricted to the slope region.
2. A flow toward northerly directions between 500 m and 1000 m depth. This flow is assumed to be a boundary current with an intensification the closer one comes to the slope from offshore.
3. A flow toward southerly directions with the general Canary Current offshore and the wind driven shelf circulation.

DISCUSSION

From the METEOR cruise in the framework of CINECA during February 1973, two north-south sections of temperature and salinity are shown in Figure 5. One section runs about 250 nautical miles offshore the continental slope; the other is further inshore and follows approximately the water depths between 2000 m and 3000 m.

The salinity distributions from winter 1973 (Figure 5) and 1983 resemble each other very much at depths. The comparatively weak vertical gradient within the layer from 600 m to 1000 m suggests intense mixing between the southward flowing Mediterranean Water (MW) and the poleward moving water masses thereabove. The tongue shaped advective/diffusive spreading of the MW produces a salinity minimum (S_{min}) at its upper boundary.

From south to north the minimum salinity increases. The thickness of the S_{min}-layer and its mean depth decreases as a consequence of the MW merging underneath the NACW. At the lower latitude the S_{min} (between 700 m and 900 m depth) results, according to Wüst (1935), from admixtures of Antarctic Intermediate Water (AIW). Farther north the S_{min} weakens, and it becomes more and more difficult to associate it with influences of the AIW. A conspicuous northward orientation of the isolines within the S_{min}-layer offshore the slope, however, suggests a northward transport of AIW concentrating as boundary phenomenon along the slope (Le Floch, 1973, 1974; Fiuza and Halpern, 1982). Great parts of this water mass come with the tropical eastward flows to the eastern margin where it contributes to the deep poleward flow at depths between 700 m and 900 m.

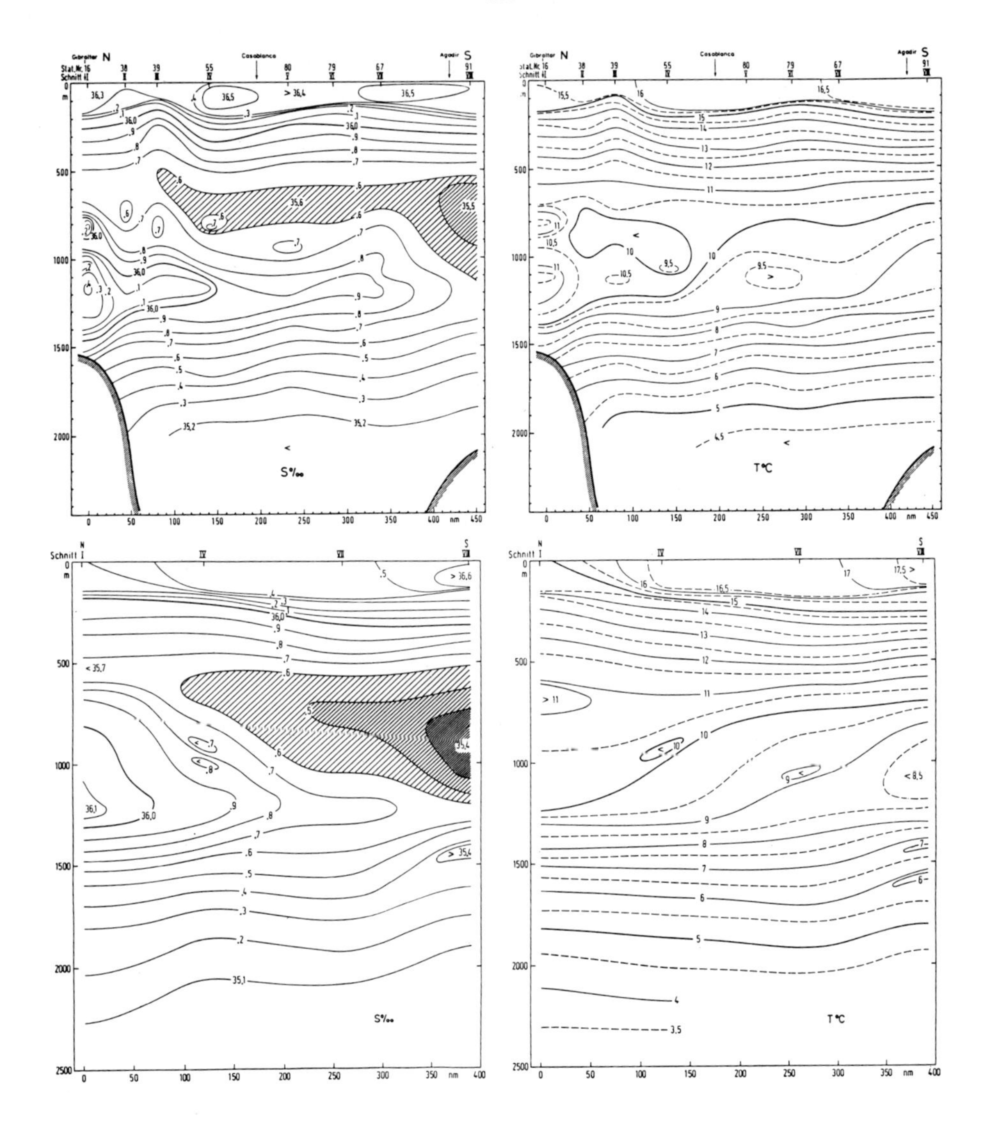

Figure 5: Meridional sections of salinity and temperature between 36°N and 30°N from February 1973 along the continental slope (above) and about 250 nautical miles offshore (Huber, Mittelstaedt, and Weichart, 1977). The salinity minimum is hatched.

Le Floch (1973, 1974) takes the view that weak admixtures of AIW are still perceivable between the Canary Islands and the African continent at about 29°N. Further north he finds that the influences of AIW disappear due to mixing with adjacent water masses. Off Mauritania, between 19°N and 22°N, Weichart (1974) finds at depths between 600 m and 1000 m maximum values of phosphate and nitrate in spring 1968. The values he gives are indicative for AIW as can be seen, for example, from the GEOSECS section (Bainbridge, 1983).

The poleward-going flow along the slope also emerges in the vertical sections from the old METEOR cruise between 14°N and 19°N (in February/ March 1927). Wüst (1957) has used the hydrographic data set from this cruise for the representation of the geostrophical currents. His results indicate that the poleward flow at these latitudes concentrates in the subsurface and surface layer, but not at the depths of the AIW. The reason might be his choice of the reference level, which coincides with the approximate depth of the AIW.

Direct current measurements in the framework of upwelling studies, during winter and early spring at latitudes between 16°N and 22°N, show a concentration of a narrow poleward undercurrent within the upper 500 m along the slope (e.g., Mittelstaedt, 1976, 1983) which may also surface. The undercurrent carries substantial admixtures of South Atlantic Central Water (SACW) toward the north, and represents the source for the upwelling water along the coast at these latitudes. At subsurface depths, SACW can be traced along the slope as far north as 26°N by its temperature/salinity relationship (Hughes and Barton, 1974; Tomczak and Hughes, 1980). Parts of the SACW come from the Gulf of Guinea, and other parts come with the tropical eastward flows as the AIW.

At 18°N, current data measured over periods of about 3 to 4 weeks offshore the continental slope clearly indicate a northward flow from the near surface layer down below 1000 m during February 1977 (Mittelstaedt and Hamann, 1981). Off Cap Bojador (26°N) Johnson et al. (1975) observed a weak northward-going flow below a depth of 400 m in summer 1972. Roemmich and Wunsch (1985) found clear evidence of a weak poleward-going flow along the slope within the layer of 400 m to 1000 m between 24°N and 28°N. This flow appears to be a broad boundary feature of a zonal extension of 1000 km. Within the layer of northward flow, below 700 m depth, oxygen is minimum and the phosphate content is maximum. In the south, the nutrient-rich water masses occur at about the same depths and are associated to AIW, while the oxygen minimum is to be found in the domain of SACW at subsurface depths within the upper 500 m (Wattenberg, 1939; Weichart, 1974; Kremling et al., 1984; Lenz et al., 1985).

Although the temperature/salinity characteristics at 32°N are not suitable to explain a deep water transport to the north, it might well be that there is still a persistent northward-going deep flow. The short-term measurements at 32°N described above seem to support this idea.

From the distribution of certain species of mesopelagic zooplankton samples during the same METEOR cruise, Weikert (1986) concludes that these species come with the poleward-going undercurrent all along the northwest

African continental slope from low latitudes up to the southwest European slope off Portugal.

The upper boundary of the flow component toward the north seems to be at depths around 500 m at 32°N. Assuming an upwelling depth of 300 to 400 m, this flow is probably no significant source for the water masses ascending up the continental slope with the wind-induced upwelling circulation.

But there might be processes such as strong internal tides or edge waves along the slope contributing to a turbulent vertical transport of deep water into shallow upwelling layers along the continental slope.

To our knowledge, there is no oceanographic evidence of the deep flow north of 32°N. Whether there is a continuous transition between the deep poleward-going flow along the African continental slope and the poleward-going flow, which occurs along the Iberian slope (see, for example, Ambar, 1982, 1985) is uncertain. The zooplankton samples by Weikert (1986) could be an indication of such a transition.

CONCLUSION

The data suggest a deep poleward-going flow within the layer of 500 m to 1000 m along the northwest African continental slope from the tropical North Atlantic to higher latitudes off Morocco. At its lower boundary this flow carries AIW toward the north along the slope. North of the Canary Islands, around 29°N, the identity of AIW seems to disappear due to mixing. But the undercurrent continues to flow toward the north at these depths.

South of 22°N the poleward-going undercurrent extends from below 1000 m depth up into the near-surface or even into the surface layer throughout the year. The upper part of the flow represents a narrow current band along the continental slope with a subsurface mean velocity maximum of 5 to 15 cm s^{-1} during the upwelling season in winter and spring. This "upwelling"-undercurrent transports substantial admixtures of SACW toward the north, which represents the major source for the upwelling water masses at lower latitudes.

On its way toward the north, the upper "floors" of the poleward-going undercurrent hit the Canary Current flowing toward the southwest at latitudes between 21°N and 24°N. This convergence produces the frontal zone between NACW and SACW. The Canary Current forces the upper poleward-

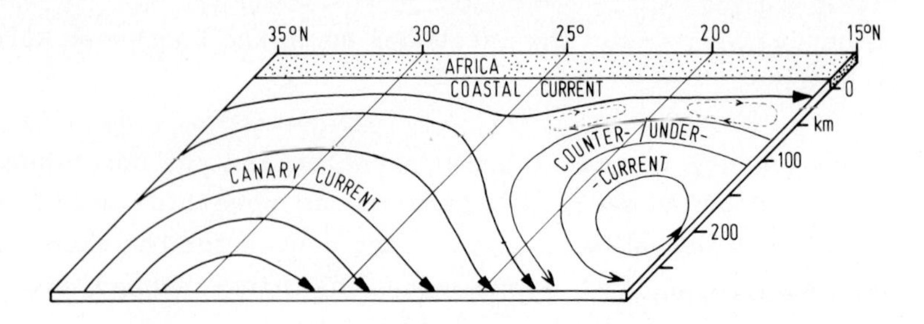

Figure 6: Schematic representation of the dominant near-surface circulation within the upper 200 m during winter and spring.

going undercurrent to turn into its own direction parallel to the frontal zone toward the southwest.

This situation is schematically represented for the winter and spring seasons in Figure 6. The circulation pattern in this figure is to describe the predominating flows within the upper 200 m. Apart from the southward flowing coastal current, this circulation scheme agrees with the representation of the general geostrophical flow (Stramma, 1984 a, b) in the eastern subtropical North Atlantic. The coarse data grid of 3° x 3° used by Stramma does not allow resolution of the narrow coastal current.

The deeper part of the undercurrent continues its way toward the north underneath the Canary Current within the layer of 400 to 1000 m depth. The small proportions of SACW which pass underneath the frontal zone at depth with the undercurrent disappear at higher latitudes around 26°N because of mixing with NACW. Due to its depth the poleward-going flow is not a major source for the upwelling waters within the surface layer off Morocco.

The question whether this flow is linked to the poleward-going subsurface flow along the Iberian Peninsula is still open and to be investigated.

ACKNOWLEDGEMENTS

The data collection off Morocco during METEOR Cruise No. 64 has been supported by the Deutsche Forschungsgemeinschaft.

From: Lothar Stromma, Inst. fur Meereskunde, Kiel University

On: Review and Commentary to paper **THE SUBSURFACE CIRCULATION ALONG THE MOROCCAN SLOPE**, by E. Mittelstaedt

Today there exists only a little knowledge regarding the oceanographic situation of the North Atlantic near Africa. This paper adds some nice additional information on the currents near the African coast. Of course we have to keep in mind that the main information is taken from some short-term observations and can't be regarded as the typical permanent flow field.

In section 2.2 the wind observations for January should also be presented for comparison with the April data. These data should be available from the ship measurements.

The schematic representation of the dominant near-surface circulation (Figure 6) is o.k., but I wonder whether it is not possible to give a schematic presentation of the subsurface circulation, which is of more interest here.

Although there is not much new information on upwelling in the ocean, this paper gives an example of how to interpret upwelling features from single sections. I also like the additional information on biological hints for the currents, which are often forgotten in oceanographic descriptions. Finally, the main point of this paper, the subsurface circulation near Africa, gives some new information on the currents in the area investigated.

POLEWARD MOTION IN THE BENGUELA AREA

G. Nelson
Sea Fisheries Research Institute,
Private Bag X2, Roggebaai 8012
South Africa

ABSTRACT

Available evidence of poleward and equatorward flow in the Benguela area is summarized and new evidence of poleward flow from direct current measurements between Cape Point (35.30°S) and Chamais Bay (27.80°S on the west coast of southern Africa is presented. Records were obtained across the shelf outwards to a depth of 3000 m. Except in the upper layers and at the shelf edge where equatorward baroclinic jets occur, these records invariably show time-averaged poleward flow along and at the base of the shelf and are indicative of cyclonic flow in the Cape Basin.

Closer to the coast, low frequency motions with periods of roughly three days display the characteristics of coastal-trapped waves. When averaged over several cycles, an ambient poleward flow of approximately 5 km d^{-1} is found for this inner shelf zone.

INTRODUCTION

In this article, poleward flow in the southern Benguela region is investigated in the broad sense of motion over the whole shelf and slope zones of southwestern Africa and over the Cape Basin. In a more restricted sense, various authors, for example Hart and Currie (1960) proposed a poleward undercurrent along the shelf edge which would compensate water removed from the 200 to 300 m level by perennial upwelling. Evidence of such a current was provided by the advection of low oxygen water as far south as the Orange River from Angola.

On a wider scale, dynamic topography (e.g., Defant, 1941) showed a substantial poleward movement of water below 2000 m in the southeastern Cape Basin. Further evidence for cyclonic motion over the Cape Basin has been obtained from marine geological surveys by Connary (1972), Embley and Morley (1980), Tucholke and Embley (1984), and Rogers (1987).

Since 1980, deployments of current meters by the Sea Fisheries Research Institute in deeper water over the shelf and continental rise of the southern Benguela region have revealed direct evidence of poleward flow. The zone of investigation is shown in Figure 1 together with geographical details, bottom topography and some of the deployment sites (see also Figure 6).

The existing data set does not allow seasonal features to be distinguished. The aim of current meter deployments was rather to identify permanent large-scale features such as bottom circulation and flow along the base of the shelf, or at the other extreme, to examine transient but repetitive phenomena such as coastal-trapped waves on time scales of a few days. The seasonal behavior of wind stress in the Benguela area has been discussed by Nelson and Hutchings (1983), Parrish et. al. (1983) and Kamstra (1985). The most significant feature is the presence of perennial southerly winds which are modulated by a three-day synoptic pressure wave, strongly so at Cape Point in the extreme south (34°S) and more weakly so as one progresses north to Möwe Point (approx 20°S). Seasonal changes in the ambient southerly wind occur as the permanent South Atlantic High pressure cell shifts or strengthens and weakens. Thus a seasonal change in Ekman transport occurs and consequently a seasonal change in a poleward compensation current would occur.

The use of current meters has revealed poleward motion near the bottom, extending to an as yet unknown height above the sea floor, but the interface between the deeper poleward moving water and the upper equatorward moving water is known to broach the surface near the coast. This has been shown both in drogue motions noted by Lamberth and Nelson (1987) and by current profiling studies reported by Nelson (1985). At one site west of Cape Columbine, the only one probed with a multiple flight of current meters, the interface showed fluctuations in height, as evidenced by low frequency switching from poleward to equatorward flow in the upper current meters, with periods of weeks.

Some details from current meter records are presented below, but first it is useful to distinguish the general physical characteristics which could influence flow over the Cape Basin and shelf zone of the Benguela area of southern Africa.

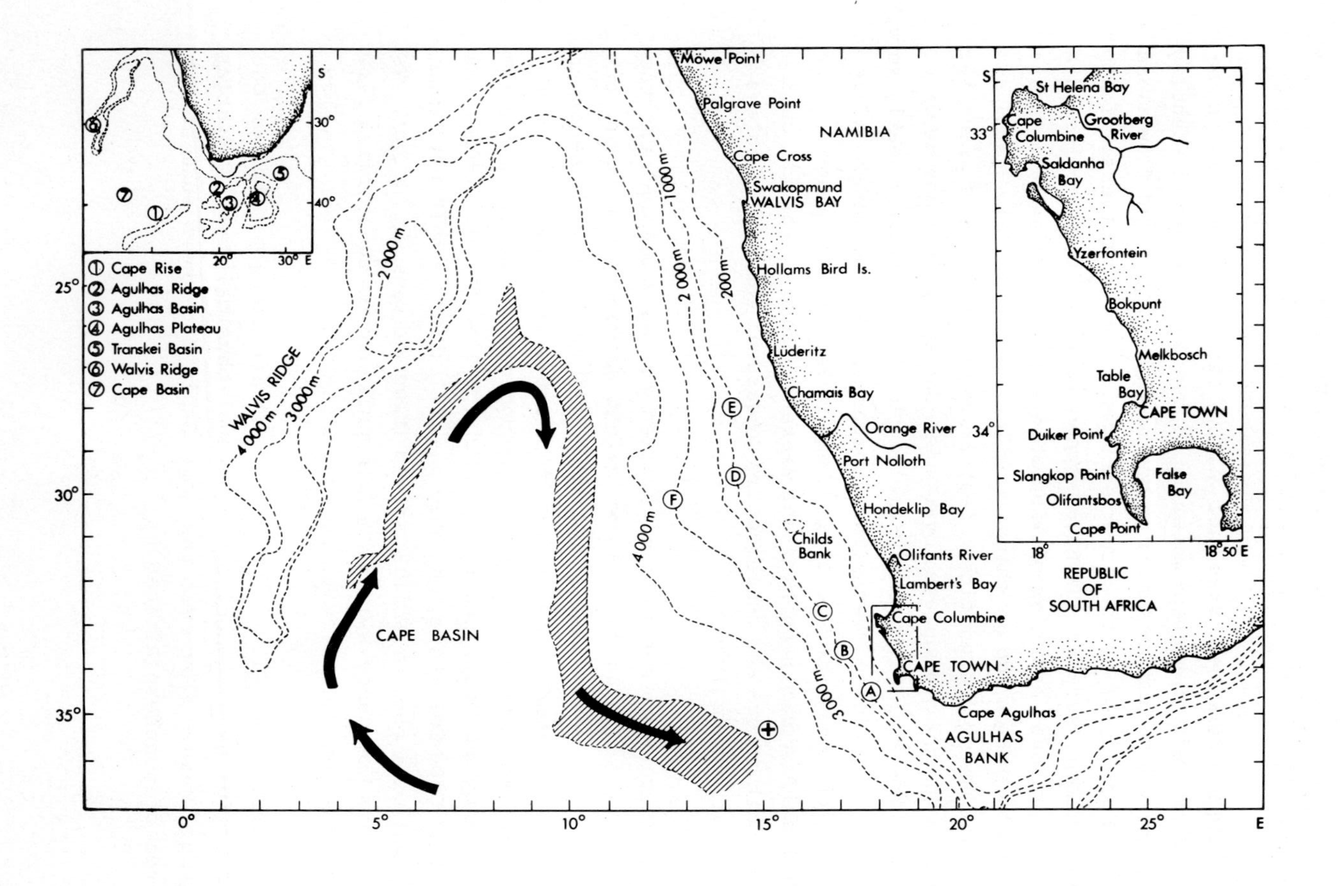

Figure 1: Geographical details, bathymetry and sites of current meter moorings (labeled A to F). The circled cross marks the site of the spot current measurement by Connary (1972) and the hatched area is the erosion zone described by Tucholke and Embley (1984). Arrows indicate the direction of bottom flow measured or inferred from various geological surveys described in the text.

The gross features of circulation based on recent current meter deployments appear to be as shown schematically in Figure 2. Although the bottom layers seem to move polewards as a uniform body of water, the physical causes of this motion are likely to be quite separate in separate domains and must be distinguished. There are numerous factors which could influence currents, some of which are as follows:

1. Density driven bottom circulation in the Cape Basin which tends to carry with it water above and water along the continental rise to the shelf-edge. The flow is cyclonic.

2. Baroclinicity over the shelf break and along the edges of terraces. An uplift in isopycnals creates equatorward baroclinic jets in these zones.

3. Net poleward motion of water over the inner shelf-zone, modulated by coastal-trapped waves. Current meter records show clear signs of wave-like currents with periodicities of three to five days and net poleward transport over periods of weeks.

4. A surface wind-driven equatorward current over the open ocean and shelf region with, additionally, a deeper geostrophic equatorward flow in the deep ocean.

5. Budgetary compensation for wind-driven Ekman divergence, presumably in the form of a poleward undercurrent along the shelf as in the classic model of Hart and Currie (1960).

6. Momentum advected onto the shelf from the deep ocean arising through changes in wind stress which generate local currents along the shelf edge, or through eddies moving against the shelf edge from the deep ocean.

Each of these will now be discussed with emphasis on items 1 and 3 and brief commentary on the others.

THE BOTTOM CIRCULATION IN THE CAPE BASIN

Of particular significance is the cyclonic motion indicated by the 2000 m relative topography of Defant (1941), based on the METEOR expedition of 1925-27 in the South Atlantic Ocean. A southeastward trend is apparent at the eastern boundary of the Cape Basin along the shelf edge and continental rise. Connary (1972) reported a southeastward set of

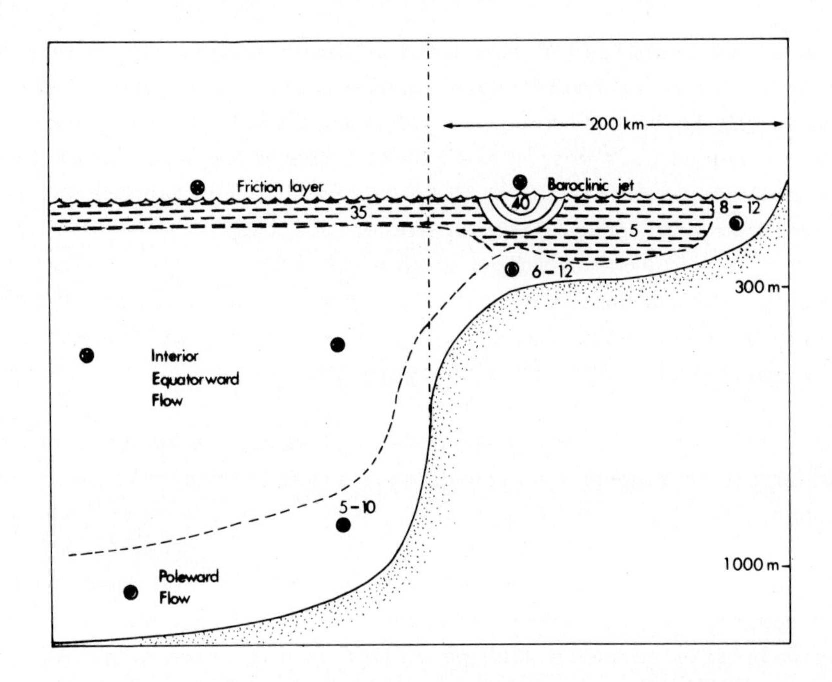

Figure 2: Schematic of poleward and equatorward flow over the shelf and eastern Cape Basin with typical current speeds (cm s^{-1}).

15 cm s^{-1} on the sea floor in this region from a spot reading instrument (see Figure 1). One should bear in mind, however, that inertial signals are strong near the bottom in this region and spot measurements of this nature could be quite misleading.

More recently, Embley and Morely (1980) have inferred cyclonic motion from bottom potential temperature. Tucholke and Embley (1984) show a band of eroded basement rock (Figure 1) claimed to be the result of scouring by strong poleward bottom currents along the eastern rise of the Cape Basin, while Rogers (1987) reported extensive erosion in a moat at about 4800 m along the northern edge of the Agulhas Ridge. Photographs of the moat show well-developed ripples. The asymmetry of the crests on these ripples indicates a westward current in excess of 20 cm s^{-1}. A camera-grab station in 4749 m of water at about 38°20'S, 17°E revealed southeastward contour currents, indicated by bending of seapens (Penatulaciae).

There seems to be little doubt that cyclonic motion occurs in the Cape Basin. Bottom currents would be deflected to the left by the Coriolis force, but contained by the topography of the Walvis Ridge to the west and north, and the continental rise to the east (Figure 1). In

Figure 3, the barrier effect of the Walvis Ridge on potential temperature is seen. Antarctic Bottom Water flows northwards from the Weddell Sea area into the basins around South Africa through fracture zone conduits (Tucholke and Embley, 1984). The Walvis Ridge, which runs obliquely from 35°S, 10°W to the African coast at about 18°S, 12°E forms an effective barrier to equatorward flow of deep water, which is consequently diverted along the continental rise, exiting the Cape Basin in the scour zone east of the Cape Rise.

Several water types are found in the Cape Basin and along the continental margin. (See Shannon, 1985, for a detailed review). South Atlantic Central Water (SACW) occurs to a depth of 600 to 800 m. This body of water is characterized by a line in TS space connecting the points (6°C; 34.5 ppt) and (16°C; 35.5 ppt) and is found along the entire continental shelf from Cape Point to Möwe Point. It is formed in the Subtropical Convergence zone by sinking and equatorward spreading into the Atlantic Ocean.

Below the SACW, Antarctic Intermediate Water (AIW) is found with a salinity minimum of 33.8 ppt at 2.2°C marking its core. With distance from the source, the salinity minimum rises to between 34.3 ppt and 34.5 ppt at temperatures of 4 to 5°C, so that a meridional change in the salinity minimum occurs.

The concept of a northward spreading of AIW along the base of the African continent which arises from TS analysis, is at variance with observed direct current measurements reported here which clearly show poleward movement. This is discussed further in Section 5.

Below 2000 m North Atlantic Deep Water and Antarctic Bottom Water are found. This is characterized by salinities lying between 34.7 ppt and 34.9 ppt at temperatures less than 3°C.

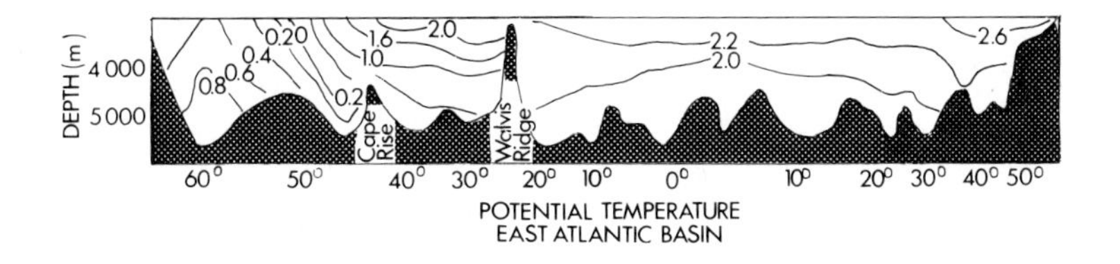

Figure 3: The barrier effect of the Walvis Ridge on potential temperature.

The recent direct current measurements in the eastern Cape Basin demonstrate that poleward flow occurs along the base of the continental shelf. Figure 1 shows the positions of five mooring sites, A to E along the upper slope in roughly 1000 m of water. Site F in the Cape Basin was at roughly 3000 m. Except at C and D which carried flights of current meters, these moorings consisted of a single current meter some 20 m off the sea floor with sufficient floatation to ensure that a tilt angle of 20° would not be exceeded in currents up to 50 cm s^{-1} .

Two types of current meter were used, Aanderaa RCM4 units, sampling over 30 or 60 minutes and Neil Brown ACM2 units vector averaging over 1, 2, or 6 minutes. The data was further vector averaged to produce hourly values.

The poleward motion along the eastern boundary is clearly seen in the form of progressive vector diagrams, some of which are shown in Figure 4. Because these records show a strong energy component at the inertial period, the data as presented here have been low-pass filtered and are displayed with supra-tidal frequencies removed. For this purpose, a Cosine-Lanczos filter with 121 weights and a half-power point at 46 hours was used. The parameters correspond to those of the Oregon State University filter as described in Woods Hole Technical Report 85-35 (1985), subsequently referred to as the OSU filter in this text.

The current at site E near Chamais Bay (Figure 4f) is seen to flow parallel to the topography over most of the period between 10 April - 5 November 1986 with two notable eastward excursions, labelled a and b in Figure 4f, that are suggestive of momentum advected onto the shelf by deep ocean disturbances. Such excursions are observed in most records, but are infrequent and not simultaneous from one site to another (See below).

Likewise, the current ran parallel to the topography at site A near Cape Point, (Figure 4a). What is clearly apparent in this record is that a periodic acceleration and deceleration of current occurred with reversal of direction, at times lasting a few days. Records from all sites exhibit this feature with periods ranging from 2.5 to 4 days. Figure 5 shows a segment of filtered time-series for mooring B in which the periodic nature of the longshelf current can be seen. In this case, 29 peaks or points of inflection can be identified over the 74 days of the record, giving a periodicity of 2.5 days. Spectra of longshelf velocity components show significant peaks between 2.5 and 3.5 days.

In Figures 4c and 4d, decreasing spatial coherence can be seen as the surface is approached. At 943 m the flow is marked simply by the

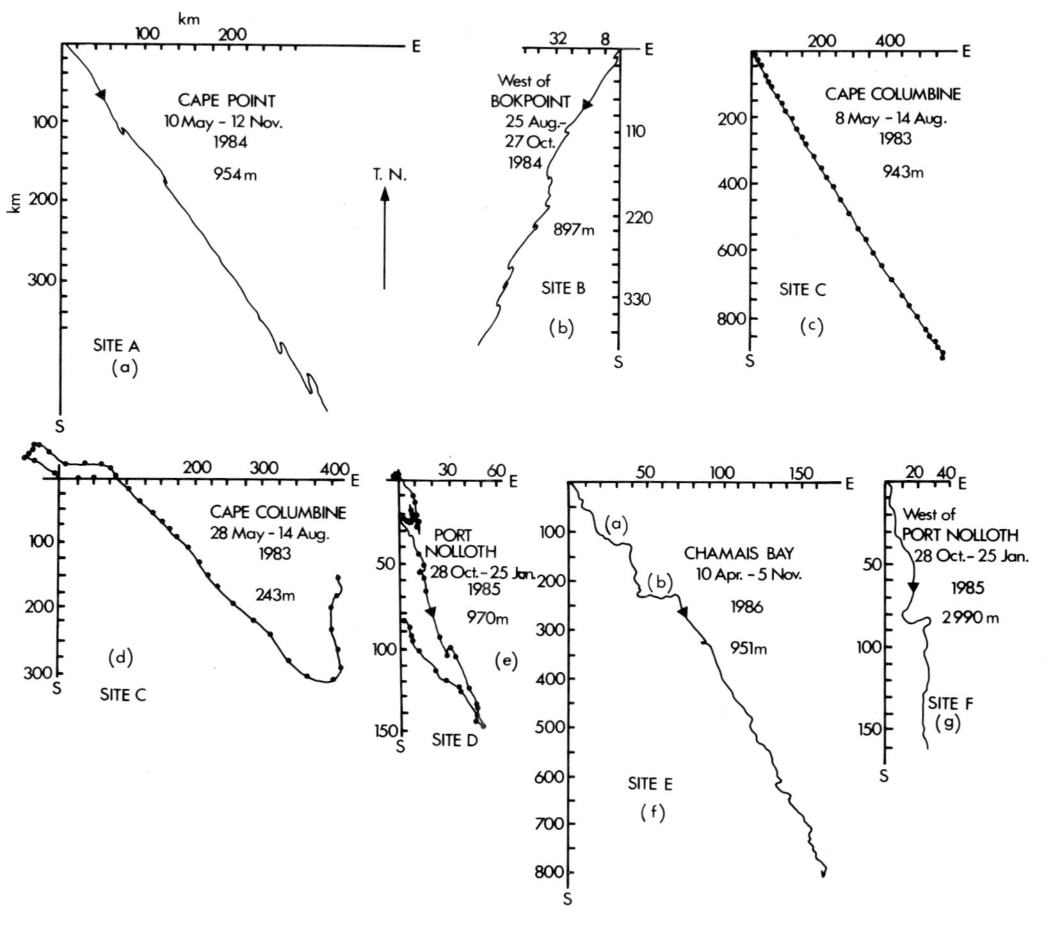

Figure 4: A selection of progressive vector displacement diagrams for sites A to F (Figure 1). Dots on (c) and (e) mark 3-day intervals and on (d) 2-day intervals. Time zero is at the origin.

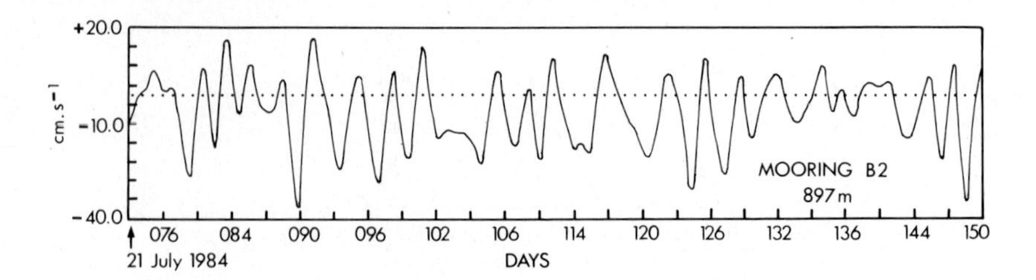

Figure 5: Segment of filtered current velocity component along the principal axis of flow (220°T) at site B (see Figure 1), showing the oscillatory nature of the flow.

characteristic periodic change in speed in the longshelf direction. At 243 m on the same mooring, two events in which reverse flow occurred over several days can be identified. Current meters at this site were deployed four times, and significant differences were noted in successive records. In particular, between May and August 1983, current speeds considerably above average were recorded.

The record from Site D near Port Nolloth shows several days of reverse flow with a slight off-shelf component. Events of this nature were seen to occur at other sites at other times, thus indicating that flow is not consistently poleward, although poleward when averaged over periods of the order of a few hundred days. Similar events were observed in records from Cape Columbine (Figure 4c is exceptional in its directional stability) and at Cape Point.

Site B lies in the Cape Canyon, a prominent incision cutting obliquely across the shelf west of Cape Columbine. The flow is observed to be uniformly downwards along the axis of the canyon, but exhibits the characteristic 3-day weakening, with slight reversal at times (Figure 4b).

That cyclonic motion occurs in the deeper water of the Cape Basin is supported by the progressive vector diagram from the record for Site F at 2990 m in 3020 m of water (Figure 4g). The flow ran parallel to the local bathymetry, with one significant eastward excursion about 40 days into the record.

Some statistical details giving vector-averaged velocities and the periodicity near 3 days for various sets of data are given in Table 1. The periodicities were obtained by counting turning points and points of inflection in OSU-filtered data and dividing by the length of record. The inclusion of points of inflection resulted in smaller variance between

Site and deployment period		Depth and sounding m		Poleward vector averaged speed		Periodicity days per peak d
				$cm\ s^{-1}$	$km\ d^{-1}$	
A	Cape Point May-Nov 84	954	983	4.3	3.7	2.5
B1	Bok Point Dec-Mar 84	1224	1253	1.6	1.4	2.1
B2	May-Oct 84	897	923	6.6	5.7	2.6
C1	Cape Columbine May-Aug 83	943	1023	16.1	13.9	3.0
C2	Sep-Nov 83	1061	1162	4.7	4.1	3.7
C3	Dec-May 83-84	889	980	5.1	4.4	3.1
C4	May-Aug 84	897	997	2.3	2.0	3.4
D1	Port Nolloth Dec-Mar 83-84	898	997	5.2	5.0	3.0
D2	Oct-Jan 84-85	970	1001	1.0	0.9	2.8
E	Chamis Bay Mar-Nov 86	951	1016	4.5	3.8	2.1
F	West of Port Nolloth Oct-Jan 84-85	2990	3001	2.1	1.8	3.4

Table 1: Statistics for poleward flow at base of the shelf edge, showing current speed vector averaged over several cycles of modulating waves and average period of waves (day).

records than was obtained from spectra in which the inflection points contributed little. Quite obviously, the acceleration of the fluid rather than its velocity is an indicator of external forcing.

BAROCLINICITY OVER THE SHELF BREAK AND TERRACE EDGES

Cross-shelf sections reveal an uplift of isopycnals at the shelf-edge or along the edge of terraces such as those west of the Cape Peninsula. Correspondingly, equatorward baroclinic jets form in these regions. Two examples of this have been discussed by Bang (1973) and Nelson (1985). The jets appear to be strongest where the topography is steepest, in particular near Cape Columbine where the bottom falls away sharply on the eastern wall of the Cape Canyon (Figure 6). Since the velocity in these jets along the shelf-edge is not uniform, local horizontal divergence must occur. This may in part account for the large offshore gyres sometimes seen on satellite images of the region west of the Cape Peninsula and particularly off Cape Columbine. Negative divergence here would result in an offshore movement of water.

The baroclinic jets off the Cape Peninsula and Cape Columbine have been studied using drogues. Notably, Harris and Shannon (1979) and Nelson and Hutchings (1983) reported accelerated motion past Cape Columbine, while Shelton and Hutchings (1982), Brown and Hutchings (1985) and Nelson (1985) have described various drogue trajectories in the vicinity of the shelf edge jets.

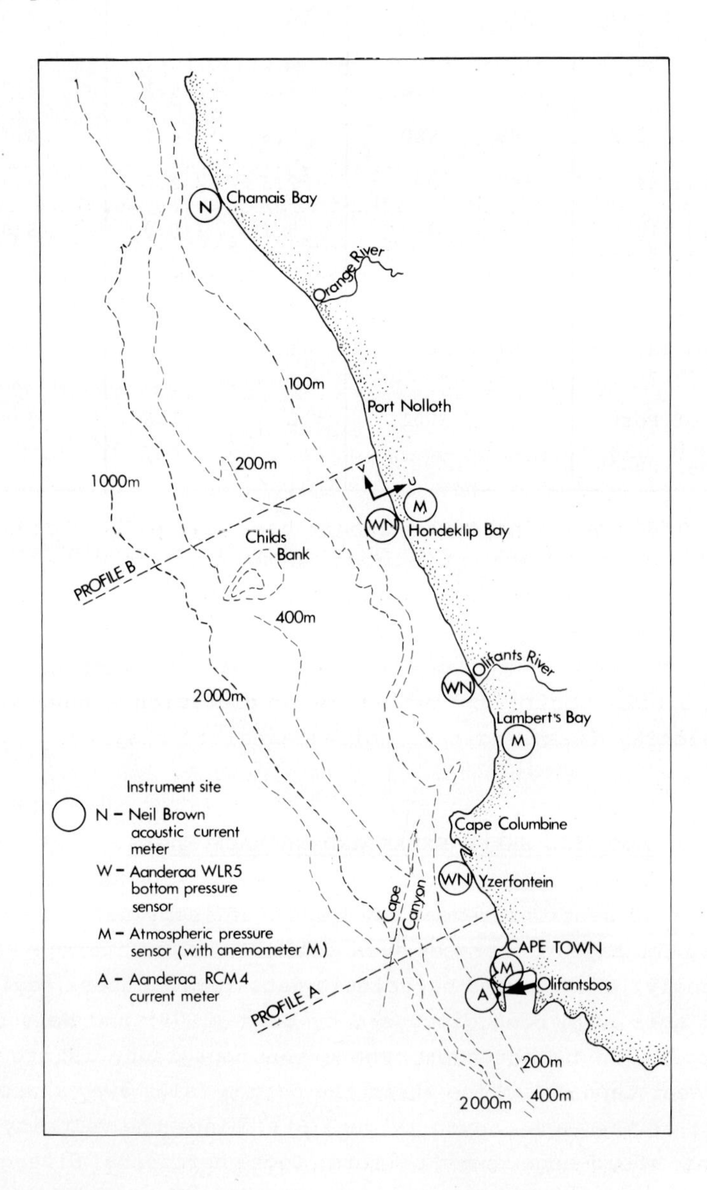

Figure 6: Geographical details for the coastal current and coastal-trapped wave experiments.

POLEWARD MOTION OF WATER OVER THE INNER SHELF

Several current meter moorings have been maintained on the shelf zone between Cape Point and Hondeklip Bay (Figure 6). Net poleward flow is observed at depths below 40 m, but the longshelf current exhibits wavelike motions with periodicities of approximately three days, similar to those of the deep shelf-edge flow. Close to the coast, poleward flow is detected in the near surface layers, and current profiling (Nelson, 1985) indicated the barotropic nature of these currents. Elsewhere, the surface layers are wind driven and show a general equatorward motion, but with poor spatial stability. A single record from Chamais Bay, north of the Orange River, showed net equatorward flow close to the coast.

A possible explanation for the observed flow over the inner shelf is that poleward-propagating coastal-trapped waves modulate an ambient poleward flow. Based on data from current meters near Cape Columbine, Holden (1987) asserted that forced modes with a periodicity of ten days occur in that area, but analysis of data from further afield, which is presented here now, indicates that free wave modes with shorter periodicities are more general for the west coast.

In an attempt to resolve such waves, three Neil Brown ACM2 current meters were deployed, together with Aanderaa WLR4, bottom pressure, and Aanderaa atmospheric pressure recorders in the vicinity. The water depth was 50 m and the current meters were approximately 25 m below the surface. The location of these moorings is shown in Figure 6. Data was obtained over the period December 1983 to January 1984.

As anticipated, the inverse barometer condition was not satisfied, and fairly large changes in bottom pressure were observed; the changes were equivalent to a variation of 20 cm in water level for a constant density water column. Figure 7a shows the OSU-filtered variation in bottom pressure in roughly 50 m of water at Hondeklip Bay and Yzerfontein. Figure 7b shows the Hondeklip Bay record with atmospheric pressure removed, and gives an indication of the magnitude of sea level variation. Large changes in sea-level such as this have been documented by Brundrit (1984) and Brundrit et. al. (1987). In these works, evidence of counter-clockwise movement of sea-level change is presented. Similar disturbances can be observed in the CODE data (Coastal Ocean Dynamics Experiment, Brown et. al., 1983) for bottom pressure. Interestingly, the periodicity for the CODE data and the Benguela data are quite different, the former being on the order of 7 days, and the latter 4 days.

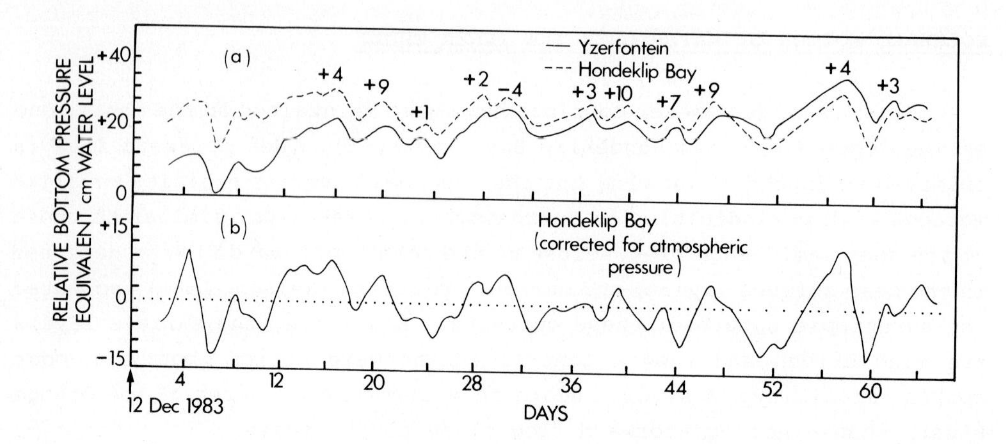

Figure 7: (a) Bottom pressure in equivalent relative cm water level variation at Yzerfontein and Hondeklip Bay showing peak-to-peak time difference in hours, Hondeklip Bay leading when positive. (b) The Hondeklip Bay record with atmospheric pressure removed showing the magnitude of sea level variation.

Changes in bottom pressure were nearly simultaneous between the northern Hondeklip Bay site and the southern Yzerfontein sites, a distance of 330 km. The peak-to-peak time difference in hours as the disturbance propagated, is shown in the annotation of Figure 7a with Hondeklip Bay leading when positive. These figures were obtained by comparing maxima in the OSU-filtered time series, as for the shelf edge data.

The range in sea level, seen here to be some 20 cm (Figure 7b), is larger than might be expected of barotropic coastal-trapped waves, as predicted by the Brink and Chapman (1985) model. Variations of a few centimeters associated with such waves would be masked by these large changes in sea level. In the classical manner, longshore wind stress would induce cross-shelf transport resulting in the generation of coastal waves (e.g., Gill and Schumann, 1974; Clarke, 1977).

Figure 8 shows low-pass OSU-filtered time-series of long-shelf currents at four nearshore sites. Three of these correspond to the synoptic shelf wave experiment done in December-January 1983-1984; net poleward transport is apparent in each of these cases, while at Chamais Bay equatorward flow was observed (Table 2). On this is superposed a fluctuating signal with an approximate three-day period, its amplitude varying from site to site.

The figure of three days is particularly significant as it is now known that three and six-day peaks occur in atmospheric pressure spectra

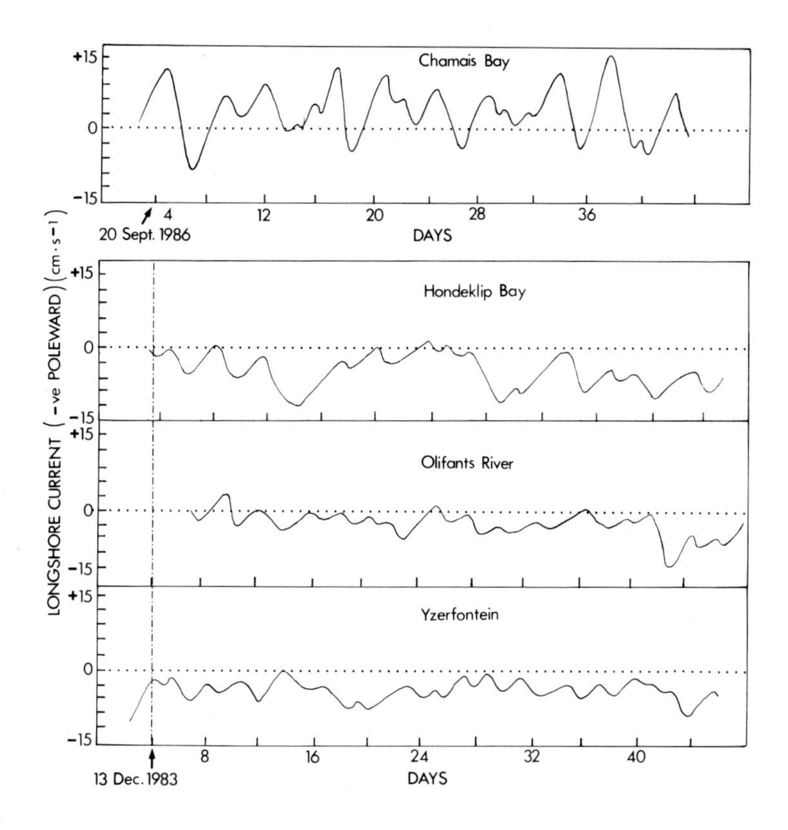

Figure 8: Filtered longshore current components along the principal axes of flow for Chamais Bay, Hondeklip Bay, Olifants River and Yzerfontein.

in the region around South Africa (Preston-Whyte and Tyson, 1974; Kamstra, 1987; see also Tyson, 1986 for more detailed references on this matter). Higher order baroclinic travelling Rossby waves in the westerlies periodically weaken the permanent high pressure cell adjacent to the west coast with a consequent periodic attenuation of ambient southeasterly winds. A short discussion of these waves is given by Tyson (1986).

Figure 9 shows segments of longshore wind time-series at Hondeklip Bay and Olifantsbos in which an approximate three-day periodicity is apparent, but with large variation in maximum velocity. The corresponding longshore currents are also shown. At times, there is a reasonable correlation between equatorward wind and poleward current, allowing for a lag of one inertial period. At other times the correlation is absent. This suggests that a mixture of free and locally forced longshore waves occurs, with local forcing and stronger poleward flow being more apparent in the narrow shelf region between Cape Columbine and Cape Point. However, one would expect the forced mode to correspond to poleward winds (Gill and Schumann, 1974) rather than equatorward winds as in the case here.

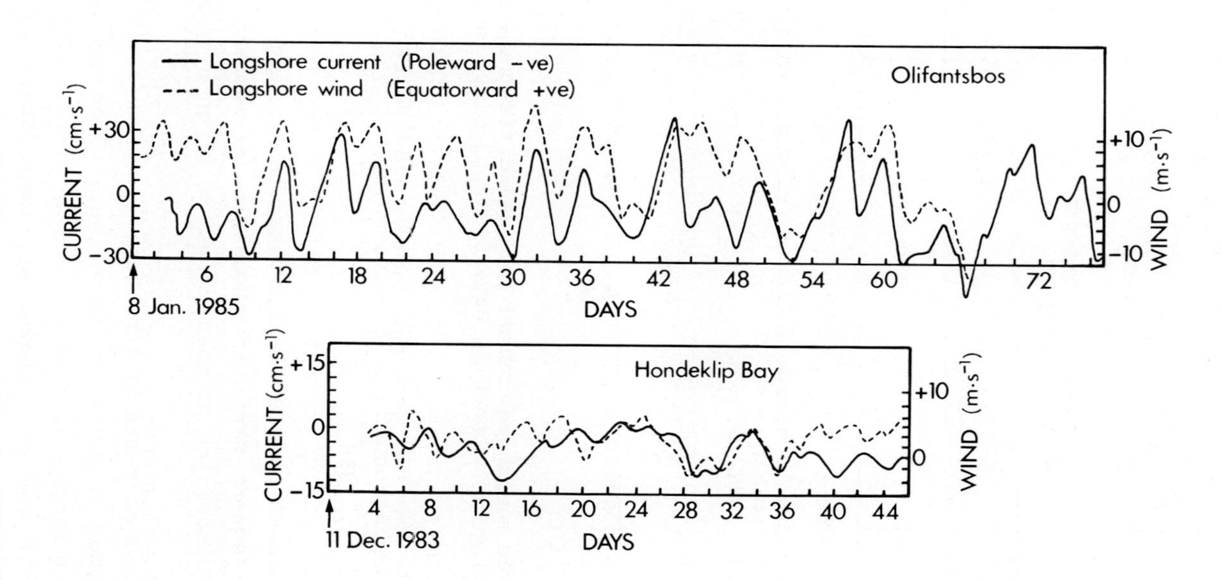

Figure 9: Filtered longshore current and wind components at (a) Olifantsbos and (b) Hondeklip Bay.

Site and deployment period (days)	Depth and sounding m		Vector averaged Poleward speed (km d^{-1})	Periodicity days per peak d
Chamais Bay Apr-Nov 86 (208)	44	66	-1.5	2.7
Hondeklip Bay Dec-Jan 83-84 (47)	27	43	3.8	2.8
Olifants River Dec-Jan 83-84 (45)	21	37	2.6	2.8
Yzerfontein Dec-Jan 83-84 (48)	25	40	4.2	2.5
Olifantsbos Jan-Feb 85 (140)	58	73	2.6	3.0

* Note weak equatorward flow at Chamais Bay.

Table 2: Statistic for inner-shelf flow showing poleward current speed (km day^{-1}) averaged over several periods of modulation by coastal waves and average period of waves (day). Note that the data is only partly synoptic.

Further, the response time of a few days is rather short for the generation of shelf waves.

Not only is the shelf geometry distinctly different in this region, but the wind forcing is more intense than it is further north. Steep atmospheric pressure gradients occur along the coast in the summer months, particularly along the Cape Peninsula (Nelson and Hutchings, 1983) giving rise to strongly modulated longshore wind stress with a three-day periodicity.

Using programs developed by Brink and Chapman (1985), dispersion relationships were calculated for free barotropic waves over two topographic sections labelled A and B in Figure 6. The results and sections are shown in Figures 10a and 10b. Notably, periods of three to five days are possible over a wide range of wave numbers corresponding to wavelengths of 200 to 500 km for the first mode. This lends support to the hypothesis of a poleward inner-shelf current modulated by coastal-trapped waves.

The different characteristics for the shelf zones north and south of Cape Columbine, in particular the stronger wind forcing in the south, may decouple these regions, producing intermittent horizontal divergence at Cape Columbine. This would provide a second possible explanation for the large gyral tongue just south of Cape Columbine, so characteristic

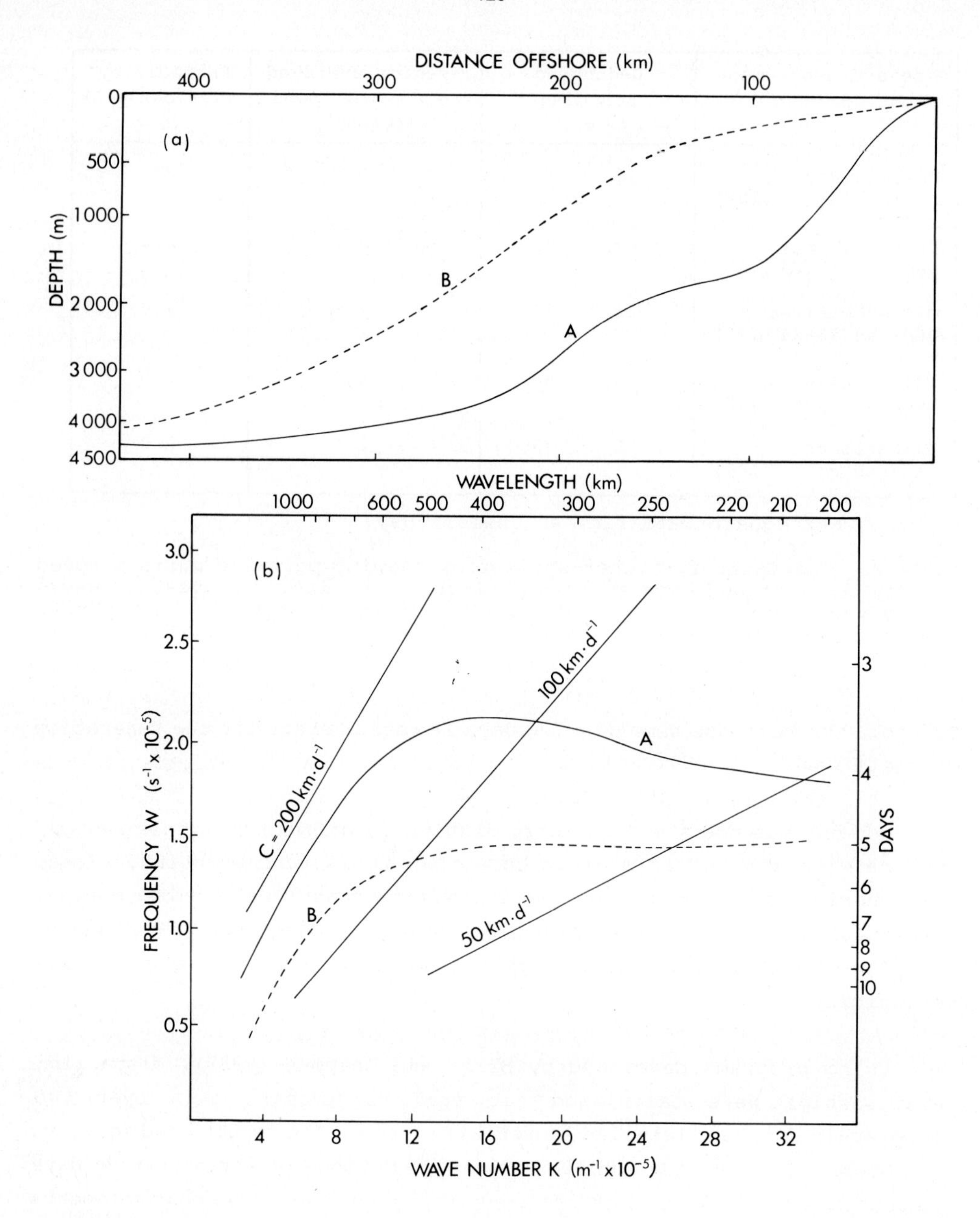

Figure 10: (a) Depth profiles for the sections A and B of Figure 6. (b) Dispersion relationships for profiles A and B for first mode free barotropic coastal-trapped waves as calculated using the Brink and Chapman (1985) routines.

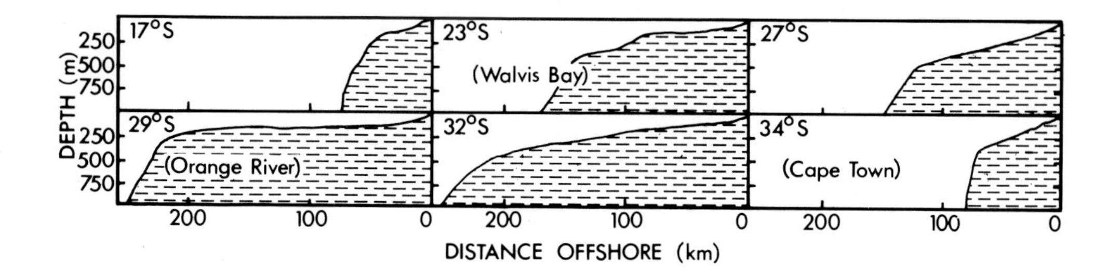

Figure 11: Zonal shelf profiles at different latitudes. (After Shannon, 1985)

of satellite images in both the visible and infra-red bands (See for example Shannon et. al., 1985).

Estimates of poleward flux are given by progressive vectors of low-pass OSU-filtered data over several wave periods. Some details are given in Table 2. Poleward flow was strong in the vicinity of the Yzerfontein mooring position, which lies on the narrow shelf above the eastern wall of the Cape Canyon. Comparing the synoptic data for the mooring near the Olifants River, where the shelf is wide, with the Yzerfontein data, one notes that the flow is weaker there. However, the width of the shelf does not appear to be a factor influencing the strength of flow, as the synoptic data for Hondeklip Bay, where the shelf is again wide, gives a figure comparable to that for Yzerfontein. The data in Table 2 is not all synoptic and further, the records are comparatively short so that considerable variation in seasonal and interannual estimates of this nature should be anticipated. A mooring placed at the same Yzerfontein site in December-January 1984 (Nelson and Polito, 1987) gave a figure of 6.2 km d^{-1} for poleward flow.

Thus, it appears that on the strength of available data, a figure of 4 to 5 km d^{-1} (4.6 to 5.7 cm s^{-1}) is a reasonable estimate for poleward flow in water 50 to 80 m deep along the coast. This is to be understood as averaged over several approximate three-day cycles of the wave-like motion observed in flow over the inner shelf zone.

The causes of the poleward flow along the inner shelf are unknown. The baroclinic structure with isopycnals lifting towards the coast from the mid-shelf region, favours equatorward flow, but both the surface Ekman and the bottom friction layers comprise a substantial fraction of the total water column, even in 100 m of water, so that geostrophy must play a small role in the inner shelf regions.

In view of the barotropic nature of currents reported by Nelson (1985) over the inner shelf, a permanent sea slope towards the equator is a possible cause of an ambient poleward current. Taking a value for the Coriolis parameter of $7.3 \times 10^{-5}\ s^{-1}$ which corresponds to latitude 30°, and an ambient flow of 5 cm s^{-1}, the gradient equation yields a slope of 3.7 cm per 100 km. There are two possible mechanisms by which such a sea slope may be maintained. One is the perennial frictional drag of southeasterly winds. Because of the orientation of winds and coastline, the Ekman transport has a small equatorward component. The other is rectification of tidal flow onto the shelf by bottom friction which would be more pronounced where the shelf is widest between a point south of Luderitz and Cape Columbine.

EQUATORWARD CURRENT

Current meters moored less than 40 m below the surface in the mid-shelf zone show equatorward motion, but with weak directional stability. Average speeds of less than 8 cm s^{-1} are characteristic of most of the mid-shelf zone north of Cape Columbine. These records are indicative of a high eddy kinetic energy density and are consistent with the convoluted patterns observed in satellite images of the west coast.

In the deeper oceanic water, equatorward geostrophic flow occurs to a depth of a few hundred meters (Defant, 1941).

The surface water over the whole southeastern Cape Basin is influenced by the permanent South Atlantic High (SAH) pressure cell and its associated southeast trade winds and thus, seasonal variations in flow can be expected.

EKMAN DIVERGENCE

The concept of poleward undercurrent along the shelf compensating and induced by Ekman divergence needs attention. In the winter months, the southeasterly wind field is disturbed by eastward propagating cyclones, but as the summer progresses, a more stable longshore wind field develops, with upwelling occurring over the whole shelf zone at discrete sites. De Decker (1970) described the poleward progression of a low oxygen tongue as the summer progressed. However, oxygen is non-conservative, particularly in the biologically and chemically active waters of the mid and inshore shelf zones in the Benguela region, and care must be exercised in using oxygen tongues to infer currents. This matter has been discussed in more detail by Chapman and Shannon (1985).

Furthermore, direct current measurements in the winter months (e.g., Nelson and Polito, 1987, their mooring H) indicate that poleward flow over the mid and inner shelf is perennial, so that a compensation current alone would not account for the observed poleward motion. One would expect the compensation current to be non-existent or considerably weaker in the winter when the wind is less favourable for continuous upwelling.

Because of the discrete zones of upwelling within the Benguela system, demands on the compensation current would be latitudinally variable. Kamstra (1985) gives figures for averaged summer wind speeds along the west coast and demarcates zones of cyclonic wind curl. The essential feature of this work is a lobe of lower wind speed north of Cape Columbine to the Olifants River, the implication being that Ekman divergence would be greater north and south of this lobe. This is suggested also in the work of Parrish et. al. (1983).

The region near Cape Columbine should thus be a source for upwelling water, with local demands on compensation current which need not necessarily be continuous along the shelf. The possibility of cross-shelf compensation from the narrow ledge at the head of the Cape Canyon should be considered.

The appearance of Antarctic Intermediate Water (AIW) below 600 m west of Cape Columbine is curious. Since the current meter records demonstrate a net poleward flow in this vicinity, the AIW cannot move in from the south and must therefore move slowly eastwards from mid-depths in the Cape Basin. A detailed TS analysis of waters in the area by Shannon and Hunter (1988), does indeed show a tongue of relatively fresh AIW touching on the continental shelf at just this point. The eastward movement could be a response to the strong upwelling in the Cape Columbine area.

ADVECTED MOMENTUM

Nothing is yet known about the scales of eddies in the Cape Basin or what effect they might have when impinging on the shelf edge. Chapman and Brink (1987), in considering the influence of the deep ocean on shelf circulation have shown that forcing in the longshore direction, which is periodic in space and time, can result in resonance with free coastal-trapped waves when the period is less than ten days.

Reference has been made above to the two eastward excursions in the progressive vector diagram obtained from data at Site E (Figures 1 and 4f). The nature of these events is unknown. They may be associated with low frequency changes in wind stress over the Cape Basin. Such excursions are common in many current meter records over the whole area explored thus far. Their frequency is on the order of 50 days and there appears to be no correlation between sites so that their spatial scales, taking into consideration their duration, would be on the order of 40 to 100 km.

SUMMARY

In conclusion, the salient features of the work on currents in the Southern Benguela zone as revealed to date are as follows:

1. Deep cyclonic motion occurs in the Cape Basin with a typical speed of 5 to 10 cm s^{-1}.

2. Wave-like currents with roughly three-day periodicities occur on the inner shelf with net poleward transport resulting from an ambient poleward current. Speeds as high as 30 cm s^{-1} are attained, with net mean flow values of typically 5 km d^{-1} (5.7 cm s^{-1}) averaged over several cycles.

3. Baroclinic jets occur along the shelf break and terrace edges with typical equatorward velocities of 30 to 60 cm s^{-1}.

4. Equatorward flow occurs in the surface layers to a depth of some 40 m over the shelf away from the coast and to depths of several hundred to thousands of meters over the deep ocean. In the surface friction layer over the open ocean, drift is typically 20 to 35 cm s^{-1}.

ACKNOWLEDGEMENTS

I am grateful to Dr. Eckart Schumann of the University of Port Elizabeth for his constructive comments in the preparation of this manuscript, to Dr. Ken Brink for suggestions and review, and to the technical staff of the Sea Fisheries Research Institute for assistance in the deployment of instrumentation and routine data processing.

THE PACIFIC OCEAN

OBSERVATIONS OF THE LOW-FREQUENCY CIRCULATION OFF THE WEST COAST OF BRITISH COLUMBIA, CANADA

Howard J. Freeland
Institute of Ocean Sciences
P.O. Box 6000
Sidney, B.C. V8L 4B2
Canada

INTRODUCTION

A variety of experiments conducted off the coast of British Columbia (BC) since 1979 have produced a general description of the circulation patterns. In this note I will present current meter data from several sources merged to form a coherent picture of the annual cycle of the current field. Figure 1 shows a diagram of the BC coast with the locations of current meter moorings indicated. Many more sites have been occupied; but to show the major features of the flow regime, I will limit discussion to two cross-shelf lines of moorings and an array deployed along the shelf edge. The two cross-shelf lines are near La Pérouse Bank, off southern Vancouver Island, and Estevan Point, halfway up the coast of the island.

Some aspects of these data bases have been discussed elsewhere, such as Freeland and Denman (1982) and Freeland et al. (1984). Data reports have been published on these and other data by Freeland (1983) and Thomson et al. (1985).

DATA REDUCTION

The stations indicated by squares on Figure 1 belonged to an experiment called CODE, Coastal-Ocean Dynamics Experiment, that extended for almost two years from 1979. The triangle locations belonged to SUPERCODE, which lasted for one and one-half years overlapping somewhat with CODE. One location was in common between the two experiments (the triangle on the shelf edge along the La Pérouse Line, we will refer to that mooring as LPB) and lasted for about three and one-half years. Initially, the time series were reduced to monthly averages for a cursory examination of the seasonal cycle; however, it immediately became obvious that the time series were very heavily dominated by the mean and annual cycle. Performing a least squares fit of a mean and 12-month cycle to the LPB data yields a simple description of that site. Subtracting the mean and annual cycle fit from the original data gives a measure of the goodness of fit. For LPB a mean and annual cycle reduces the variance of the time series by 77 percent, 69 percent, and 80 percent for time series at depths 50 m, 100 m and 150 m, respectively. (In terms of standard deviations that is 88 percent, 83 percent, and 89 percent, respectively.) This is an impressive data compression; so in the remainder of this note, the data will be represented in this form. Furthermore, for the 3 meters discussed, LPB at 50, 100, and 150 m, we find 96 percent, 98 percent, and 96 percent, respectively, of the total variance in the along-shore component. Hence, we will display only that part of the velocity field.

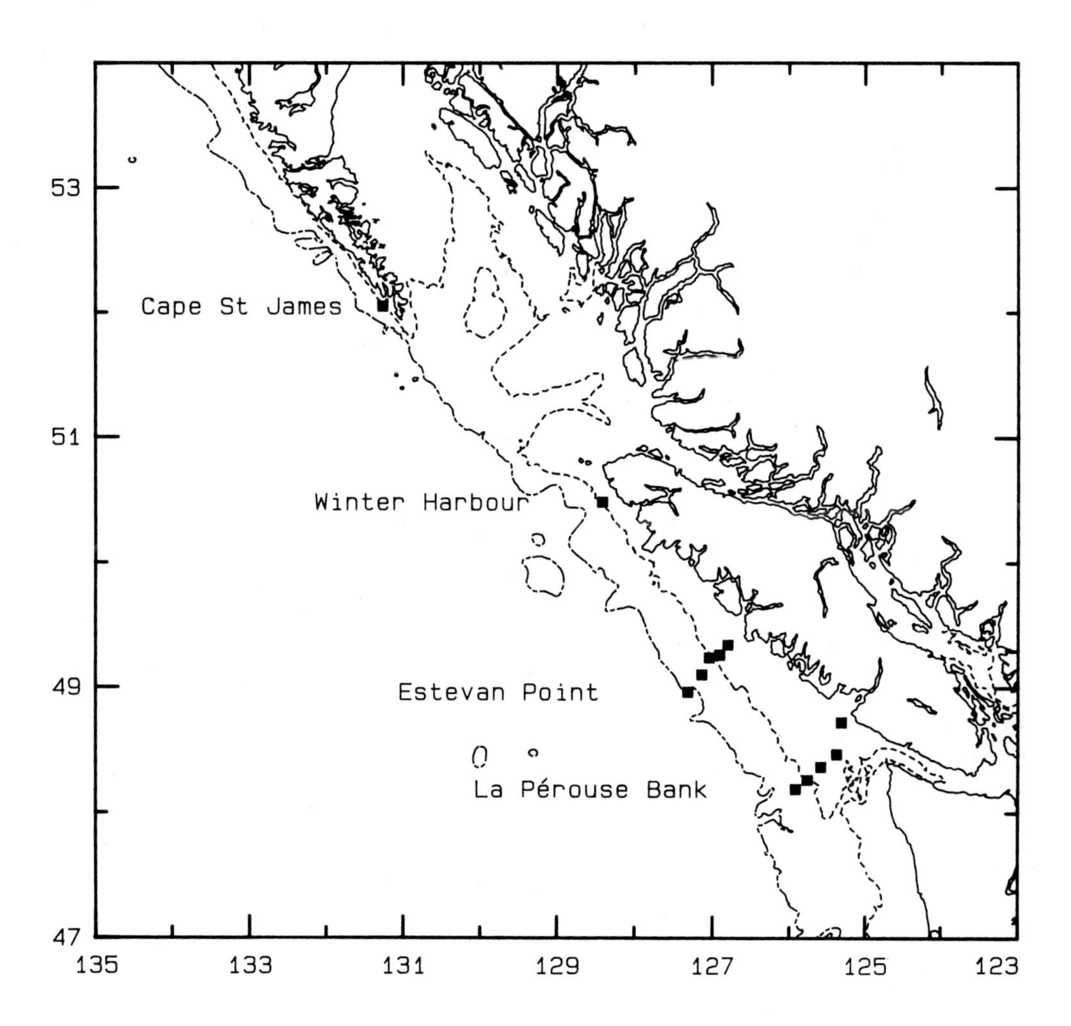

Figure 1: Map of the coast of British Columbia showing the locations of the current meter moorings used in this paper. The dotted and dashed lines indicate depths of 100 f and 1000 f, respectively.

LA PÉROUSE BANK CURRENTS

Figure 2 shows the annual cycle of currents along the La Pérouse line of current meters in the perspective plot. Each of the panels shows the distribution of currents for one month across the continental shelf and part way down the continental slope. For the January panel a scale of speed is indicated. The direction sense is such that flow to the northwest is represented by arrows to the left on each panel.

In winter, October through February, surface currents show an intense northward flow over the shelf edge and slope and intense northward flow very close in to shore. Generally current speeds decrease with depth. However, at the shelf edge mooring (fourth from the coast), we see a strong northward flow, at 150 m depth, near the bottom. In the case of the January flow regime, this appears to be the strongest flow anywhere in the water column. At the coastal site the vertical shear is particularly strong with a very large northward near-surface flow and an extremely weak (reversed but essentially zero) deep flow at 100 m.

March appears to be a period of rather unsettled flow; all vectors are small and close to zero except for the northward flows at the shelf edge, near-bottom, and coastal regime, near-surface. This is the time of the "spring transition."

In April we see the development of what appears to be the characteristic summer pattern of currents on the Vancouver Island shelf, a pattern that is fully developed in June. This pattern is characterized by southward flow over the outer continental shelf and slope except for the persistent deep northward flow at LPB at 150 m depth. In the inner shelf region, the surface flow remains strongly northward and the deep flow weak and indeterminate. Through the rest of the summer and fall a transition is gradually effected back to the winter conditions, the time of the fall transition being late September or early October.

In a region that at first sight seems to be dominated by seasonal variability, the most striking feature is the lack of variability at two sites, near surface, close inshore and the deep northward flow near the shelf edge. The former is a fresh water current named the Vancouver Island Coastal Current. This originates at the mouth of the Juan de Fuca Strait and is maintained by buoyancy sources distributed homogeneously along the coast of Vancouver Island. The second strong northward flow is of more direct interest to this conference, and appears to be a northern part of the California Undercurrent. The northward flow has temperature and salinity of 6.7°C and 33.9 ppt giving a sigma-t of 26.6. This is close

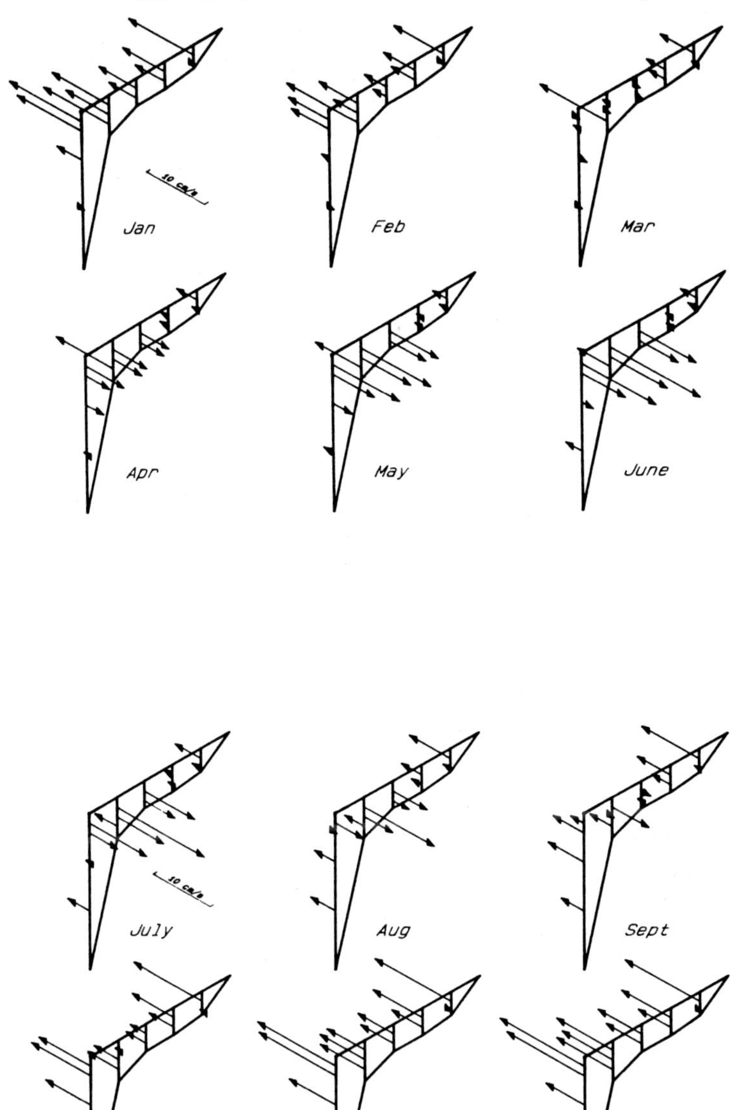

Figure 2: The annual cycle of along-shore currents off the La Pérouse Bank. The actual depths are proportional to the height of arrows on the individual panels. The top two arrows at each mooring site are at 50 m and 100 m depth; the water depth at each site from the coast is 120 m, 150 m, 140 m, 210 m, and 800 m, respectively.

to the standard values described by Hickey, 1979. The northward flow also has a high nutrient content and very low dissolved oxygen content, properties that are all standard for the California Undercurrent.

ESTEVAN POINT CURRENTS

Figure 3 shows a profile of currents for the Estevan Point line of moorings displayed in a form analogous to Figure 2. The shelf profile is a little different. The shelf is only about 40 km wide off Estevan Point, compared with 60 km near La Pérouse Bank. However, Figure 3 shows a profile of currents that is remarkably similar to the description of the previous section. Persistent northward flows are evident that can be identified as the Vancouver Island Coastal Current and the California Undercurrent. The shelf edge flows are otherwise northward in the winter and southward in the summer. During the winter all flows are northward, but a distinct mid-shelf minimum in flow speed is evident as it is on the La Pérouse section.

ALONG-SHORE STRUCTURE

Lest anyone should think that the circulation off the BC coast is simple and well determined, we present in Figure 4 a sketch of the annual cycle of currents at three depths at four sites distributed along the shelf edge. The edge of the continental shelf is, in this case, defined rather arbitrarily as the 210 meter depth contour. This contour lies about 60 km, 40 km, 20 km, and 4 km offshore from the coast at the 4 sites near La Pérouse Bank (LPB), Estevan Point (EP), Winter Harbour (WH) and Cape St. James (CSJ), respectively. In Figure 4 northward flow is denoted by leftward arrows, and the scale to the right of each month name indicates 10 cm/sec.

As indicated by earlier discussion, the currents at EP and LPB are very similar. Both show northward flow at all depths during the winter, with an indeterminate or transitionary behavior in March, followed by strong southward near-surface flow and a maintenance of a deep northward undercurrent year round. At WH, northern Vancouver Island, the pattern is similar but differences are appearing. The near-surface flow does become southward during the summer, but the period of southward flow is relatively brief, and the flow speeds relatively weak. At 150 m the differences are more striking; apparently a bias towards northward flow does exist but it is weak. The evidence for continuation of the California Undercurrent to the northern tip of Vancouver Island is marginal.

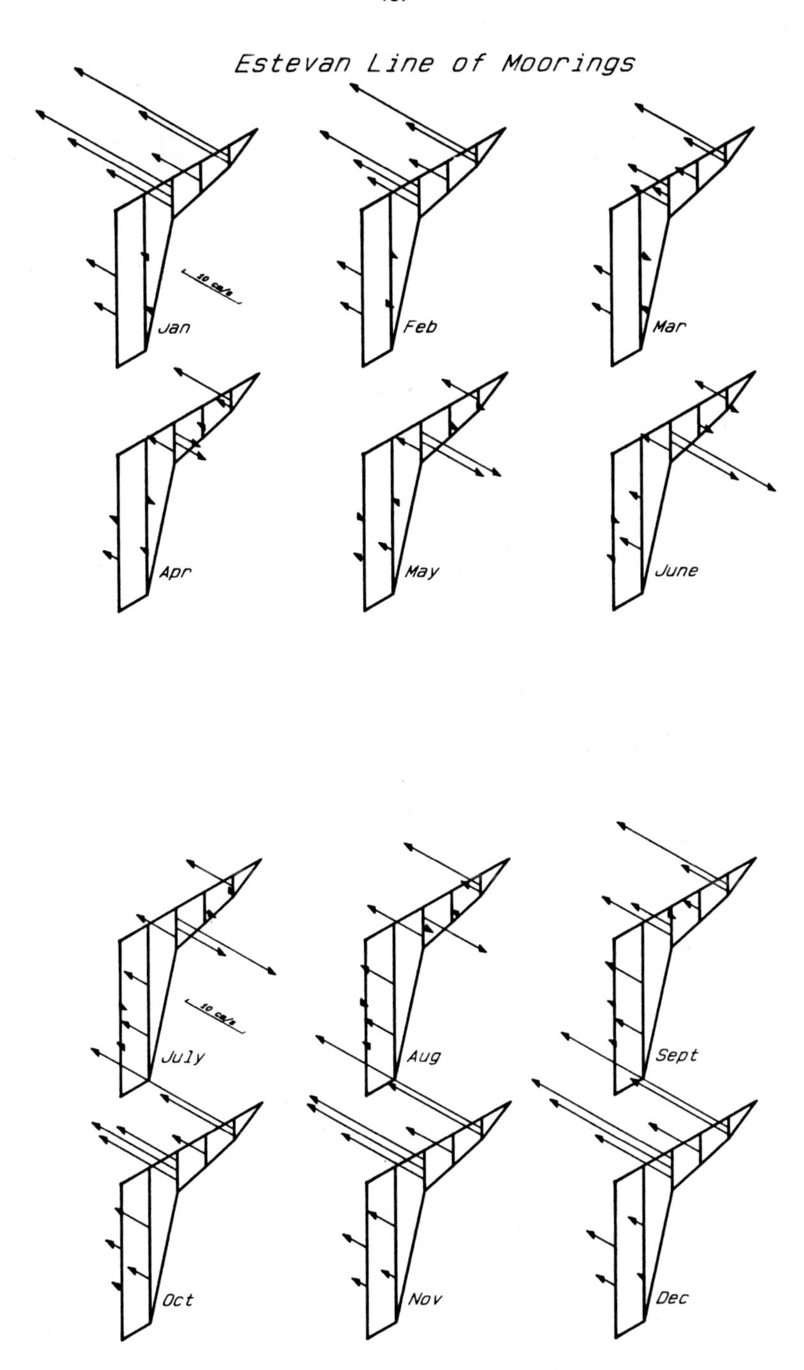

Figure 3: The annual cycle of along-shore currents off Estevan Point. The water depth at each site from the coast is 110 m, 160 m, 210 m, 800 m, and 2000 m.

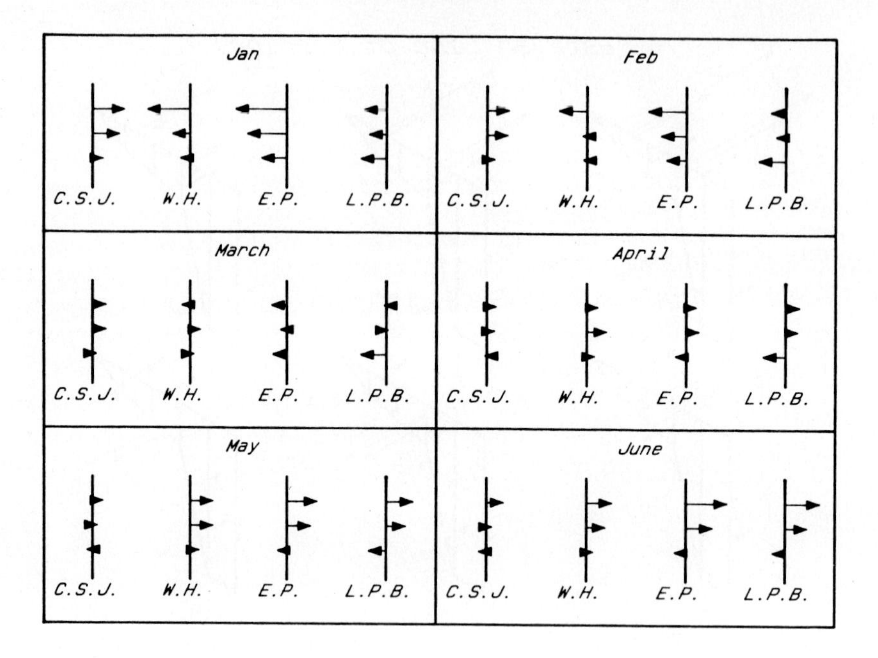

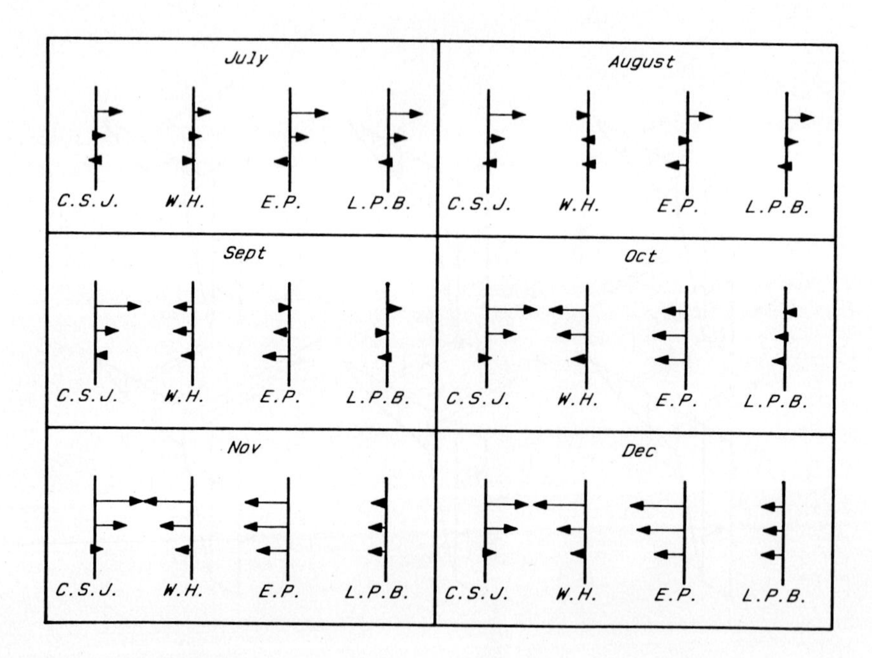

Figure 4: The annual cycle of along-shore currents at four sites distributed along the edge of the continental shelf. The sites are indicated by the solid triangles in Figure 1.

The flow patterns exhibited off the west coast of the Queen Charlotte Islands, exemplified by the mooring CSJ, have no apparent relationship to the other current regimes. The currents at 50 m and 100 m are southward year round. Furthermore, the phase of the annual cycle is different from that off Vancouver Island by about 180°, the strongest flows to the south appearing in mid-winter. At the deep meter, 150 m depth, there is no evidence of a preferred sense of flow, and so no evidence of any deep undercurrent. The general sense of flows off CSJ is baffling. The period of strongest southward flow also corresponds to the period of strongest winds to the north. A clue may exist in the detailed structure of the currents. A strong oscillation in the along-shore currents exists at the tidal period MSf. This suggests that the currents off CSJ may result from tidal rectification and so may be a very local feature not typical of the general circulation off the west coast of the Queen Charlotte Islands.

From: John Church, CSIRO Marine Laboratories, Hobart, Tasmania

On: Review and Commentary to paper **OBSERVATIONS OF THE LOW-FREQUENCY CIRCULATION OFF THE WEST COAST OF BRITISH COLUMBIA** by Howard J. Freeland

The author has assembled an impressive series of current meter observations for the continental shelf and slope adjacent to British Columbia, Canada. This data set is probably the only current meter data set for which a seasonal cycle can be determined on an eastern boundary for two cross-shelf arrays of five moorings (each with between two and four current meters) plus two additional shelf edge moorings.

In regards to the topic of this volume, the stability of the poleward undercurrent seen near the bottom at the shelf edge at latitudes of about 48°N and 49°N is quite remarkable. In fact, the poleward undercurrent is present during all seasons and appears to be the "continuous ever-flowing river" referred to by Smith in the Introduction of this volume. However, the undercurrent does not seem to be present at locations further north (at about 50.5°N and 52°N) along the coast of British Columbia. However, this may be a problem of sampling a narrow current with only a few current meter moorings.

These observations immediately raise some further questions. How continuous is the poleward flow off British Columbia with the California Undercurrent? How stable is the California Undercurrent? What happens to the poleward undercurrent between 49°N and 52°N? These are questions which could be addressed with a series of CTD/ADCP sections spaced along the eastern boundary of the North Pacific."

How the water properties varied along the core of any continuous poleward flow would be a useful diagnostic tool for examining mixing processes in this ecologically important region.

I feel the paper would be far more complete if the current meter data was supplemented by CTD data.

Author's reply:

> "[Reviewer Church was] interested in the observations at the lone current meter off the northern end of Vancouver Island and felt that if we had CTD sections suitable for geostrophic calculations, then the structures [revealed] would form a

useful extension of the report. Unfortunately, when we deployed the moorings off Winter Harbour and north, we were exploring the range of flow fields in an area that has to date had scant attention. No suitable CTD sections are available. CTD sections made in the past have extremely coarse resolution, entirely unsuitable for exploring coastal currents. Past practice has been to sample with one station on the continental shelf, one on the slope, and then a string of stations in the deep ocean. Only very recently have CTD sections had the resolution needed to examine the structure of shelf and slope flows."

POLEWARD FLOW IN THE CALIFORNIA CURRENT SYSTEM

Adriana Huyer
P. Michael Kosro
College of Oceanography
Oregon State University
Corvallis, OR 97331

Steven J. Lentz
Robert C. Beardsley
Woods Hole Oceanographic Institution
Woods Hole, MA 02543

INTRODUCTION

In the California Current System, poleward flow has been observed at different levels and in different seasons. In winter, there is broad poleward flow at the surface along most of the coast along Washington, Oregon, and northern California. During the upwelling season, in spring and summer, there is a narrow, inshore poleward surface current which appears off northern California whenever upwelling-favorable winds weaken; off southern and central California, this inshore poleward flow seems to persist through most of the upwelling season. During the upwelling season, there is also subsurface poleward flow, both near the bottom over much of the continental shelf and at depths of a few hundred meters along the upper continental slope. Whether and how these "branches" of poleward flow are related to each other is still unknown. We have some ideas of how the surface poleward flows are driven, but still very little information on the continuity and the driving of the subsurface undercurrents. In this note, we shall summarize briefly the main characteristics of each of these "branches," and present some recent observations of the California Undercurrent in the CODE region near San Francisco.

POLEWARD FLOW AT THE SURFACE: THE DAVIDSON CURRENT

Broad Northward Flow in Winter

The first evidence of a northward surface current along the west coast of North America was the use of redwood logs by North Coast Indian tribes in making dugout canoes and totem poles. This inferred northward flow was called the Davidson Current in honor of Professor George Davidson who was one of the early surveyors of the west coast. Systematic observations of the surface currents at lightships between San Francisco and the Strait of Juan de Fuca (Marmer, 1926) showed predominantly northward flow from October through March at all locations, with the strongest monthly-mean northward flow (20-30 cm/sec) occurring in January or February each year. Drift bottle releases (Wyatt et al., 1972), near-surface drogues (Reid and Schwartzlose, 1962), repeated hydrographic sections (Huyer, 1977; Chelton, 1984), and current measurements over the midshelf and shelfbreak (Hickey, 1981; Strub et al., 1987) all indicate that this winter northward surface current is considerably wider than the continental shelf, at least at latitudes north of Point Conception. The strongest northward flow (with a typical monthly mean of 15 cm/sec at the surface) occurs over the inner shelf, adjacent to the coast (Huyer et al., 1978). There seems to be considerable year-to-year variability in the strength of this current; during the El Nino winter of 1982-1983, the northward current was twice as strong as in the preceding and subsequent "normal" years (Huyer and Smith, 1985).

This winter northward surface current seems to be forced by the winter southeasterly winds which cause downwelling of the surface waters at the coast. This downwelling, combined with the increased coastal runoff caused by winter rains, results in an offshore density gradient which is in approximate geostrophic balance with the vertical shear of the northward current (e.g., Huyer et al., 1979).

Inshore Northward Flow

Poleward surface currents have also been observed during the upwelling season, with particular frequency and persistence along the coast of southern and central California. These northward flows apparently occur inshore of the coastal upwelling jet which flows equatorward along the front between dense, freshly-upwelled coastal waters and lighter oceanic surface waters. In the Southern California Bight, this inshore northward flow prevails most of the year, disappearing for only a month or so each year in March or April (Tsuchiya, 1980). The inshore poleward flow here may be as wide as 100 km or more; it seems to form the inshore limb of a large cyclonic eddy that fills most of the Bight (Reid et al., 1958). Further north, in the CODE region near San Francisco, an inshore countercurrent appears whenever upwelling-favorable winds relax (Send et al., 1987) and disappears when they become unusually intense (Winant et al., 1987). This inshore countercurrent is only 10-20 km wide and can have velocities of > 30 cm/sec (Kosro, 1987); it seems to become more persistent as the upwelling season progresses (Winant et al., 1987). Off central Oregon, a similarly narrow, inshore countercurrent is frequently observed in mid- and late summer (Stevenson et al., 1974; Kundu and Allen, 1976).

These inshore countercurrents may be driven partly by along-shore pressure gradients which are set up to balance the strong equatorward wind stress: when the winds relax, the residual pressure gradient causes a northward acceleration. This mechanism has been invoked to explain the rapid reversals of flow over the Pacific Northwest shelf that occur whenever previously strong winds relax (Hickey, 1984). However, the details of the dynamics involved are not yet clearly understood: in particular, the along-shore scales of the relevant pressure gradients are unknown, and both the sign and magnitude of estimates of the pressure gradient are sensitive to the choice of scale. There may well be other processes and dynamics involved.

UNDERCURRENTS OVER THE SHELF

Moored current measurements over the continental shelf off Washington and central Oregon have consistently shown a poleward undercurrent along the bottom over the mid- and outer-shelf during the summer upwelling season (Mooers et al., 1976; Huyer et al., 1975a, b). This undercurrent has its maximum strength within 20-30 m of the bottom (Huyer et al., 1978). Maximum velocities are relatively weak: monthly means of 2-5 cm/sec are typical. Similar undercurrents have been observed over the shelf in the CODE region just north of Point Reyes (Winant et al., 1987; their Figure 19b), and off Half Moon Bay at 37.4°N and Purisima Point at 34.8°N (Strub et al., 1987; their Figure 7). This undercurrent is absent immediately after the spring transition which marks the onset of seasonal upwelling (Huyer et al., 1979; Lentz, 1987), but it appears within a few weeks and slowly intensifies through the summer. In fall, the depth of the current maximum seems to shoal (Reid, 1987), and the shelf undercurrent becomes indistinguishable from the northward surface current which extends across the entire shelf in late autumn and winter.

Why the maximum velocities are observed so near the bottom and how this shelf undercurrent is driven remain uncertain. It seems increasingly likely that what we are observing here is merely the upper and inshore portion of the larger undercurrent that flows poleward along the upper continental slope, which is often called the California Undercurrent.

POLEWARD FLOW ALONG THE CONTINENTAL SLOPE: THE CALIFORNIA UNDERCURRENT

A poleward undercurrent over and along the continental slope has been observed at several latitudes between Baja, California and Vancouver Island. Indirect evidence for this flow is clearly visible in the large-scale temperature-salinity characteristics of coastal waters as northward-tending tongues of relatively warm, saline water (Tibby, 1941; Reid et al., 1958). More detailed studies of particular regions (e.g., Wickham, 1975; Reed and Halpern, 1976; Freitag and Halpern, 1981) also show a concentration of waters of more southerly origins along the continental margin. The undercurrent can also be clearly seen in the dynamic topography (ΔD) of isobaric surfaces of 150-300 dbar (relative to a deeper isobaric surface): in the maps of $\Delta D_{200/500}$ from repeated CalCOFI surveys off Baja, California and southern California (Wyllie, 1966); in a map of $\Delta D_{300/900}$ covering slope waters off northern California in May 1977 (Freitag and Halpern, 1981); and in maps of $\Delta D_{150/1000}$ off Washington and Oregon in September 1973 (Reed and Halpern, 1976) and July 1975 (Halpern et al., 1978). Freitag and Halpern (1981) estimate a northward transport of 1-3 Sv along the slope. Analysis of repeated CalCOFI sections shows that the undercurrent off Point Conception (Chelton, 1984; his Figure 3) persists year-round, has a width of 50-100 km, is strong (> 7 cm/sec) and shallow (core at 100 m) in winter, and weak (< 3 cm/sec) and deep (core at 250 m) in early spring; the lateral and vertical shears along the undercurrent are strongest from May through August. Similar analysis of sections off Point Sur (Chelton, 1984; his Figure 2) shows only a weak (3 cm/sec) undercurrent in late summer (July to September) that seems to shoal, intensify, and merge with the Davidson Current in winter, and to disappear in spring and early summer (March to June).

Direct measurements of this northward flow were first made by deploying parachute drogues and tracking them for several days. Reid (1962) found a 20 cm/sec northward current, about 75 km wide, at a depth of 250 m off Monterey (36°N) in November 1961. Wooster and Jones (1970) found a very narrow (20 km) undercurrent with a speed of about 30 cm/sec at about 31°N off Baja, California in August 1966. However, Stevenson et al. (1969) failed to find a northward undercurrent off Oregon in most of their 15 drogue deployments between January 1962 and September 1965.

Longer-term direct measurements of the northward flow required the installation of moored current meters. Prior to the CODE experiment conducted in 1981 and 1982, moorings of at least a few weeks duration were successfully deployed at several locations along the continental slope (Figure 1). The most complete sections, showing the most detailed vertical and offshore structure, were obtained off southern Washington in the

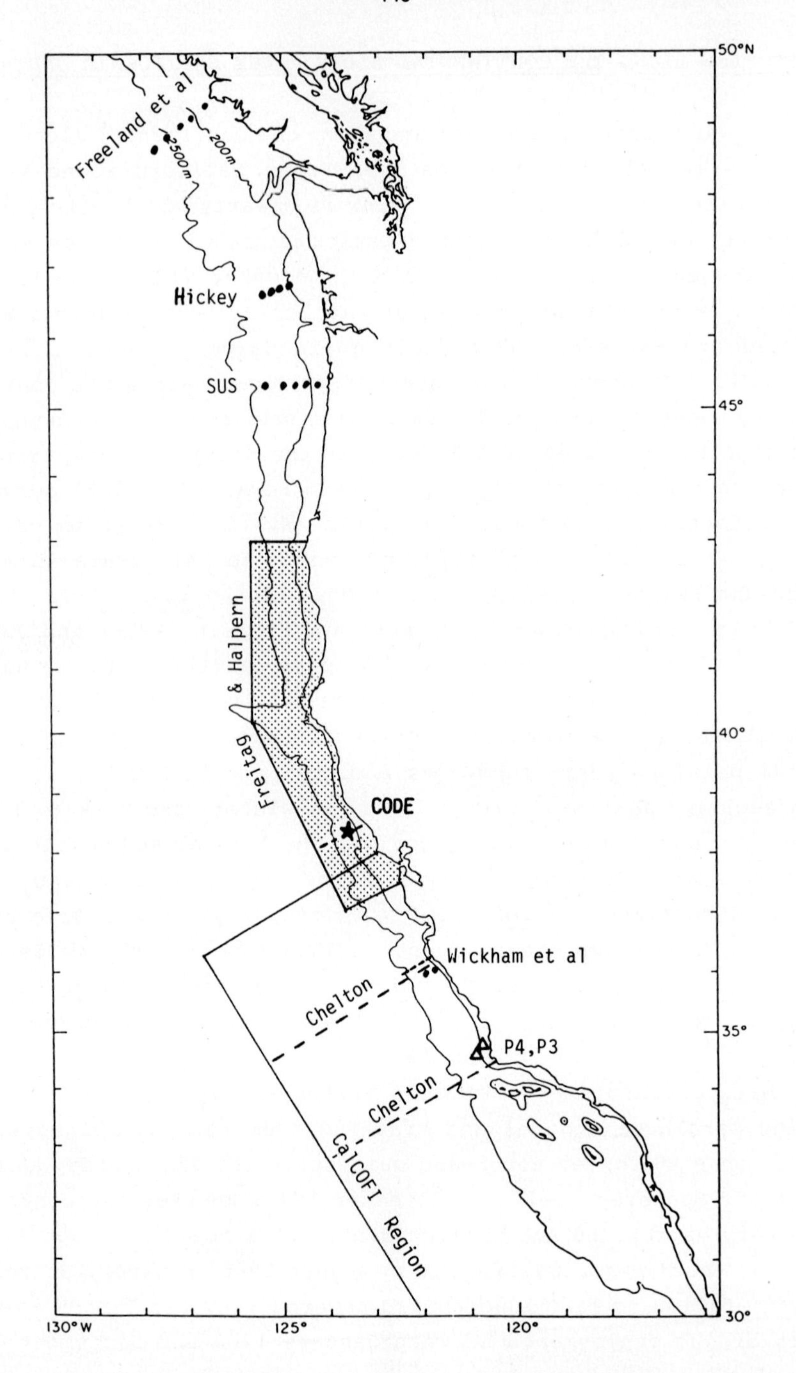

Figure 1: Map of a portion of the west coast of North America, showing the location of the principal historical studies of the poleward undercurrent along the continental slope. Repeated hydrographic sections are indicated by dashed lines, and moorings of at least a few weeks duration are indicated by dots. The CODE C-5 mooring is indicated by a star and the shelf moorings off Purisima are indicated by triangles.

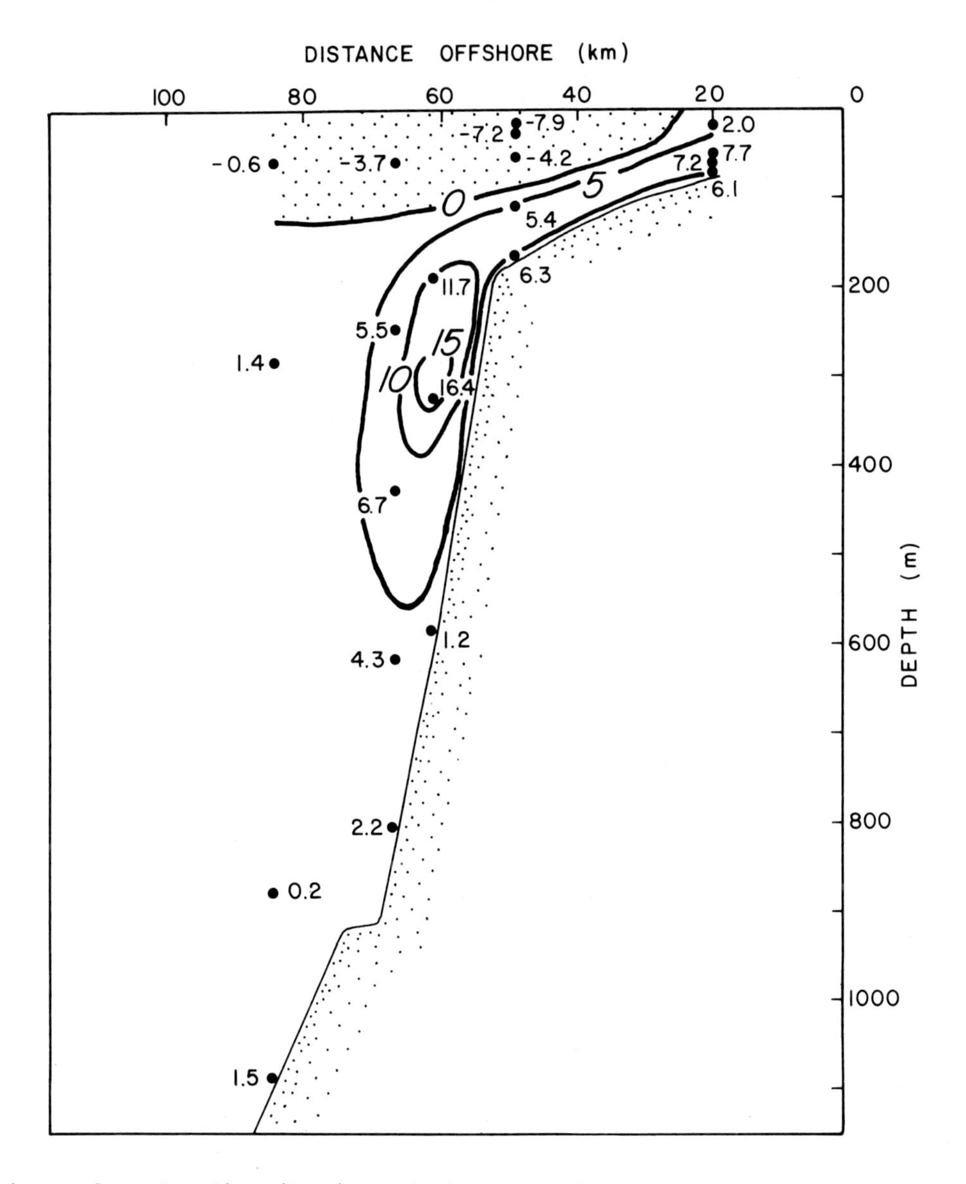

Figure 2: The distribution of the mean along-shore flow off southern Washington, 21 July to 28 August 1972. Adapted from Hickey (1979).

summer of 1972 (Hickey, 1979) and off Vancouver Island in the summer of July 1980 (Freeland et al., 1984). Average along-shore velocity sections from these arrays are shown in Figures 2 and 3; in both cases, the average was calculated over a period of about a month, in summer. Both show that the poleward velocity has a definite maximum adjacent to the continental slope, but the depth and strength of this core are rather different. Off Washington in 1972, the 15 cm/sec maximum occurs at a depth of about 300 m; off Vancouver Island in 1980, the 4 cm/sec maximum occurs at a depth

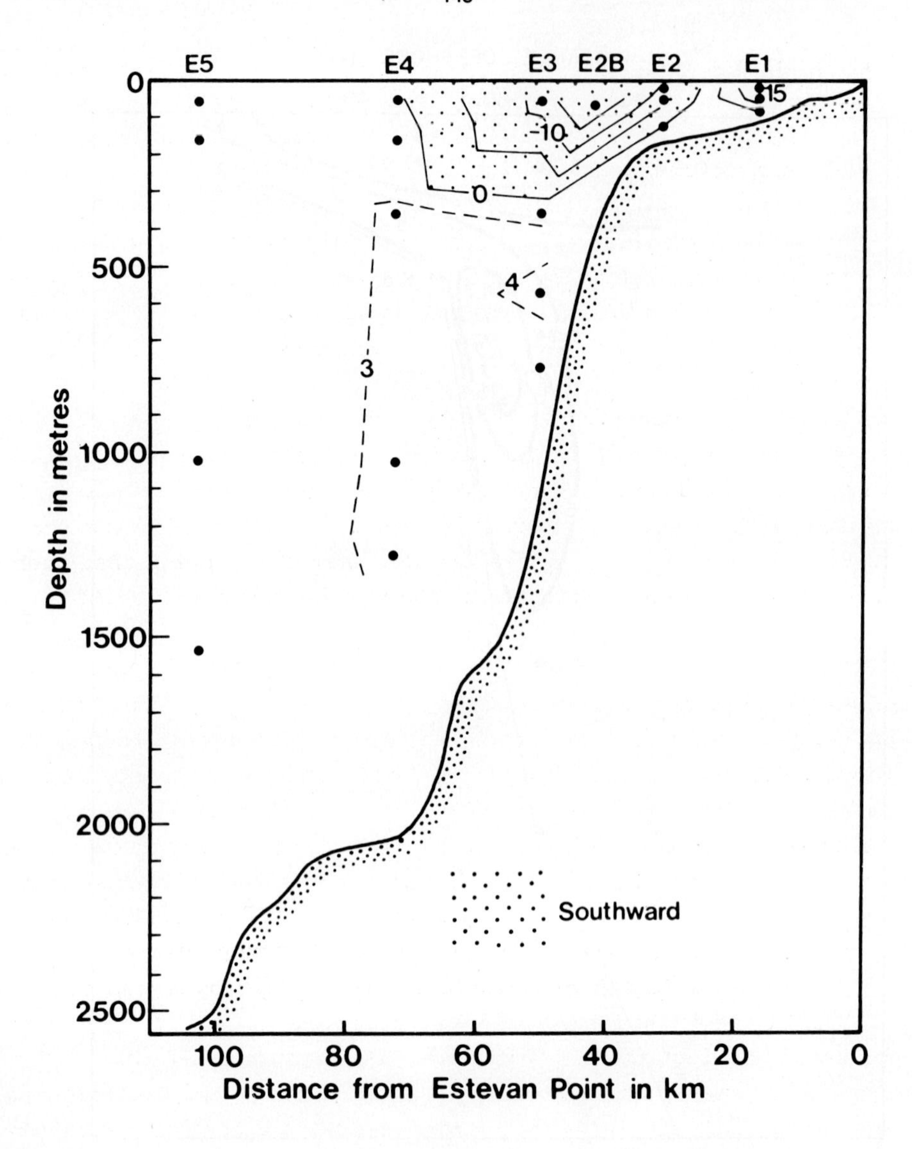

Figure 3: The distribution of the mean along-shore flow off Vancouver Island, July 1980. From Freeland et al. (1984).

of 600 m. We do not know whether these differences are associated with the difference in location or with the difference in years; i.e., whether they reflect spatial or interannual variations. The difference is probably not seasonal, since the data sets were taken at very nearly the same time of year.

A Slope Undercurrent Study (SUS) designed to observe the seasonal variation of the undercurrent off central Oregon in 1977 and 1978 met with limited success. An earlier mooring there had shown a definite undercurrent over the upper slope (500 m isobath), with the strongest northward current (5 cm/sec) near the bottom (Halpern et al., 1978). There were five moorings in this array, spanning the entire continental margin; unfortunately, there were more instrument failures and losses to fishing than usual, and a strong warm-core eddy remained over the offshore moorings for more than two months in January and February 1978 (Huyer et al., 1984). Nevertheless, the results do indicate some seasonal differences (Figure 4): in winter, there is broad northward flow (presumably the Davidson Current) over the entire margin; in spring, there is predominantly southward flow with a weak (5 cm/sec) undercurrent near the shelf-break; and in summer, there is still a weak undercurrent over the shelf-break and an apparently separate broad poleward flow over the outer margin. These sections look very different from those from the southern Washington and Vancouver Island sites (Figures 2 and 3). The data from the upper slope during summer 1978 is very similar to the summer 1975 data (Halpern et al., 1978) taken at a nearby location. We, therefore, think that much of the difference should be attributed to the shape of the continental margin, which is much wider here than farther north.

Current meter moorings were deployed for several two-month intervals at three sites over the upper slope near Point Sur between July 1978 and July 1980 (Wickham et al., 1987). Although there was little temporal overlap in the data from different moorings, the results indicate that the strongest poleward flow is adjacent to the upper slope. The strongest average flow (15 cm/sec) was obtained at the shallowest current meter, at a depth of about 100 m, only 10 km from shore; geostrophic velocity sections indicate there was southward flow at shallower depths. If these observations are assumed to be representative of the flow regime off Point Sur, we would infer that the undercurrent is much shallower there than off Oregon, Washington, and Vancouver Island. This shoaling of the undercurrent to the south is supported by the long-term shelf moorings deployed along the west coast in 1981 and 1982; the upper instruments from these moorings show an increasing tendency for northward flow at more southerly sites (Strub et al., 1987a). In particular, the moorings off Point Purisima at 34.8°N show persistently northward flow (Figure 5) which is interrupted for only a few weeks in early spring (in March 1981 and April 1982).

Data from the CODE C-5 Mooring

During the Coastal Ocean Dynamics Experiment conducted over the continental shelf and slope between Point Arena and Point Reyes, a pair of current meter moorings were maintained at 38.5°N, 123.7°W over the 400 m isobath on the upper slope (Winant et al., 1987). One of the moorings

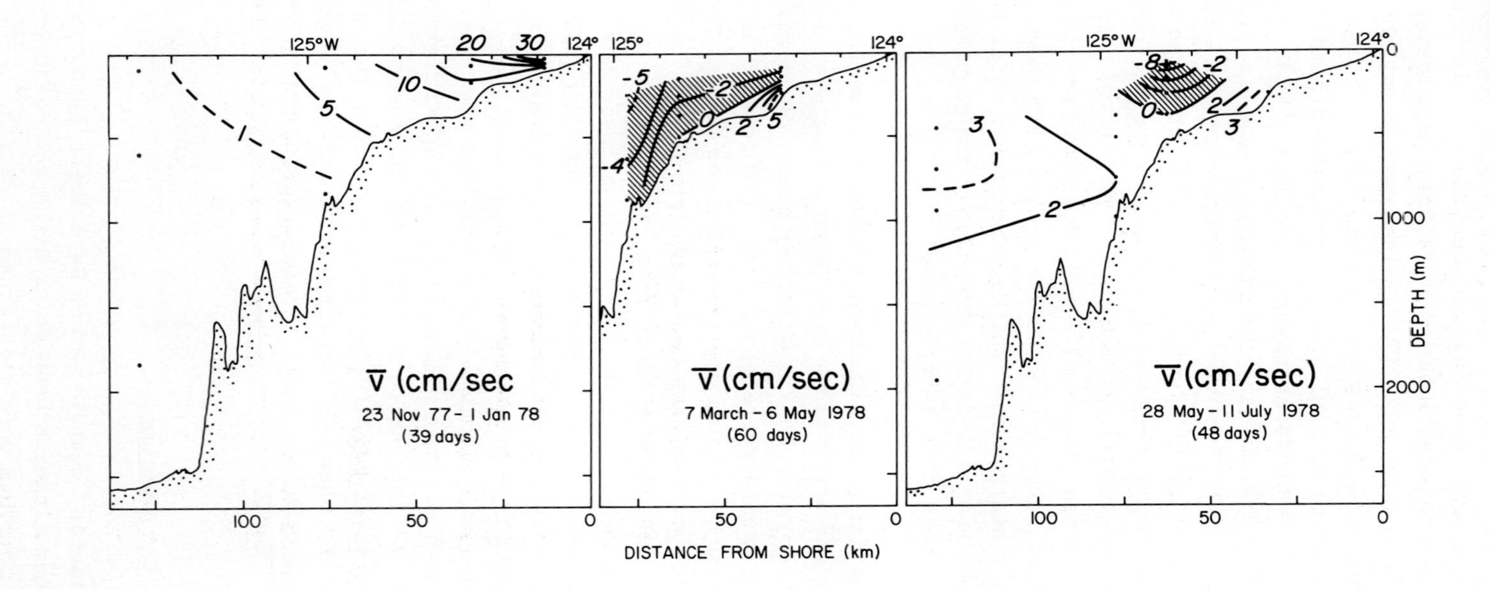

Figure 4: The distribution of average along-shore flow off central Oregon for three periods between November 1977 and July 1978. From Huyer et at. (1984).

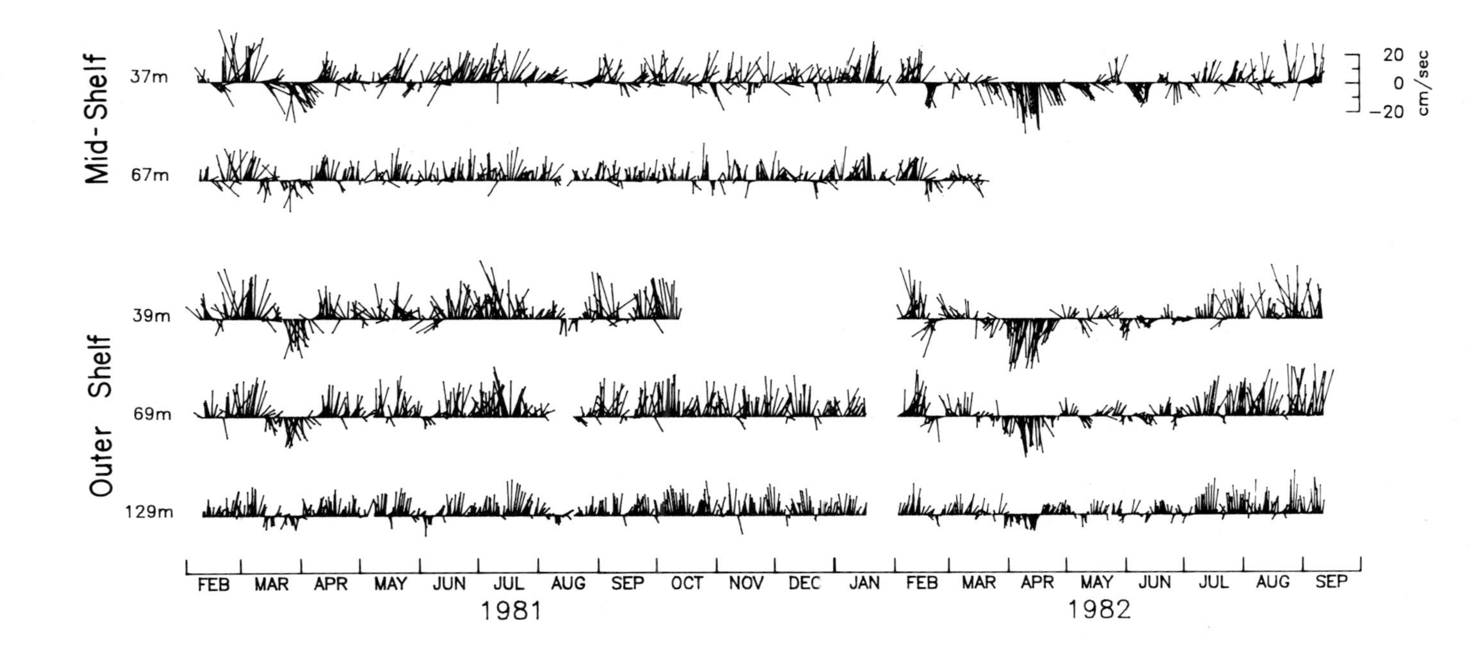

Figure 5: Time series of daily low-passed (<0.6 cpd) current vectors over the mid-shelf and outer-shelf (P3 and P4, over the 90 and 155 m isobath, respectively), off Purisima Point at 34.8°N, February 1981 to September 1982. Vectors are oriented so that true North is at the top of the page. Replotted from Denbo et al. (1984).

had surface flotation and a vector-averaging current meter (VACM) at 9 or 10 m; the other mooring was entirely subsurface, with instruments at or very near 75, 150, 250, and 350 m. The moorings were installed in April 1981, serviced in August, December, and April, and finally recovered in August 1982; these observations preceded the onset of El Nino conditions in October 1982 (Huyer and Smith, 1985). Current records at the same (or similar) depth from consecutive installations were joined by applying a predictive filter (Ulrych et al., 1973) to fill the short gaps in the hourly data. The 9 m record has a two-month gap due to an instrument failure in October 1981, but the records at the other depths are complete. The joined records were low-pass filtered (half-power at 0.6 cpd) to suppress tidal and inertial fluctuations; a coordinate rotation yielded onshore (toward 65°T) and along-shore (toward 355°T) components of the current. Time series of twelve-hourly low-passed current vectors (Figure 6) show considerable variability on time scales of days and weeks; nevertheless, it is clear that equatorward flow predominates at the surface and poleward flow predominates at depths greater than 100 m (Figure 6, Table 1). Both the equatorward surface current and the poleward undercurrent seem to be strongest in the season when the local wind stress (calculated from winds measured at nearby NDBC Buoy 13; Beardsley et al., 1987) is persistently favorable for upwelling. This is seen more clearly in the vertical profiles of three-month averages of the along-shore current (Figure 7, Table 2) which show that the undercurrent is strongest (averaging about 10 cm/sec) in spring and summer, and absent or very weak in winter. Since the spring profiles from the two years are very similar (both show strong mean vertical shear and an undercurrent core-depth of 250 m), it seems likely that they are representative for this location. However, we do not know how large a region the C-5 mooring site represents—it may be only a few kilometers across. Repeated current measurements by means of a shipborne doppler acoustic profiler along sections at 38.5°N and 39°N indicate that there are significant differences in the strength and structure of the under-current between lines separated by only 50 km (Huyer and Kosro, 1987; their Figure 27). Preliminary analysis of T-S characteristics along a section through C-5 suggests that the undercurrent core-depth is more than 100 m deeper over the mid-slope than at C-5, a separation of only a few kilometers; more work will need to be done to verify this result and to grasp its implications. On the other hand, preliminary results from recent moorings over the upper slope at other locations between Monterey and Point Conception (Chelton et al., 1987) seem to indicate vertical profiles that are similar to those observed at C-5; again, more work will be needed to verify this.

As yet, we still have a very incomplete picture of the California Undercurrent that flows poleward along the continental slope. We think

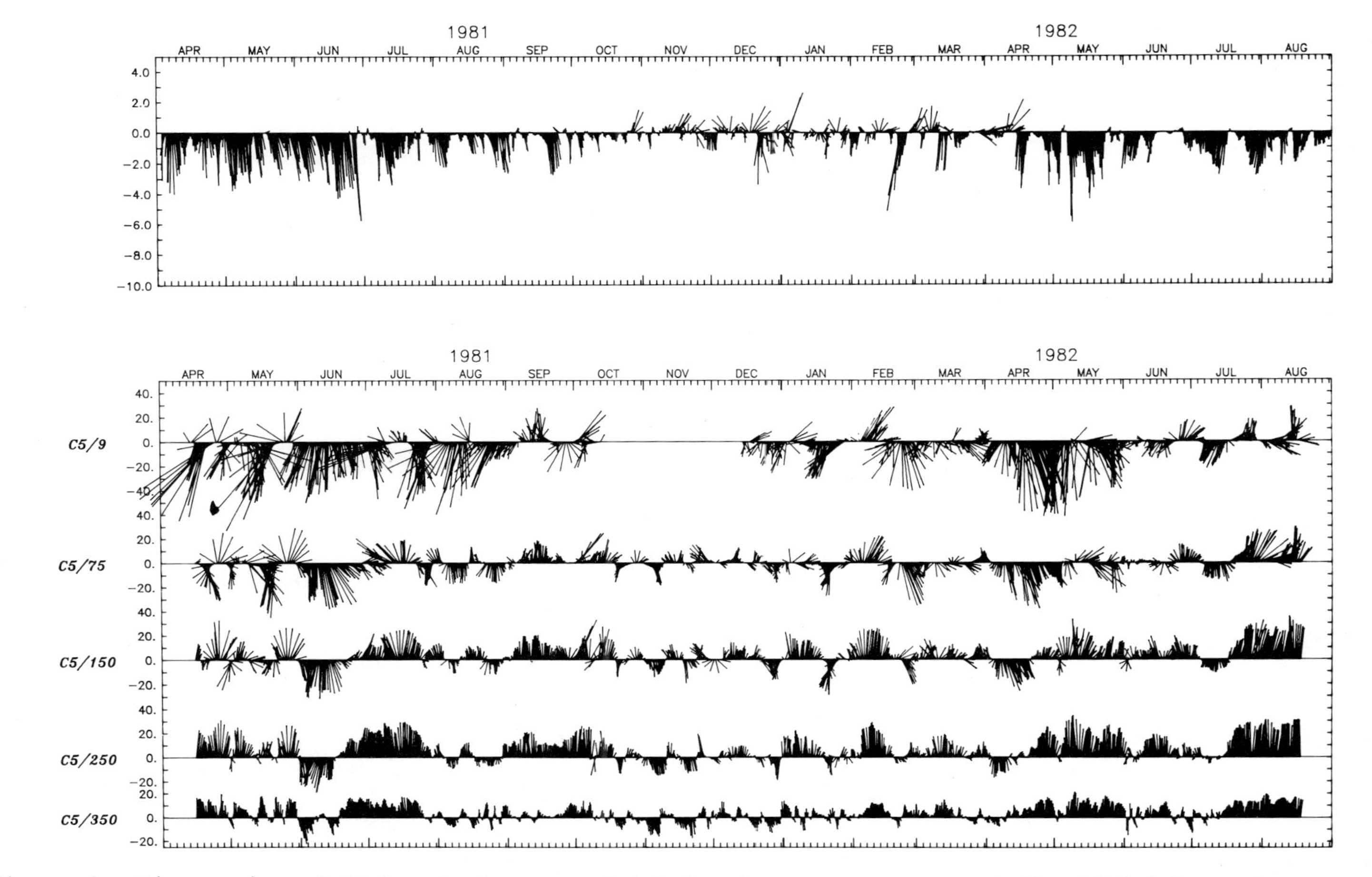

Figure 6: Time series of 12-hourly low-passed (<0.6 cpd current vectors at the CODE C-5 mooring over the 400 m isobath at 38°31'N, 123°40'W from April 1981 to August 1982. Low-passed vectors of wind stress are shown at the top. Vectors are oriented so that 335°T is at the top of the page and 65°T is to the right.

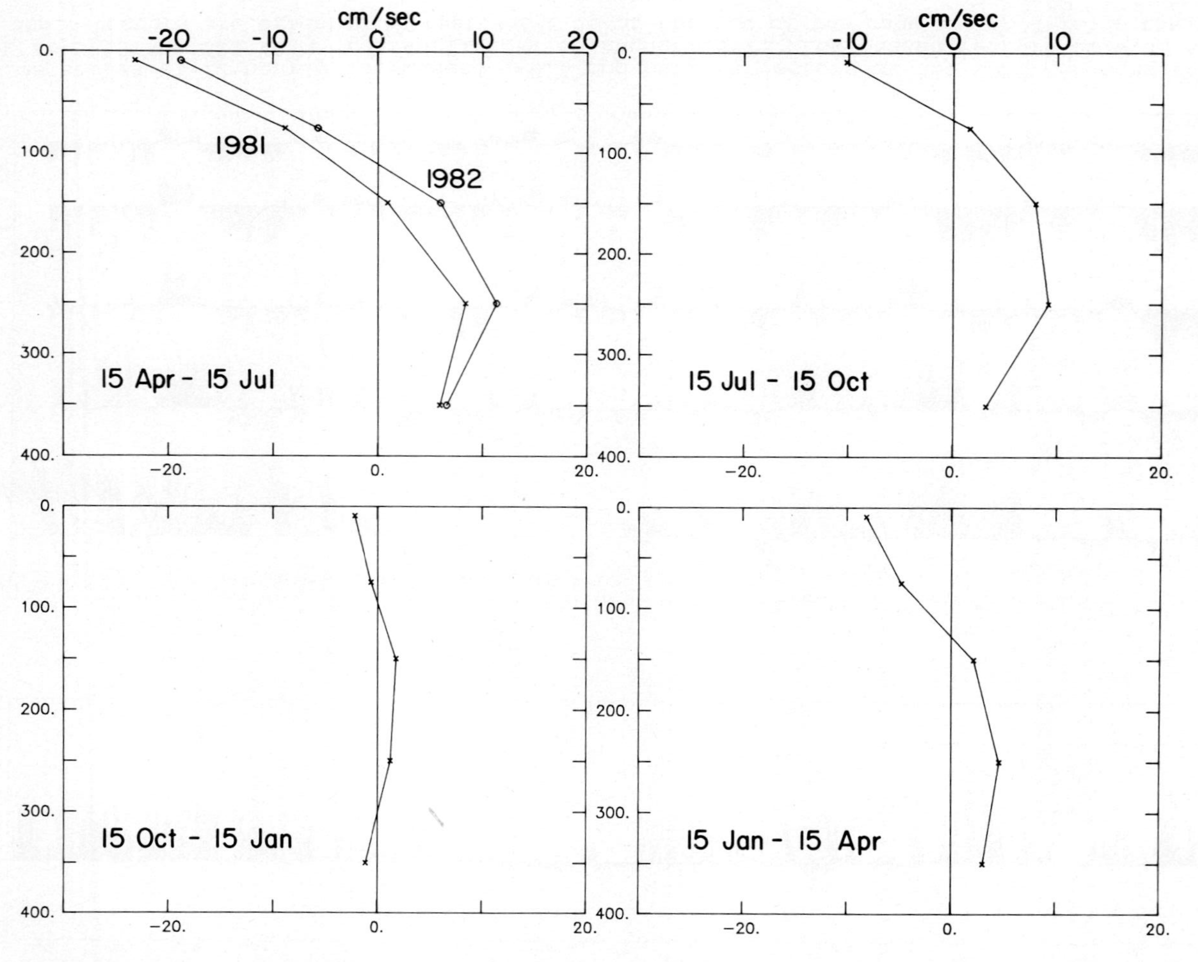

Figure 7: Vertical profiles of the three-month average along-shore current (positive toward 335°T) at C-5.

Table 1. Overall means and standard deviations of the current and temperature at the C-5 mooring.

Direction of Major Axis			Onshore Current (65°T)		Along-shore Current (335°T)		Temperature		
Depth	N	(°T)	mean	std dev	mean	std dev	mean	std dev	N
9	1717	317	-1.1	13.1	-12.8	18.8	11.30	1.18	1968
75	1960	337	2.9	7.3	-2.6	12.7	9.75	0.81	1960
150	1959	333	1.2	4.4	5.0	12.3	8.48	0.13	1959
250	1960	335	0.4	2.0	8.1	11.9	7.84	0.11	1960
350	1959	338	0.5	0.3	4.2	7.5	7.08	0.07	1959

Table 2. Three-month means and standard deviations of the onshore (65°T) and along-shore (335°T) components of the current at C-5.

	15 Apr- 15 Jul 1981		15 Jul- 15 Oct 1981		15 Oct- 15 Jan 1982		15 Jan- 15 Apr 1982		15 Apr- 15 Aug 1982	
depth	mean	std dev	mean	std dev	mean	std dev	mean	std dev	mean	std dev
u 9	-6.0	18.8	-4.9	10.0	(2.7)	(9.1)	-0.6	11.3	2.5	9.8
75	+4.2	10.5	1.2	4.8	1.3	5.4	1.5	7.5	3.6	5.9
150	1.6	5.2	0.4	3.4	0.6	4.4	1.6	5.2	1.0	3.8
250	0.3	2.8	0.0	1.4	0.7	1.8	0.7	2.0	0.3	1.7
350	0.0	1.1	0.1	0.9	0.6	0.7	0.5	0.8	0.6	0.8
v 9	-22.8	18.3	-10.2	19.3	(-6.3)	(8.3)	-8.0	13.8	-18.9	20.0
75	-9.2	15.5	1.6	10.5	-0.6	7.5	-4.6	11.5	-6.0	11.0
150	0.7	14.6	7.9	9.4	1.8	9.4	2.3	12.4	5.9	10.0
250	8.2	14.6	9.3	9.9	1.2	3.0	4.6	3.0	11.4	9.7
350	5.9	8.7	3.2	6.1	-1.2	2.1	3.0	1.6	6.5	6.6

it is strongest during the upwelling season and that it disappears or merges with the Davidson Current in winter. We also think that its core occurs at different depths (and densities) at different latitudes, but this is really only a guess since simultaneous observations from different latitudes have only recently become available. We do not know what determines the undercurrent variability on time-scales of weeks which does not seem to be directly related to variations in the wind. With so little known about the structure and variability of the undercurrent, we obviously do not know how it is forced.

CONCLUDING REMARKS

There is a great deal we still do not understand about the undercurrent that flows poleward along the continental slope off California and the Pacific Northwest. Many questions remain: Is the undercurrent continuous along the coast? What determines the width, strength, and the core-depth of the undercurrent? How and why do these vary with location and from season to season? Are the northward-tending tongues seen in water property distributions due primarily to advection, or are mixing and interleaving more important? Is the poleward flow observed near the bottom over the outer shelf merely a manifestation of the slope undercurrent, or is it an independent phenomenon with its own characteristics and forcing mechanism? We hope that these and other questions about undercurrents will be addressed by an integrated program of additional observations and improved models in the decade to come.

ACKNOWLEDGEMENTS

We wish to express our appreciation for many fruitful discussions with colleagues over the years, especially to Bob Smith, John Allen, and Barbara Hickey. Thanks also to Steve Ramp for his thorough review of this article. This paper was completed with support from the National Science Foundation through Grants OCE-8410546 and OCE-8709930 to Oregon State University.

From: Steven Ramp, Naval Postgraduate School, Monterey, CA

On: Review and commentary to paper **POLEWARD FLOWS IN THE CALIFORNIA CURRENT SYSTEM** by A.Huyer, P.M. Kosro, S.J. Lentz, and R.C. Beardsley

I feel that I must tread lightly when commenting on a review paper written by a group of authors whose combined experience in the California Current exceeds my own by many years. Indeed, the authors have done a very complete job of summarizing the available observations in the California Current System. (They found some I had not even heard of!). My primary criticism would be needless heavy emphasis on the CODE experiment, which occupies fifty percent of the figures and a good portion of the text. This is perhaps understandable in view of the very close association of all the authors with that experiment; but being primarily a shelf experiment, it was no more illuminating than many other experiments when it comes to the California Undercurrent. While recognizing the need for brevity, I would include a few more figures. There are no figures illustrating the section on "undercurrents over the shelf." This is where CODE shines, and I would include perhaps Figure 19b from Winant et al. (1987), or Figure 7 from Strub et al. (1987), although there may be better ones somewhere. I would also include Figure 2 from Chelton (1984), which shows the mean seasonal distribution of along-shore currents off Point Sur, from the CalCOFI data and represents a very nice summary of what is known about that section of the coast. This paper could also be strengthened by including some comments on the paper by Lynn and Simpson which just appeared in the most recent issue of JGR (Vol. 92, No. C12, Nov. 1987). This paper is perhaps the most complete analysis so far of the CalCOFI data, and thus, covers a spatial area far greater than the earlier work, which is useful for things like examining the continuity and seasonal variability of the coastal countercurrent and the undercurrent. Two of the more interesting results of Lynn and Simpson (1987), as far as the poleward flows are concerned, are: 1) observational evidence that the coastal countercurrent represents a surfacing of the UC, and 2) that the time of maximum flow in the UC is in fall and winter, rather than when the surface equatorward winds and currents are strongest as suggested by the CODE data.

I suspect that dynamics are being covered more completely in other papers being submitted to the volume, but on page 3, paragraph 2, I would at least list the possibilities for forcing of the poleward surface flow in the face of equatorward wind stress. McCreary (1987) lists four possibilities: remote forcing, relaxation of equatorward wind, the positive curl of equatorward wind stress, and thermohaline forcing, and comes down heavily in favor of the wind stress curl for the California Current case. Likewise, at the bottom of page 9, we do not know how the undercurrent is forced, but some possibilities could be mentioned. Two which immediately come to mind are the alongshore pressure gradient established by the equatorward wind stress at long time scales, and Kelvin waves excited by periodic forcing at periods of 20 days or less (Philander and Yoon, 1982).

To summarize, this paper makes a nice contribution to the "lecture notes" series. I feel it could be strengthened by a slightly more balanced approach that focuses less on the CODE experiment, and includes some comments on the most recent paper by Lynn and Simpson, which are very relevant to the subject at hand.

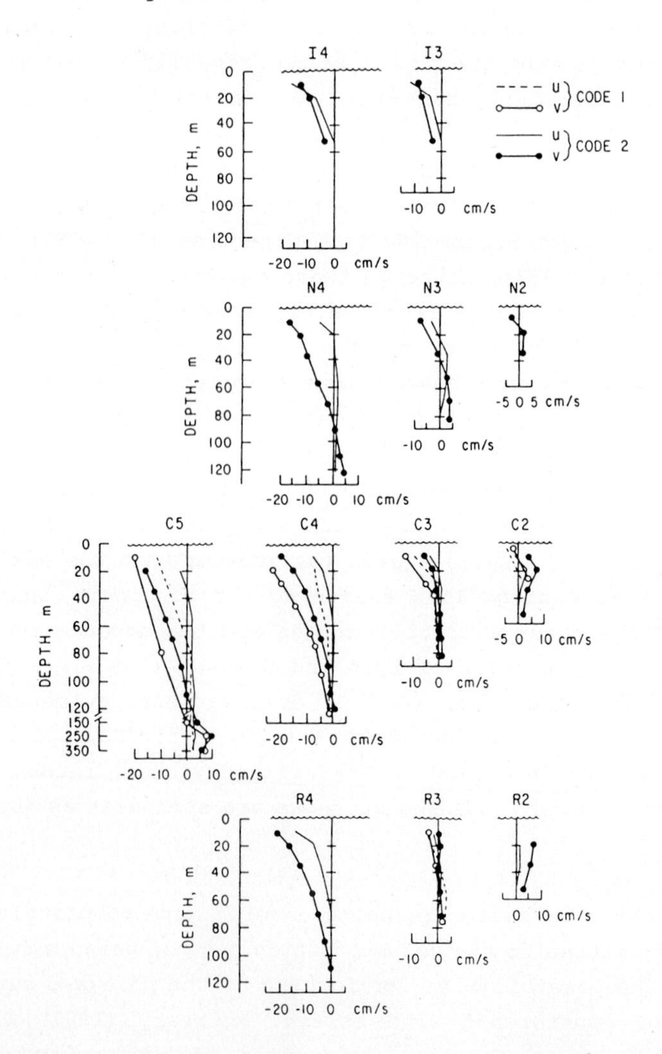

Spatial structure of time average currents during the CODE 1 and CODE 2 common analysis periods in vertical profile. (from Winant, et al., 1987; their Fig. 19b).

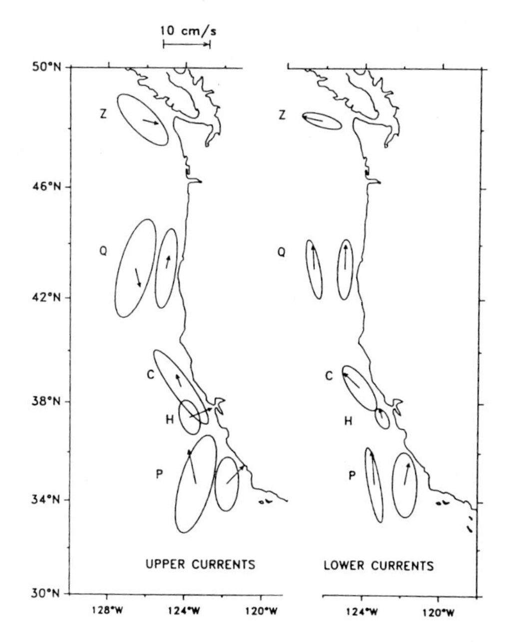

Annual vector means (arrows and principal axis ellipses, calculated from one complete year of 6-hourly current data (May 1, 1981 to May 1, 1982), for the top and bottom instruments on each mooring. The centers of the ellipses are at the latitudes of the moorings, but the longitudinal positions have been displaced for better viewing. (from Strub, et al., 1987a; their Fig. 7).

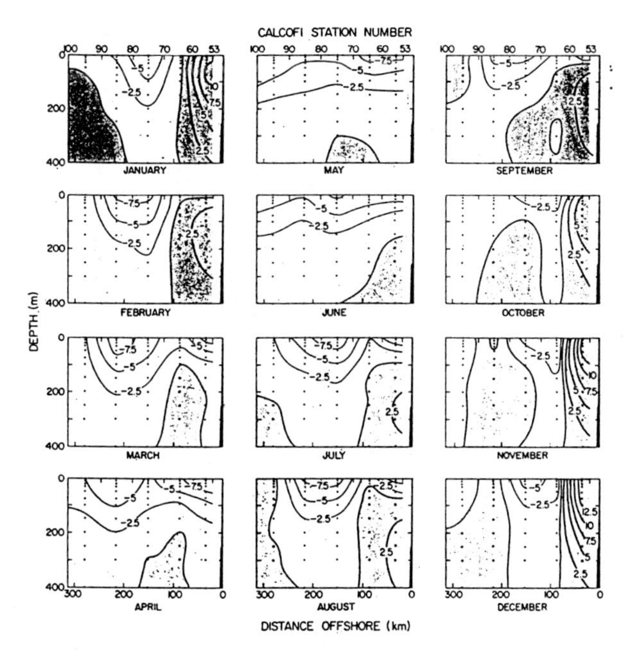

Vertical sections of seasonal alongshore geostrophic velocity relative to 500 db along CALCOFI line 70 off Point Sur. Shaded regions correspond to poleward flow (from Chelton, 1984; his Fig. 2).

POLEWARD FLOW NEAR THE NORTHERN AND SOUTHERN BOUNDARIES OF THE U.S. WEST COAST

Barbara Hickey
School of Oceanography
University of Washington
Seattle, WA 98195

ABSTRACT

Poleward flow over the shelf and slope is a common feature of west coast circulation. This paper describes the present state of knowledge of poleward flow near the end points of the U.S. West Coast, Washington State, and southern California, and compares the occurrence of poleward flow in the two regions. Attention is given to the occurrence of equatorward flow, which is now becoming the anomaly rather than the norm in west coast current observations. In particular, the occurrence of equatorward flow over the shelf during winter in relation to the Columbia River plume, and an equatorward undercurrent over the slope during winter off the Washington coast are described.

INTRODUCTION

Poleward flow over the shelf and slope is a common feature of west coast circulation. This paper uses existing data to describe the present state of knowledge of poleward flow near the end points of the U.S. West Coast, namely, Washington State and southern California. Attention is given to the occurrence of equatorward flow, which is now becoming the anomaly rather than the norm in west coast current observations. For a complete description of the seasonal flow off the Washington coast, the reader is referred to Hickey (1989a). For a complete description of the seasonal flow in the Santa Monica/San Pedro basin and shelf area off southern California, the reader is referred to Hickey (1989b).

DATA AND DATA SOURCES

Results in this paper are based primarily on current meter records. The meters were generally Aanderaa's and moorings were generally a subsurface taut wire type. Nearsurface and nearshore data were collected with EG&G Vector Measuring Current Meters suspended from surface toroids. Washington data were collected over a 15-year period from 1971 to 1984. Southern California Bight data were collected over the period 1985 to 1987. The sampling interval was generally 10 or 20 minutes. Data were filtered to obtain hourly values of current speed and direction. Because the flow generally follows the isobaths, the data were rotated into a local isobath system for most of the figures. The majority of the data are presented as monthly means. Since typical time scales are on the order of 10 days, only records whose length exceeded nine days were included in these averages.

RESULTS

Washington Shelf

Poleward flow is the rule rather than the exception on the Washington shelf for the below-pycnocline (20 m) water column. This is illustrated in Figure 1, which depicts monthly mean along-isobath velocity for selected depth intervals at inner-, mid-, and outer-shelf locations. The data available off Washington are insufficient to define the seasonal cycle in the upper 20 m of the water column. The meager data that do exist (two short summer data sets over the mid- and inner-shelf) indicate that considerable vertical shear and even flow reversals occur over the upper 20 m (Figure 2) (Hickey, 1989a). Equatorward flow occurs at all depths below the surface layer and at all locations only during spring. Equatorward flow is more persistent on the outer-shelf than at mid-shelf, occurring from spring through late summer on the outer-shelf, but only through early summer at mid-shelf. This differs from the Central Oregon shelf (~45°N), where measurements indicate that flow at mid-shelf is primarily equatorward throughout the summer (Kundu and Allen, 1976; Huyer et al., 1978). Thus, during late summer, the equatorward coastal jet is located farther offshore off Washington than off Oregon. The apparent offshore migration of the location of the equatorward coastal jet as the summer advances is illustrated in Figure 2, which gives monthly mean velocity across the Washington shelf during early (June) and late (August) summer. A second mid/late summer (July/August) example is given in Figure 5, which is presented in the next section. The shelf edge, upper slope, and location of the equatorward jet during mid-to-late summer off Washington is similar to that reported for the coast of Vancouver Island (Freeland et al., 1984), off southern Oregon (Strub et al., 1987) and off northern California (Winant et al., 1987). The occurrence of stronger monthly mean mid-shelf poleward flow off Washington more than off Oregon during mid-to-late summer may be related to the poleward decrease in

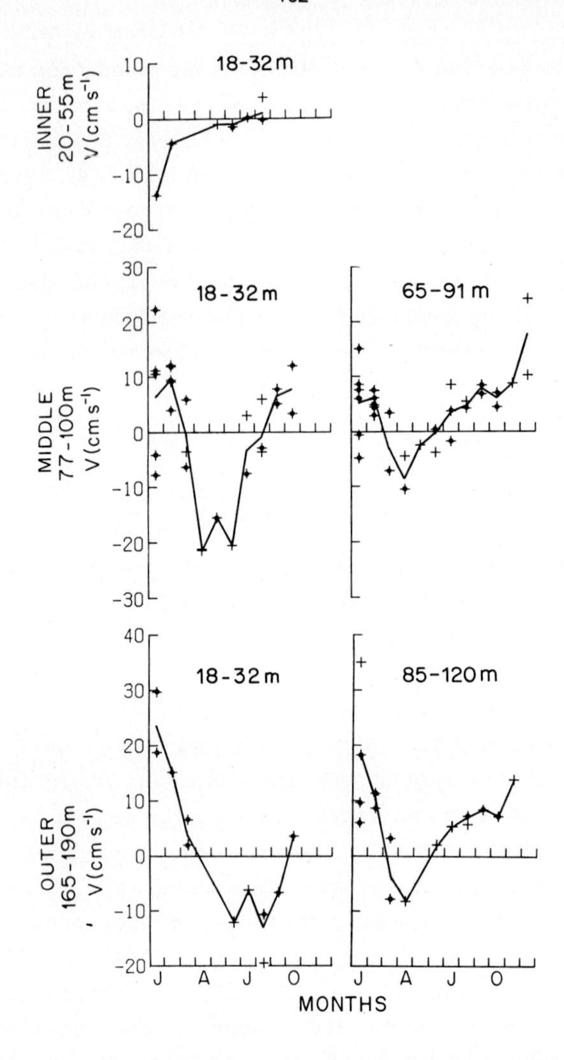

Figure 1: Monthly mean along-isobath velocity in several depth intervals at inner-, mid-, and outer-shelf locations off the Washington coast. The weighted monthly mean is shown as a solid line. Values calculated using less than 15 days of data are indicated with simple crosses.

equatorward wind stress that occurs between Oregon and Washington during the summer season (Huyer, 1983).

Poleward flow occurs at most shelf locations along the west coast during the winter season (Strub et al., 1987). An exception to this rule appears to occur off the Washington coast on the inner-shelf during late winter (Figure 3). This equatorward flow may be related to the Columbia River plume, which has a seasonal maximum in winter, as well as in spring. In support of this hypothesis, we note that geostrophic velocities

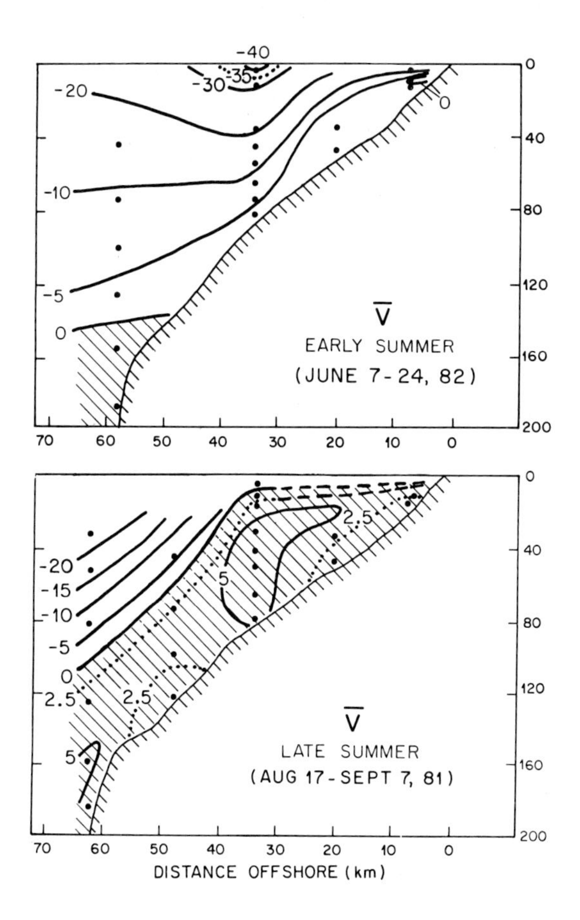

Figure 2: Cross-shelf sections of mean along-isobath velocity off the Washington coast during early and late summer. Units are cm s^{-1}. Regions of poleward flow are shaded. Dots indicate locations of current meters.

computed from seasonal mean hydrographic data also indicate equatorward flow (below a thin, poleward flowing surface layer) on the inner-shelf off Washington where the plume occurs, but not off Oregon where the plume is generally absent during this season (Figure 4).

Monthly mean poleward flow during winter often increases from the mid- to outer-shelf or upperslope off the Washington coast. This result is confirmed in mean geostrophic data (Figure 4) and in individual data sets 1975, Figure 6; 1977 and 1973, Figure 3; 1980, Figure 7), as well as in the statistical mean (Figure 1). Off Oregon, on the other hand, both mean geostrophic data (Figure 4) and direct measurements (1975, Huyer et al., 1978; 1978, Huyer et al., 1984) indicate that monthly mean poleward flow decreases from the mid- to the outer-shelf.

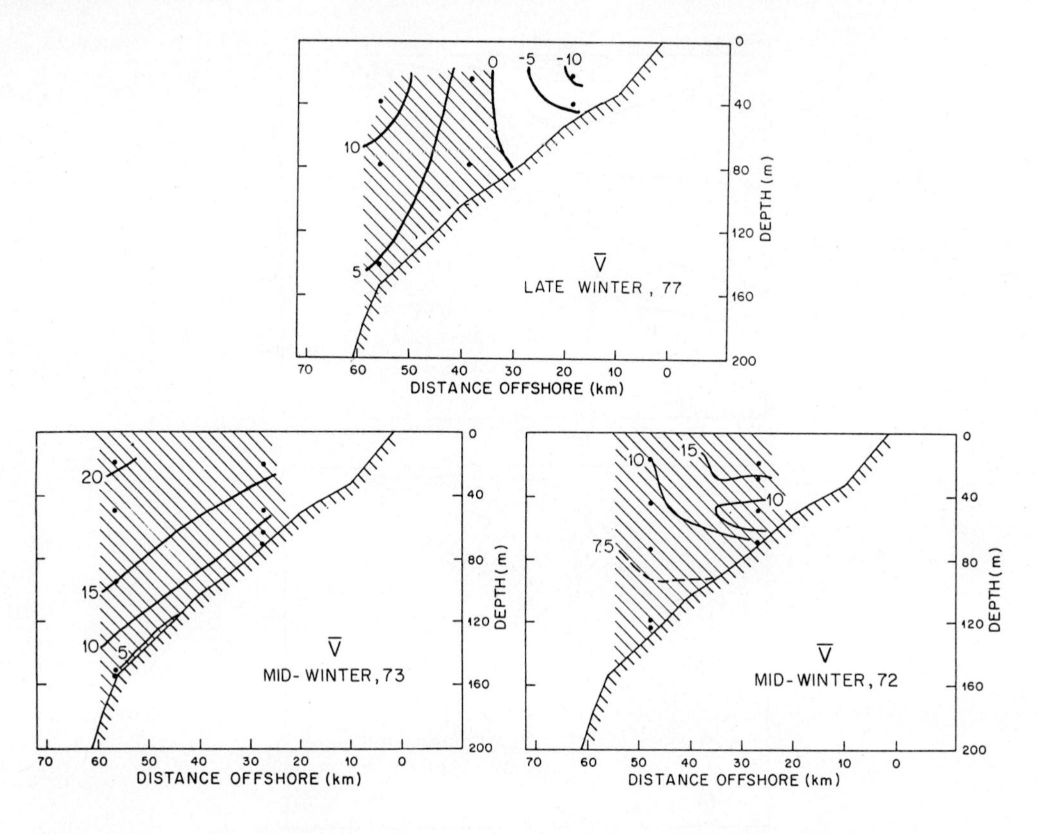

Figure 3: Cross-shelf sections of mean along-isobath velocity off the Washington coast during winter. Units are cm s^{-1}. Regions of poleward flow are shaded. Dots indicate locations of current meters.

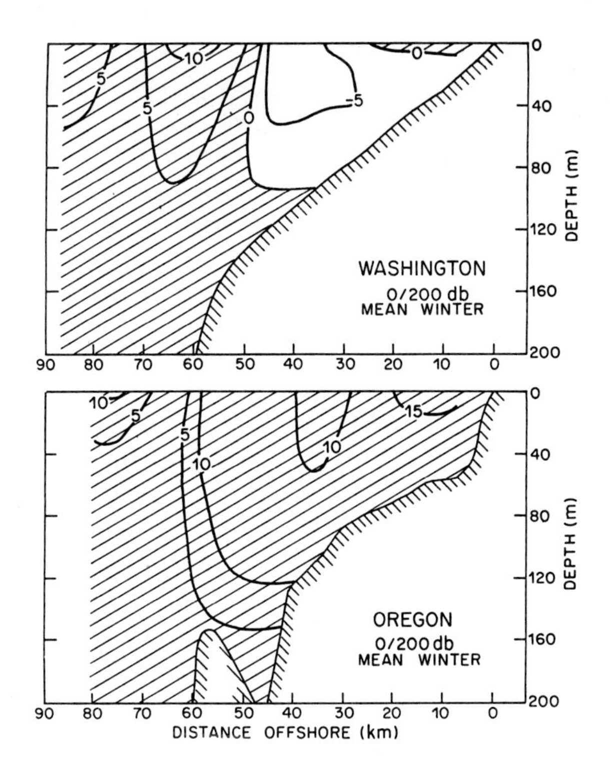

Figure 4: Geostrophic along-isobath velocity (0/200 dB) corresponding to seasonally averaged sigma-t sections for mid-Washington and mid-Oregon during winter. Units are cm s^{-1}. Data have been extrapolated into regions shallower than the reference level according to the method of Montgomery (1941). Regions of poleward flow are shaded. (Figure 2.13 in Hickey, 1989a.)

Washington Slope

Measurements have indicated that poleward flow occurs over the Washington slope below the surface layer (usually below ~100 m) throughout most of the year Figure 5, summer; Figure 6, winter). No data are available over the Washington slope during spring. However, data from the Oregon slope suggest that during spring, poleward flow is replaced with equatorward flow to depths of about 400 m (Huyer and Smith, 1976). During summer the flow over the Washington slope has the structure of a poleward jet, with maximum speeds occurring at shelf break depths (Figure 5). The jet appears to be trapped over the slope. During winter, there is no evidence to suggest that a poleward undercurrent persists; that is, that a subsurface maximum occurs (Figure 6). Rather, the limited data that are available off the Washington coast suggest that poleward flow below the

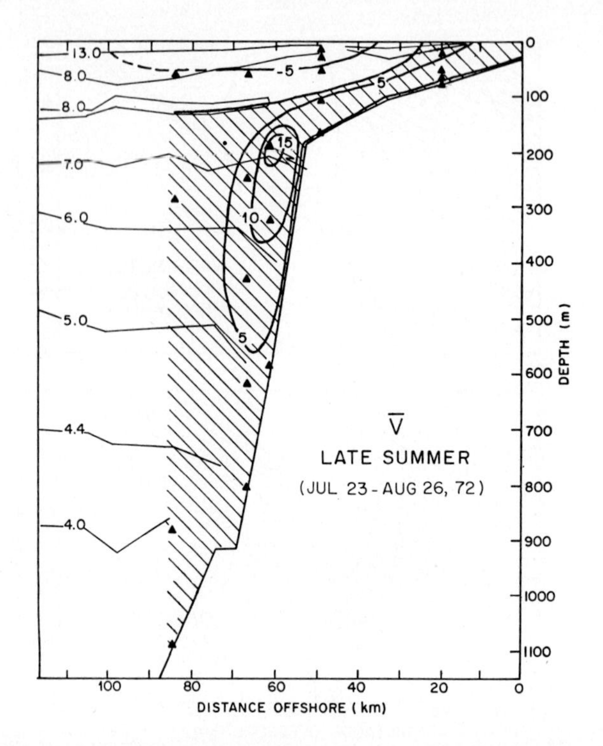

Figure 5: Cross-shelf section of seasonal mean along-isobath velocity off Washington during late summer 1972 and of temperature on 27 July 1972. Units of velocity and temperature are cm s^{-1} and °C, respectively. The data illustrate the cross-shelf structure of the poleward undercurrent. Regions of poleward flow are shaded. Locations of current meters are indicated by triangles. (Adapted from Figure 10 in Hickey, 1979.)

shelf break has a relative minimum near ~400 m (Figure 6) or that an equatorward undercurrent may even occur (Figure 7). This result seems consistent with available information on wind and along-shore pressure gradient (as measured by adjusted sea level slope); namely, in winter off Washington, along-shore wind stress and along-shore pressure gradient force are opposite in direction, and both are reversed in direction from the summer case (Figure 8). If a simple along-isobath balance of momentum is considered, it seems reasonable to expect the resulting current distributions to be opposite during winter and summer, as observed. Data from the Oregon slope during October and November 1977 clearly demonstrate the existence of persistent vertical shear in the water column below the shelf break (Figure 9). The shear is such that the tendency toward equatorward flow increases with depth. Maximum equatorward flow appears to occur at least as deep as 325 m (from the Oregon data) or 392 m (from the Washington data). Vertical shear of the same sign is also evident in mid-shelf data (Figure 9). However, the shear on the shelf is insufficient to produce persistent equatorward flow. Although one example of

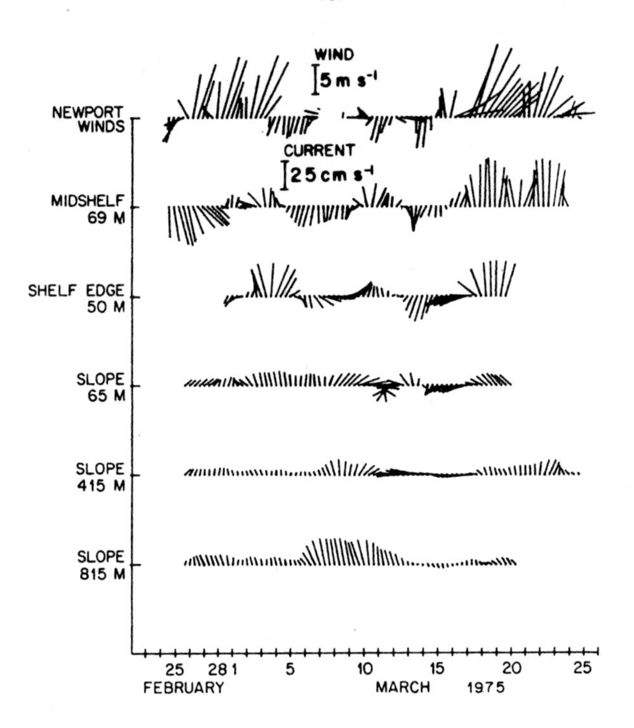

Figure 6: Selected sub-tidal (8-hourly) velocity vectors over the Washington shelf and slope during late winter 1975. Mooring depths are 94,182, and 1000 m for mid-shelf, shelf edge and slope locations, respectively. Winds at Newport, Oregon are also shown. (Adapted from Figure 12 in Hickey, 1979.)

equatorward flow over the slope during late winter has been related to the presence of an eddy (Huyer et al., 1984), we note that in this case the equatorward flow decreased rather than increased with depth. Since each of the three data sets available for the Pacific Northwest indicate the presence of persistent vertical shear over the slope below the shelf break during winter, it seems reasonable to expect that this phenomenon may be related to a seasonal reversal of the wind stress/sea level slope relationship rather than to the random occurrence of eddies.

In the above discussion concerning the origin of the poleward flow, the assumption has been made that poleward flow occurs as a result of an applied along-shore pressure gradient force in the presence of bottom stress (as in Csanady, 1978). The ultimate driving mechanism of the poleward flow is whatever is creating the along-shore sea level slope. Hickey and Pola (1983) demonstrated that both the amplitude and the phase of the seasonal and long-term mean signal in along-shore sea level slope in the Pacific Northwest could be reproduced using seasonal and mean along-shore wind stress data and the simple physics of the arrested topographic wave model (Csanady, 1978). The driving mechanism is a combination of local

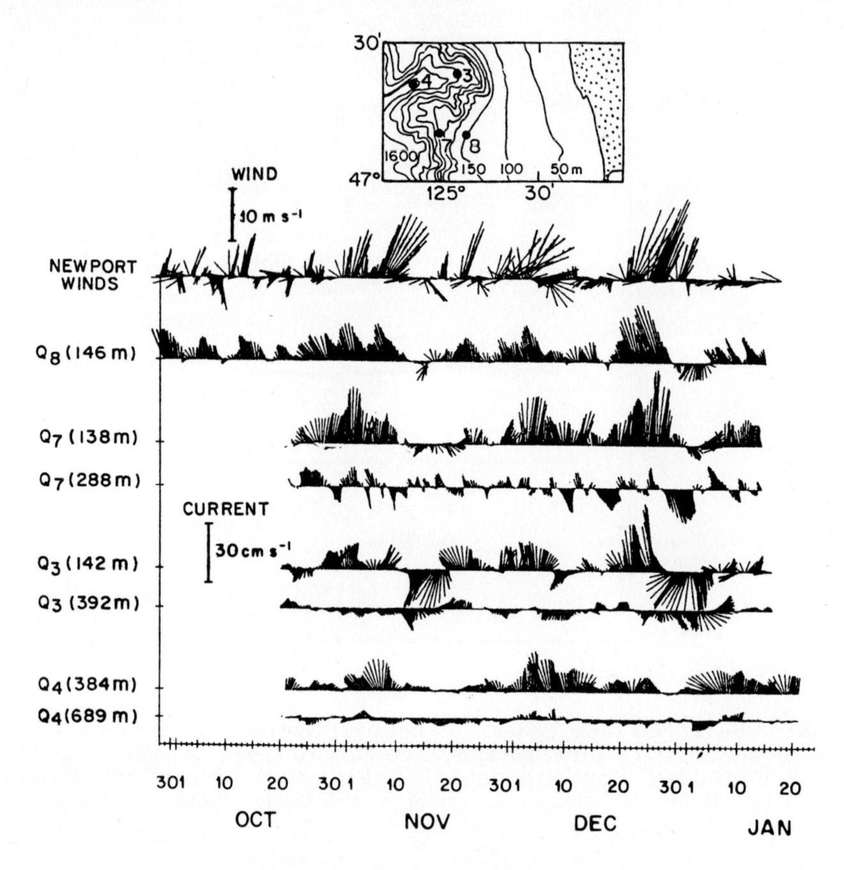

Figure 7: Selected sub-tidal velocity vectors in a north-south reference frame at locations on a section across the Washington shelf and slope during fall and winter 1980. Mooring locations are shown on the inset map. Winds at Newport, Oregon are also shown. (Adapted from Figure 9 in Werner and Hickey, 1983.)

and remote wind stress forcing. The model was not very successful off southern California where, nevertheless, a large along-shore sea level slope occurs. Recently, several other theories for poleward flow have been presented. These theories indicate that poleward flow can be generated by a variety of mechanisms; e.g., the interaction of nonlinear shelf waves or eddies with along-shore variations in topography (Holloway et al., this volume) or topographic form drag (Samelson and Allen, 1987). It is unclear at present what fraction of the signal can be attributed to each mechanism. However, intuitively, it seems that the availability of regions of distinct types of topography and distinct seasonal forcing should allow the relative importance of the proposed mechanisms to be separated.

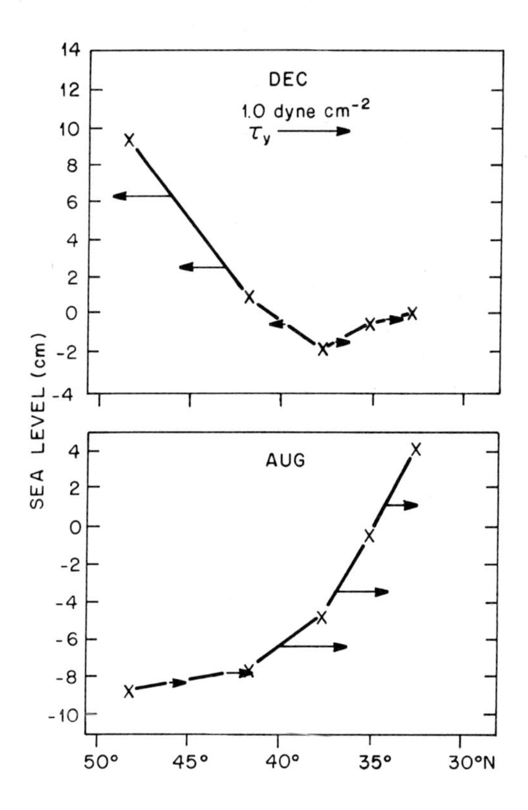

Figure 8: The north-south component of selected sub-tidal velocity time series on a section across the Oregon shelf and slope at 45°20'N during fall and winter 1977. Note scale change for current time series. The north-south component of wind on the Columbia River Lightship (46°10'N) is also displayed. Shelf (E) and slope (O) moorings are in water depths of 110 and 600 m, respectively. At the 50 m depth over the slope only direction was measured. Time series were constructed for that meter using a uniform speed of 10 cm s^{-1}.

Southern California Bight Slope

Poleward flow in the California Bight over the upper slope inshore of the Channel Islands is remarkably more persistent than that over the Washington slope. Poleward flow occurs at 40 m and 100 m depths everywhere except on the seaward side of the Santa Monica basin during winter and over the slope off San Onofre in the upper water column during late fall and spring (compare Figure 10 for southern California with Figures 5 and 6 for Washington). Off Washington, on the other hand, equatorward flow occurs from April through September in the upper 100 m over the slope. We note that flow over the open slope off San Onofre is less poleward than that off Santa Monica, but still more poleward than that off the Washington slope. Results similar to those for Santa Monica have been reported just north of the Santa Monica basin in another constricted region; namely,

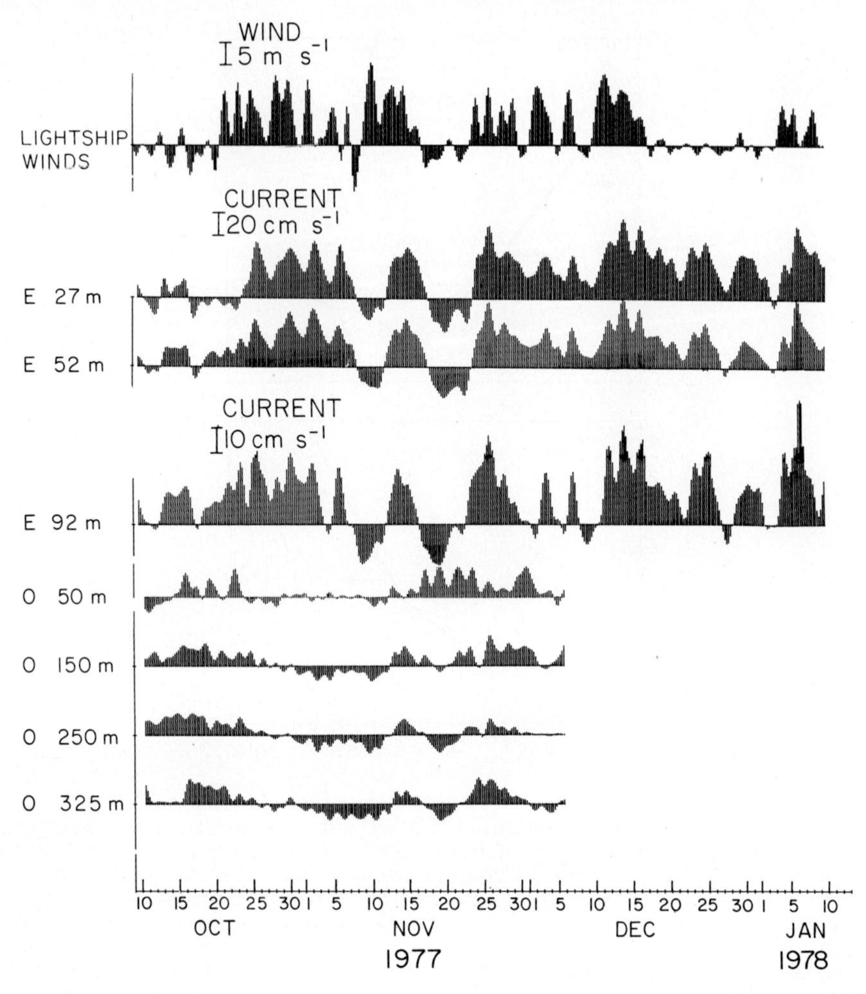

Figure 9: Monthly mean adjusted sea level height and the north-south component of wind stress along the west coast during typical winter and summer months. Wind stress data are from Nelson (1977). Sea level height data are from Hickey and Pola (1983).

the Santa Barbara Channel (Brink and Muench, 1985). Only a single ~one week reversal to equatorward flow was observed along the north side and eastern entrance to the channel at a depth ~30 m during the observation period of April to June 1983. Data from year-long (1984) current records in the channel also indicate poleward monthly means throughout the year at the ~30 m depth. Only infrequent reversals to equatorward flow occur even at this relatively shallow (30 m) depth (Gunn et al., 1986). At least at the eastern entrance to the channel, reversals are more common during spring (several occurred during that period) and almost never occur after July (only two occurred between July and February). The reversals almost never reached the deeper measurement depths of 80 m and 126 m.

ALONG-ISOBATH VELOCITY

x 40m

• 100m

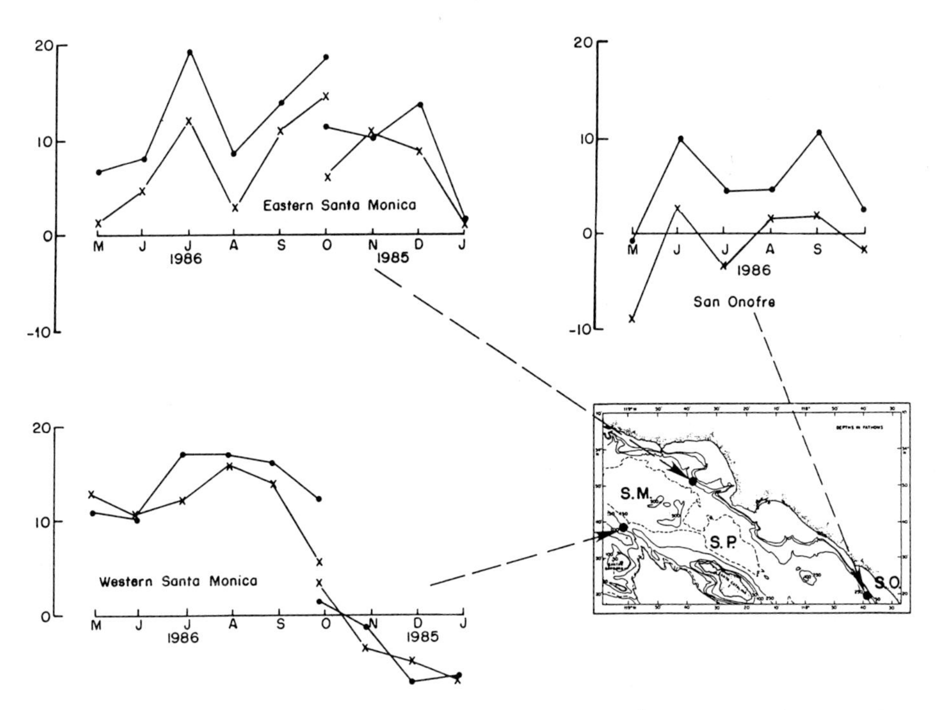

Figure 10: Monthly mean speed at selected locations over the continental slope (Z_b = 800 m) in Santa Monica/San Pedro basin and off San Onofre. Units are cm s^{-1}. Mooring locations are shown on the inset map. The data years are presented out of order to allow the semblance of a continuous record. A positive speed indicates flow directed poleward; a negative speed indicates flow directed equatorward. Since the flow is roughly paralled to local isobaths, positive and negative speeds conform almost exactly to poleward and equatorward components of along-isobath velocity.

No directly measured data are yet available over the Santa Monica slope from February to April, and no data are available above the pycnocline over the slope during any season. While it is possible that equatorward flow may occur in a thin (<20 m) surface layer during spring and summer, available satellite data usually suggest that the surface layers move poleward through the region during the spring season (not shown). Instruments presently in the water will resolve the question of the direction of the surface layer flow over the slope. Deployments during 1988 will determine whether flow reversals occur at deeper depths during the spring, as reported for <u>shelf</u> locations just outside the Bight (Strub et al., 1987). We note, however, that large-scale geostrophic data presented by Hickey (1979) suggests that no reversals occur in the seasonal averages at depths 200 m in the northern half of the Bight, even during spring.

The amplitude of the seasonal variation is similar to that reported for the shelf outside the Bight (Figure 10) (Strub et al., 1987). The seasonal variation of the amplitude is not inconsistent with the existence of a semi-annual signal in the poleward flow. A semi-annual signal in this region, with maxima in late summer/early fall and winter, has been observed in large scale geostrophic data (Hickey, 1979). These large scale results have recently been substantiated with direct measurements on the Point Conception Shelf (Strub et al., 1987). In the present data set, the seasonal maxima along the eastern boundary of the basin occur in July (in the Santa Monica-San Pedro Basin) or June (south of the basin) and between September and December. The scatter in the data makes the evidence barely convincing. The flow on the western side of the basin is directed equatorward after October. Note that the December maximum is obtained from data in 1985. During 1986 poleward flow stronger than the winter maximum in 1985 occurs as early as September. Clearly inter-annual variability is significant in the seasonal cycle in the Bight, as it is off the Washington coast.

The poleward flow in the Bight appears to have a subsurface maximum during much of the year (Figure 10); that is, it has the structure of an undercurrent. This is in contrast to the case off Washington, where the poleward undercurrent structure disappears or perhaps even reverses during fall and winter. However, as off Washington, the vertical shear is generally a maximum during summer and early fall and a minimum in late fall and winter. The difference between the two locations is qualitatively consistent with the observed differences between the wind and sea level slope distributions. Off Washington both along-shore wind stress and along-shore sea level slope reverse direction seasonally (Figure 8); off southern California they do not reverse except perhaps during January; rather they are always directed the same as occurs off Washington during summer, so that the tendency for the existence of a poleward undercurrent persists throughout the year. Note that the amplitude of the sea level slope has a seasonal cycle that is similar to that of the seasonal cycle of vertical shear: the maximum poleward decreasing sea level slope occurs in late summer and early fall in both regions, corresponding to the maximum in vertical shear in both regions; the minimum sea level slope occurs in winter (near zero off southern California, large, but decreasing equatorward off Washington) corresponding to the minimum in vertical shear off southern California and a minimum or a reversal in shear direction off Washington (Hickey and Pola, 1983).

The flow in the Bight does not appear to be trapped over the slope as is the case off Washington, except perhaps during winter (Figure 10). During the period of strongest poleward flow, the flow extends at least

40 km to the other side of Santa Monica Basin, with almost uniform strength across the basin. The strength of the seasonal mean flow is of the same order as that off Washington.

Southern California Bight Shelf

The shelf in the Southern California Bight is not continuous like the Washington shelf; rather, several bays cut into the coastline in the Northern Bight (Figure 10). The shelves in these bays are relatively wide (~20 km), although still only half the width of most of the Washington shelf. Between the San Pedro shelf and Northern Baja, the shelf width is relatively uniform, but narrow (~3-10 km). Over both types of shelves, flow appears to be equatorward, at least in the near-surface layers, during much of the year, in contrast to the flow over the Washington shelf. For example, on the wide Santa Monica shelf, measurements to date suggest that the seasonal mean flow is equatorward at 5 m and 10 m from the surface in 30 m of water from October to February, and offshore or equatorward at a depth of 25 m from May to October (Hickey, 1989b). Similar results have been reported for the narrow (~3 km) shelf just north of San Diego (Winant and Bratkovitch, 1981). Equatorward flow was observed 5 m from the surface in 15, 30, and 60 m water depths in every season, and in spring and fall as deep as 10 m from the surface at the deeper measurement sites. The strongest equatorward flow occurred during winter, when the equatorward flow extended throughout the water column at each of the three sites. The magnitude of the monthly mean equatorward flow on both the narrow and wide shelves (~5-10 cm s^{-1}) is much less than that of the poleward flow over the upper slope. The only significant poleward flow on the narrow shelf occurs during summer (August data, Winant and Bratkovitch, 1981) in the lower half of the water column at the 30 m and 60 m bottom depths: the typical undercurrent that we expect on this eastern boundary. No undercurrent was observed in the monthly means at 5 m above the bottom in a depth of 30 m on the wider Santa Monica shelf.

On the Santa Monica shelf, equatorward flow during fall and winter decreases offshore: flow even in the surface layers at the outer shelf monthly mean flow is poleward rather than equatorward (Hickey, 1989b). On the narrower San Diego shelf, equatorward flow during winter and summer increases between the mid- and outer-shelf measurement sites (Winant and Bratkovitch, 1981). The data are insufficient to determine the location of the flow maximum.

SUMMARY

In this paper differences and anomalies in west coast seasonal flow structures have been highlighted. In particular, it was emphasized that flow off Washington has significant differences from that of Oregon; namely, during summer, flow at mid-shelf is more poleward off Washington, and during winter, flow on the inner-shelf is more equatorward off Washington than off Oregon. The former result may be related to the poleward decrease in the longshelf wind stress; the latter result may be related to the presence of the Columbia River plume. Off southern California the near-surface flow over the shelf is more persistently equatorward than that off Washington. Conversely, the flow over the slope in the upper 100 m of the water column is more persistently poleward than that off Washington. Also, the undercurrent structure, that is, a subsurface maximum, is maintained at least from summer to early winter off southern California (no data are yet available from spring), but only during summer and early fall off Washington. We note that the seasonal cycle of vertical shear in the two locations is similar, although a reversal in sign sometimes occurs off Washington.

ACKNOWLEDGEMENTS

This work was supported by the Department of Energy under Grant DE-FG05-85ER60333#4 and by the National Science Foundation under Grant OCE 86-01058#1.

From: Adriana Huyer, College of Oceanography, Oregon State University, Corvallis, OR.

On: Review and Commentary to paper **POLEWARD FLOW NEAR THE NORTHERN AND SOUTHERN BOUNDARIES OF THE U.S. WEST COAST**, by Barbara Hickey.

Although this paper covers some of the same ground (or water) as my own, I think the two are not redundant. This paper gives a rather detailed view of the structure and variability of the poleward flow at the "ends" of the California Current System as it flows along the U.S. West Coast. Although the data sets are rather fragmentary, Hickey does manage to draw conclusions about local differences in structure. She clearly shows the variability of the poleward flow on time scales of days-to-weeks by including several time-series plots of current vectors.

POLEWARD FLOWS OFF MEXICO'S PACIFIC COAST

A. Badan-Dangon
J.M. Robles
J. García

CICESE. Apdo. 2732
Ensenada, B.C., Mexico

ABSTRACT

Poleward flows are predominant off the west coast of Mexico. Off the peninsula of Baja California, the northern third of the Mexican Pacific coastline, one finds the southern extension of the California Current System. Poleward flows in this region occupy a broad domain of about 200 km width, adjacent to the coast, and include an intensified countercurrent that can reach 50 cm s^{-1}, often within 20 km of the shelf and slope. From late winter to early summer, the upwelling favorable northwesterly winds oppose the poleward flow at the surface and the current proceeds as an undercurrent that hugs the continental slope. This poleward flow advects Subtropical Subsurface Water from the tip of Baja California into the Southern California Bight and farther north. The coast of mainland Mexico is exposed, on the contrary, to a broad extension of the equatorial return flow that feeds the North Equatorial Current from the Equatorial Countercurrent. This flow is known as the Costa Rica Coastal Current; it extends around the Costa Rica Dome, and northwestwards to the mouth of the Gulf of California, where the Eastern Pacific Transition Zone usually abutts the continent. This poleward flow is most prominent during the fall and may reverse directions in the spring, appearing as an intensified California Current extension. The strength of the large scale drifts, the surface expression of the counterflows, and the latitude at which the currents meet, all vary seasonally and depend on the intensity of the dominant winds. Direct observations of the poleward flows off Mexico are few and restricted to northern Baja California. They confirm a counterflow is probably present throughout the year, mostly over the shelf and slope regions, but with its surface expression obliterated by the upwelling favorable northwesterly winds that blow strongly in the late winter and spring. There are no direct measurements to provide a detailed description of the Costa Rica Coastal Current and, in particular, to know whether it possesses a counterflow.

MOTIVATION

Poleward flows are important, maybe even dominant, components of eastern boundary current systems. They are noteworthy for a number of reasons, the principal one being that their direction of motion often opposes that of the dominant winds and wind-driven currents, whence they must be propelled by rather robust sources of momentum. These flows are known as countercurrents if they flow at the surface, or undercurrents when they do not (Hickey, 1979; this volume). Counterflows are effective at distributing water properties over long distances (Tibby, 1941; Wooster and Reid, 1963; Cannon, Laird, and Ryan, 1975; Reed and Halpern, 1976) and compete against the wind-driven advection to determine the hydrographical composition of nearshore waters (Bernal, 1981; Bernal and McGowan, 1981; Chelton, 1981). Counterflows often provide the source water for the Ekman pumping offshore in eastern boundary currents (Chelton, 1982) and for upwelling at the coast (Yoshida and Tsuchiya, 1957; Smith, 1968,1981; Huyer, 1983); they thus contribute importantly to coastal ocean frontogenesis and to the nature of the jets and squirts that result from cross-shore exchanges with the open ocean (Mooers and Robinson, 1984; Davis, 1985; Huyer and Kosro, 1987). They also influence the propagation of internal tides and inertial waves (Mooers, 1975), and alter the exchanges with the atmosphere in the coastal zone (Tont, 1981). All this, of course, has substantial implications in matters such as the chemistry of nearshore waters (Codispoti, 1981) and variations in the structure of the coastal ecosystem as manifested, for example, by the alterations of the salmon migration routes (McLain and Thomas, 1983) or by the modification of plankton distribution patterns (Peláez and McGowan, 1986; Chelton, Bernal, and McGowan, 1982). The west coast of Mexico constitutes the southeastern boundary of the North Pacific Anticyclonic Gyre, and the northeastern limit of the eastern equatorial circulation, which contributes water from the Equatorial Countercurrent into the North Equatorial Current. As a result, poleward flows are prevalent in the entire region, in the first case as a classical example of a counterflow to midlatitude eastern boundary currents (Wooster and Reid, 1963), and in the second because the equatorial return flow constitutes the Costa Rica Coastal Current. The Costa Rica Coastal Current is a broad poleward flow that originates around the Costa Rica Dome, proceeds along the coast of the Mexican mainland, and merges in various proportions with the southern extension of the California Current to feed the North Equatorial Current (Wyrtki, 1965, 1966).

The present report summarizes the configuration and scales of variation of poleward oceanic motions off the Pacific seaboard of Mexico. It is organized as follows: Section 2 provides a broad geographical description, based mostly on large scale hydrographical surveys and ship drift compilations reported in the literature; Section 3 describes observations that detail the nearshore scales of the counterflows off northern Baja California, the only region off Mexico where direct measurements have been conducted; Section 4 summarizes some of the efforts that have been made to explain the origins of counterflows, and hypotheses from which future studies in the region might draw support.

THE POLEWARD FLOWS ON THE LARGE SCALE

Most of the Mexican west coast (Figure 1) is found in North America, extending from the border with the United States at 32°N, to the Gulf of Tehuantepec at 16°N; only a portion of about 300 km, the eastern side of the Gulf of Tehuantepec, is in Central America (Urrutia-Fucugauchi, 1988). The continental shelf is seldom wide, with the notable exceptions of some portions off Baja California, where it exceeds 100 km in width, and on the eastern side of the Gulf of Tehuantepec. This otherwise regular coastline, with a northwest to southeast orientation, is interrupted between Cabo San Lucas (23°N) and Cabo Corrientes (21°N) by the broad entrance to the Gulf of California, a marginal sea over 1500 km in length that lies to the east of the peninsula of Baja California. Two major wind-driven current systems compose the oceanographic makeup off the west coast of Mexico. These are the North Pacific Anticyclonic Gyre, whose southeastern limit constitutes the California Current System, and the extension of the Equatorial System into the Eastern Tropical Pacific, bounded by the coasts of Central and North America. The North and South Equatorial Currents flow to the west, separated by the eastward flowing Equatorial Countercurrent, located just north of the equator, between 2°N and 12°N. This major current penetrates seasonally into the Eastern Tropical Pacific, and a branch of it turns around the Costa Rica Dome, close to 8°N; 89°W, proceeds along the coast of Mexico, and feeds the North Equatorial Current.

The large scale configuration of the wind and current systems that encompass oceanic motions of the Eastern Tropical Pacific Ocean have been described abundantly. Early maps of these flows were offered by Puls (1895), and more detailed studies have been reported by Cromwell and Bennett (1959), Austin (1960), Reid (1961), and by Bennett (1963). Extensive data compilations and reviews of the circulation and water masses of the region have been prepared by Wyrtki (1964a, 1964b, 1965, 1966, 1967) and Tsuchiya (1974). Our description is taken largely from these latter works, and we reproduce the convenient summary made by Baumgartner and Christensen (1985) for their study of interannual

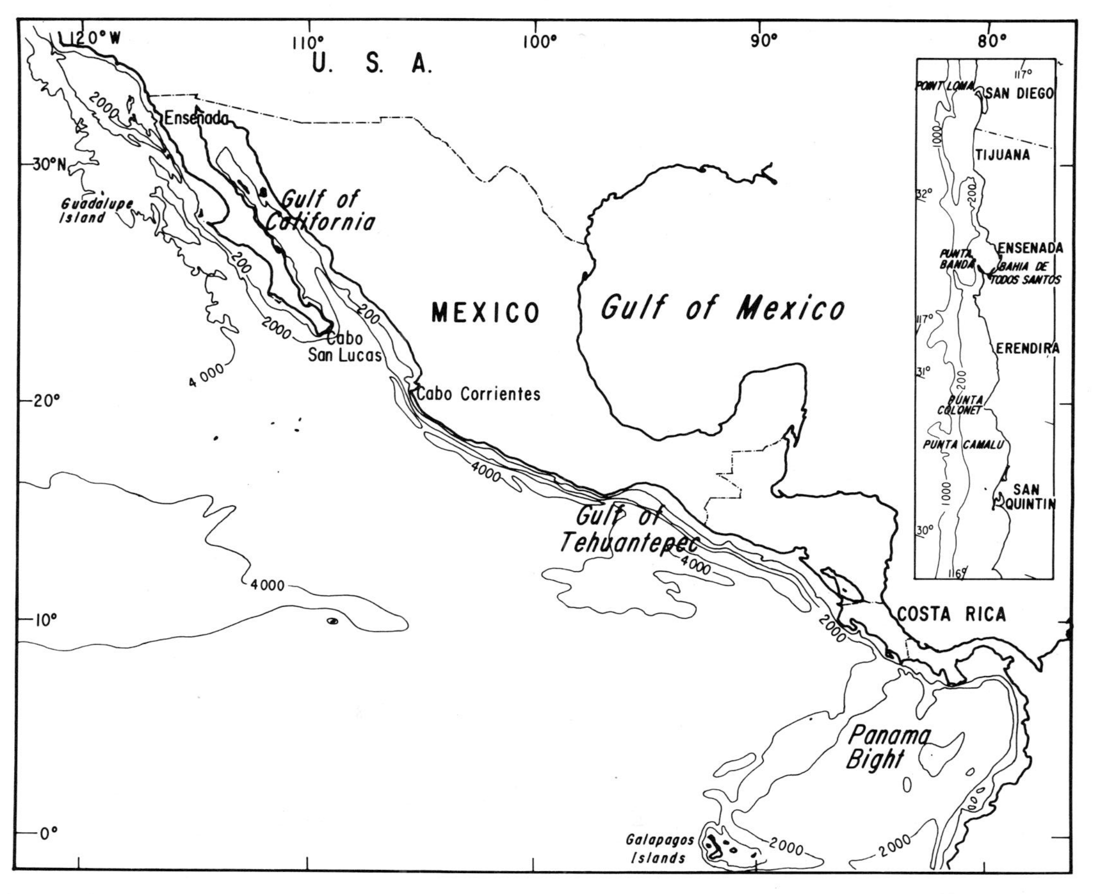

Figure 1: Configuration of the coast and of the continental shelf in the California Current and Eastern Tropical Pacific Regions off Mexico. The inset details the portions of the coast where direct measurements of poleward flows have been conducted off northern Baja California.

fluctuations in the Gulf of California (Figure 2). Other studies have included the Eastern Tropical Pacific because of the interest in the role of the Equatorial region in the generation and propagation of the El Niño signal (e.g. see Journal of Geophysical Research, Vol. 92C13).

The major pressure cells that determine the conditions of the wind field are the North and South Pacific centers of high pressure, and the Aleutian and Indonesian centers of low pressure. The positions and strengths of these large wind systems induce the fluctuations of major current systems in the equatorial and eastern portions of the Pacific Ocean. In addition, in a more local context, the Mexican Low is an important generator of fluctuations in the California Current System and in the Gulf of California (Baumgartner and Christensen, 1985). Eastern boundary currents in the subtropical gyre are characterized by equatorward winds that shift progressively into the Trade Winds. These are separated close to the equator by the doldrums, centered about the Intertropical Convergence Zone, whose position fluctuates considerably with the seasons; it is found roughly coinciding with the northern limit of the Equatorial Countercurrent at 12°N in the northern summer, and close to 2°N in the northern winter. These two extremes correspond to the times when the Equatorial Countercurrent is most developed in the northern hemisphere summer, but essentially non-existant in winter east of 130°W. Therefore, the equatorial climate is shifted to the north by about 10°; subtropical conditions reach only to about 20°N in the northern hemisphere, but can be found reaching the equator at the Galapagos Islands in the southern hemisphere.

Oceanic flows in the Eastern Tropical Pacific are well related to the local winds (Reid, 1948); they are typified by a pair of counter-rotating gyres centered at about 8°N; 89°W, and 5°N; 88°W. The northernmost is cyclonic and projects a well known hydrographical structure, the Costa Rica Dome. The circulation in the Eastern Tropical Pacific is intimately linked to these structures; an eastward current proceeding from the Equatorial Countercurrent flows between the two and splits into two branches (Roden, 1962). The southern one turns west, and eventually joins the South Equatorial Current. The northern branch flows around the Costa Rica Dome, whence it becomes the Costa Rica Coastal Current, a flow along the coast of Central America and off mainland Mexico. It reaches a latitude of confluence with the California Current, which is the eastern terminal of the North Pacific Transition Zone (Roden, 1970, 1971, 1972), and separates from the coast to feed the North Equatorial Current. The Costa Rica Coastal Current is of extreme importance in the context of the present discussion, for not only is it a swift poleward current in its own right, it is also the most likely process by which the Tropical Surface Water

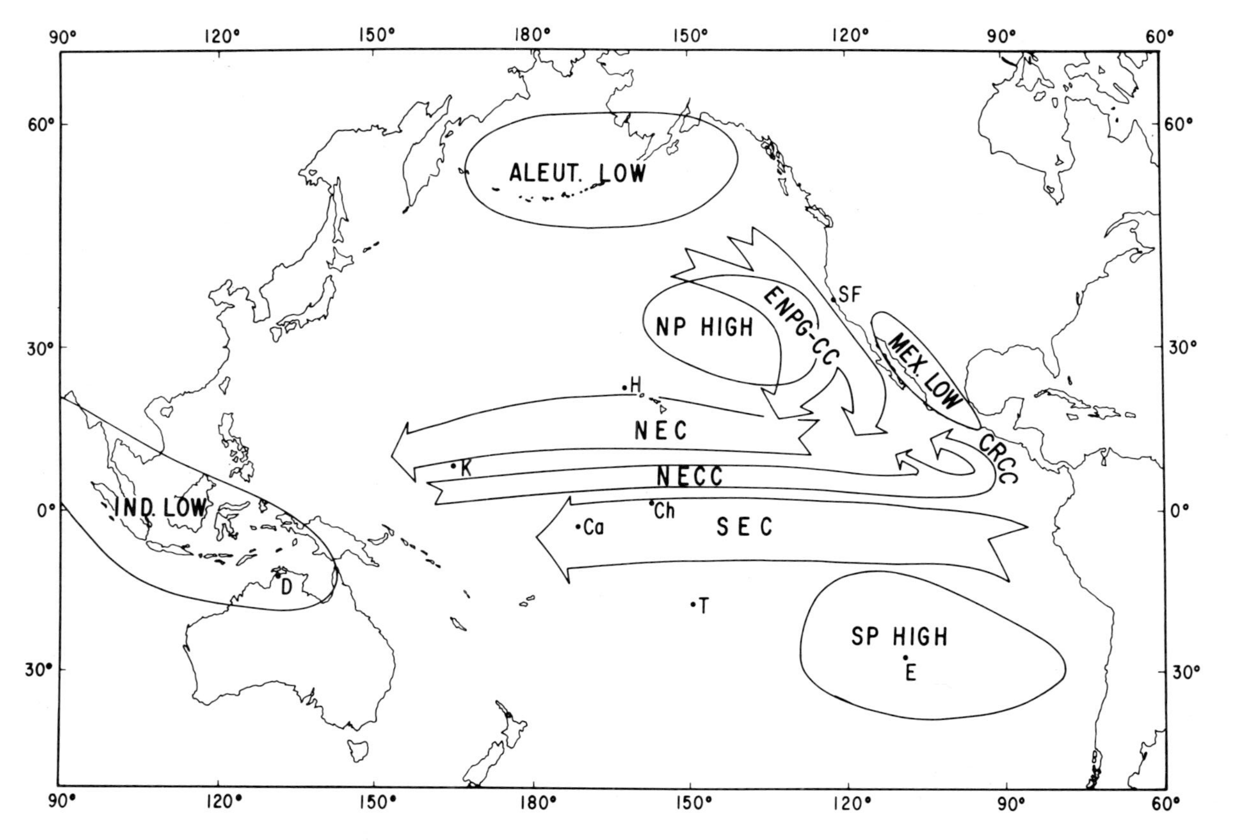

Figure 2: Locations of atmospheric pressure centers (Aleutian, Mexican, and Indonesian Lows; North Pacific and South Pacific Highs) and ocean surface current systems (Eastern North Pacific Gyre-California Current, North Equatorial Current, North Equatorial Countercurrent, South Equatorial Current, and Costa Rica Coastal Current) that contribute to the circulation off the Pacific coast of Mexico. The locations of some coastal (San Francisco) and mid-Pacific island sea level stations are also shown (H: Honolulu; K: Kwajalein; Ch: Christmas; Ca: Canton; T: Tahiti). From Baumgartner and Christensen (1985).

from the equatorial region and the Subtropical Subsurface Water, which originates in the Central Pacific, are recirculated and brought to the mouth of the Gulf of California (Wyrtki, 1967, Reid and Mantyla, 1978); there, a portion of the Subtropical Water feeds the California Countercurrent through a path that remains unknown (Simpson, 1983; Baumgartner and Christensen, 1985).

This entire scheme of circulation varies markedly with the seasons. Wyrtki (1965, 1966) distinguishes three periods throughout the year, summarized in Figure 3, which are characterized by varying intensities and changing configurations of the currents. The most stable and longest period lasts from August through December and is a well intensified version of the basic configuration described above. The Intertropical Convergence Zone is at its northernmost position, close to 10°N, and the Equatorial Countercurrent is fully developed; it flows past and around the Costa Rica Dome, and feeds the equally strong Costa Rica Coastal Current, which reaches past the Gulf of Tehuantepec and to the mouth of the Gulf of California. From the north, the California Current is inferred to leave the coast at about 25°N, and supplies the North Equatorial Current only north of 20°N. By January, the Intertropical Convergence Zone moves south, the Equatorial Countercurrent weakens, and the California Current System intensifies accordingly. The second period lasts from February through April. The Intertropical Convergence Zone is at its southernmost position, about 3°N, the Equatorial Countercurrent is essentially absent, but the two eddies off Central America remain (Tsuchiya, 1974). The Costa Rica Coastal Current is very weak, and the flow north of the Gulf of Tehuantepec is to the southeast, providing an impression of an extremely developed California Current System, which supplies most of the water for the North Equatorial Current. However, neither drift bottles (Crowe and Schwartzlose, 1972; Badan-Dangon, 1972) nor drogues (Poulain, Illeman, and Niiler, 1987) have traced the surface flow from the California Current System along the coast past Cabo Corrientes, at 21°N; the transition zone off the mouth of the Gulf of California appears unaffected by the intensification of the California Current System, and the composition of the southeastward flow off the mainland of Mexico in late winter and early spring remains unexplored. During the third period, from May to July, the Intertropical Convergence Zone returns to its northern position, and the Equatorial Countercurrent forms again, flows past the Costa Rica Dome and drives the Costa Rica Coastal Current to Cabo Corrientes. The California Current remains strong and contributes still a considerable fraction of the North Equatorial Current, although the southeastward flow does not penetrate much into the Eastern Tropical Pacific. From July to August the California Current weakens progressively.

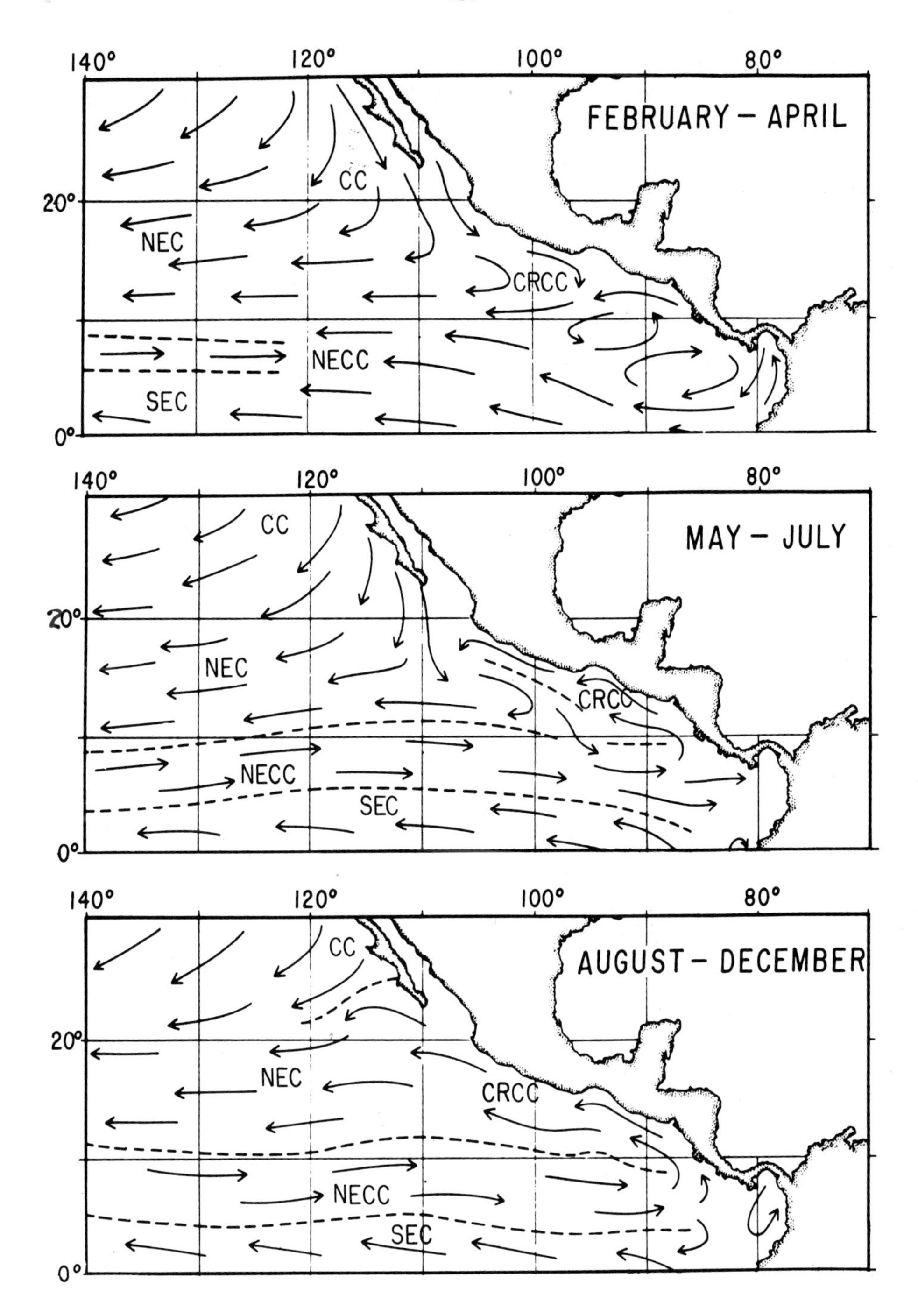

Figure 3: Principal seasonal circulation patterns of wind driven surface currents in the Eastern Tropical Pacific. Abbreviations are as in Figure 2. Redrawn from Wyrtki (1965).

In the northern portion, off Baja California, the circulation consists of the southern extension of the California Current and conforms to the classical scheme of an eastern boundary current, as described by Wooster and Reid (1963). Careful summaries of earlier knowledge on the California Current System have been prepared by Reid, Roden, and Wyllie (1958), and Hickey (1979) later integrated additional data to an extensive literature review, which remains the most complete report on this current system, although various studies have added to its recent understanding (Chelton, Bernstein, Bratkovitch, and Kosro, 1987; Lynn and Simpson, 1987; Kosro, 1987; Winant, Beardsley, and Davis, 1987). A poleward undercurrent appears to be present essentially everywhere along the peninsula of Baja California (Chelton, 1982), with its maximum usually at about 20 km from the slope, but which can reach 100 to 150 km in width (Chelton, Bratkovitch, Bernstein, and Kosro, 1988), and at depths probably not exceeding 600 m. This flow must originate at least as far south as the southernmost extension of the California Current itself, since the common hydrographical signature of the undercurrent is that of the Subtropical Subsurface Water, of relatively warm, saline, of high nutrient but low dissolved oxygen contents; the hydrographic distribution suggest this flow becomes narrower north of 30° (Figure 4). The typical T-S values in the undercurrent change remarkably little along the peninsula of Baja California, indicating that lateral mixing is probably not very important along its path and due mostly to the impingement of eddies offshore (Wickham, 1975; Lynn and Simpson, 1987); close to 9.5°C and 34.6 ppt near Cabo San Lucas (Hickey, 1979), they remain above 9.0°C and 34.4 ppt in the core of the undercurrent off northern Baja California (Wooster and Jones, 1970; Barton and Argote, 1980). By comparison, Hickey (1979) reports values of 7°C and 33.0 ppt in the counterflow off Vancouver Island.

The geostrophic computations of Wyllie (1966), some of which (Figure 5) have been recalculated with closer detail by Hickey (1979), and the compilations of historical ship drift data kindly provided to us by Mr. A. Bakun (Figure 6), illustrate the seasonal variations of the principal features of the California Current. Poleward flow is present at depth off Baja California most of the year, manifested mostly as a sequence of cyclonic gyres along the coast, of a scale comparable to the supposed width of the counterflow. The Southern California Countercurrent (Sverdrup and Fleming, 1941) or eddy (Schwartzlose, 1963) is the most important of these gyres. It occurs off northern Baja California and southern California as the flow of the California Current, partially shielded from the dominant northerly winds by the change in direction of the coast in the Southern California Bight (Figure 1), moves onshore and a portion branches north in association with the countercurrent (Reid, Schwartzlose, and Brown, 1963). Oddly, this onshore tendency of the California Current and its

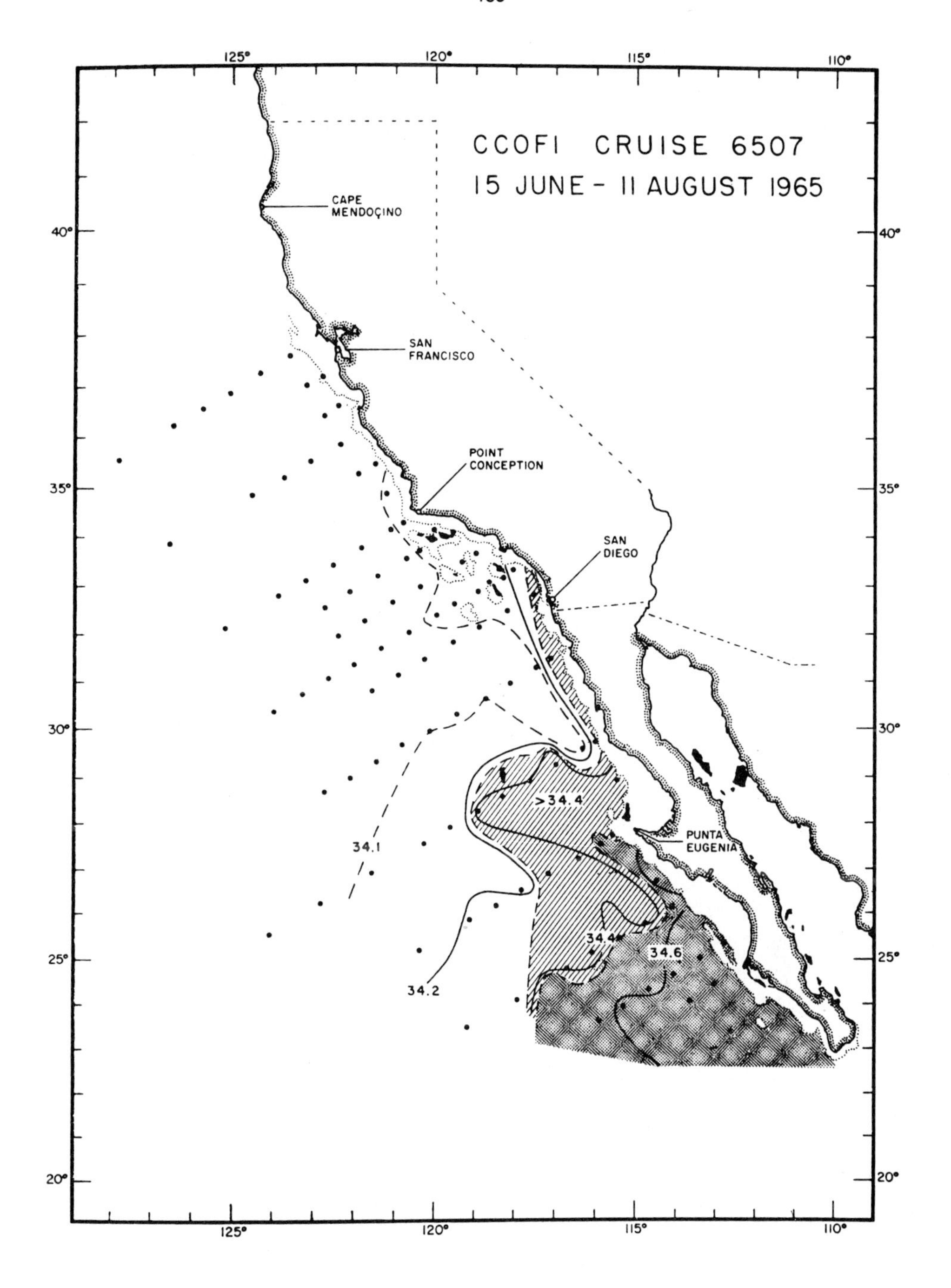

Figure 4: The signature of the California Undercurrent in the salinity field on the isanosteric surface 150 cl t^{-1}. From Wooster and Jones (1970).

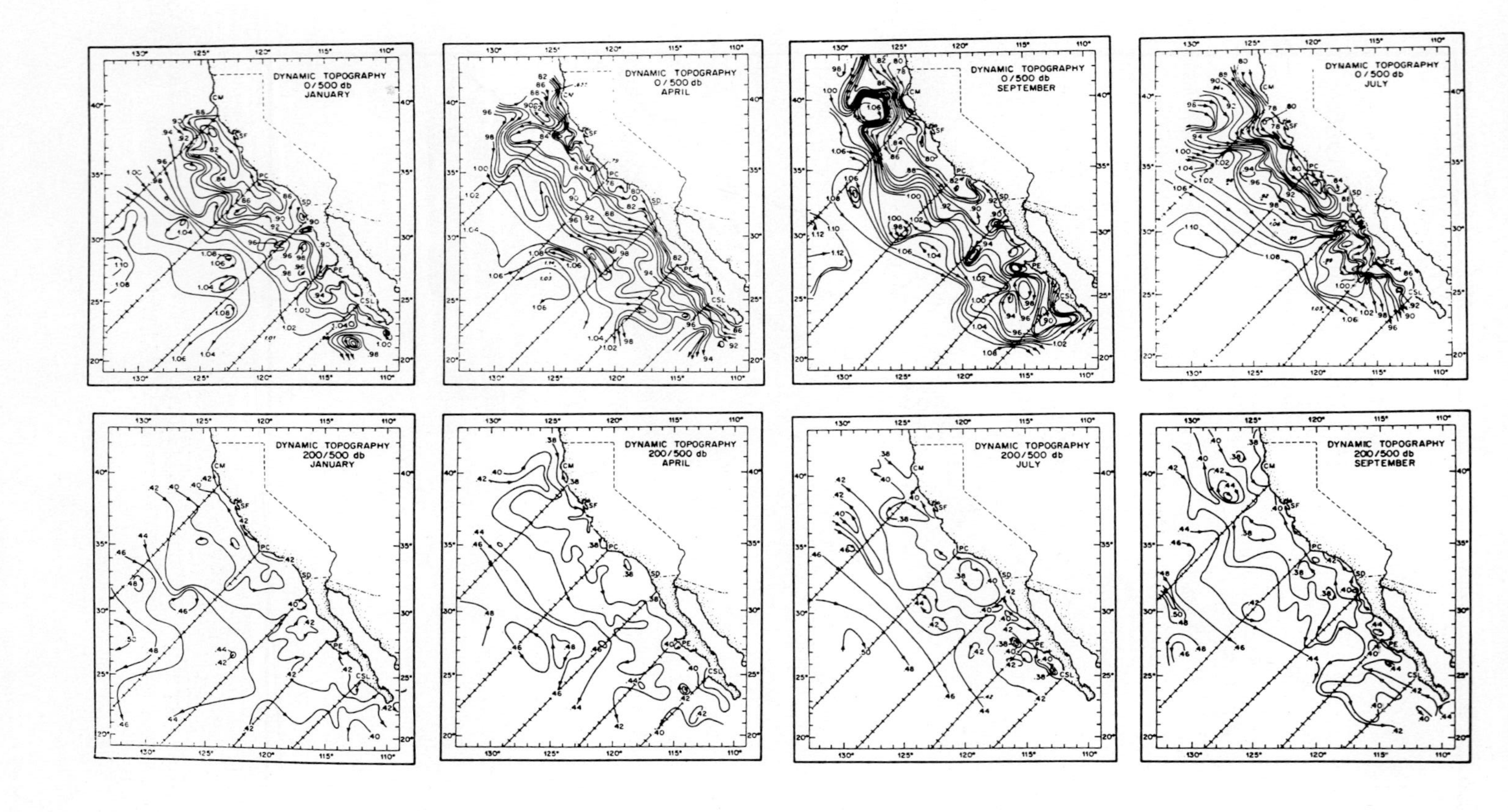

Figure 5: The seasonal variations of the California Current system at the surface and at 200 m depth. From Hickey (1979).

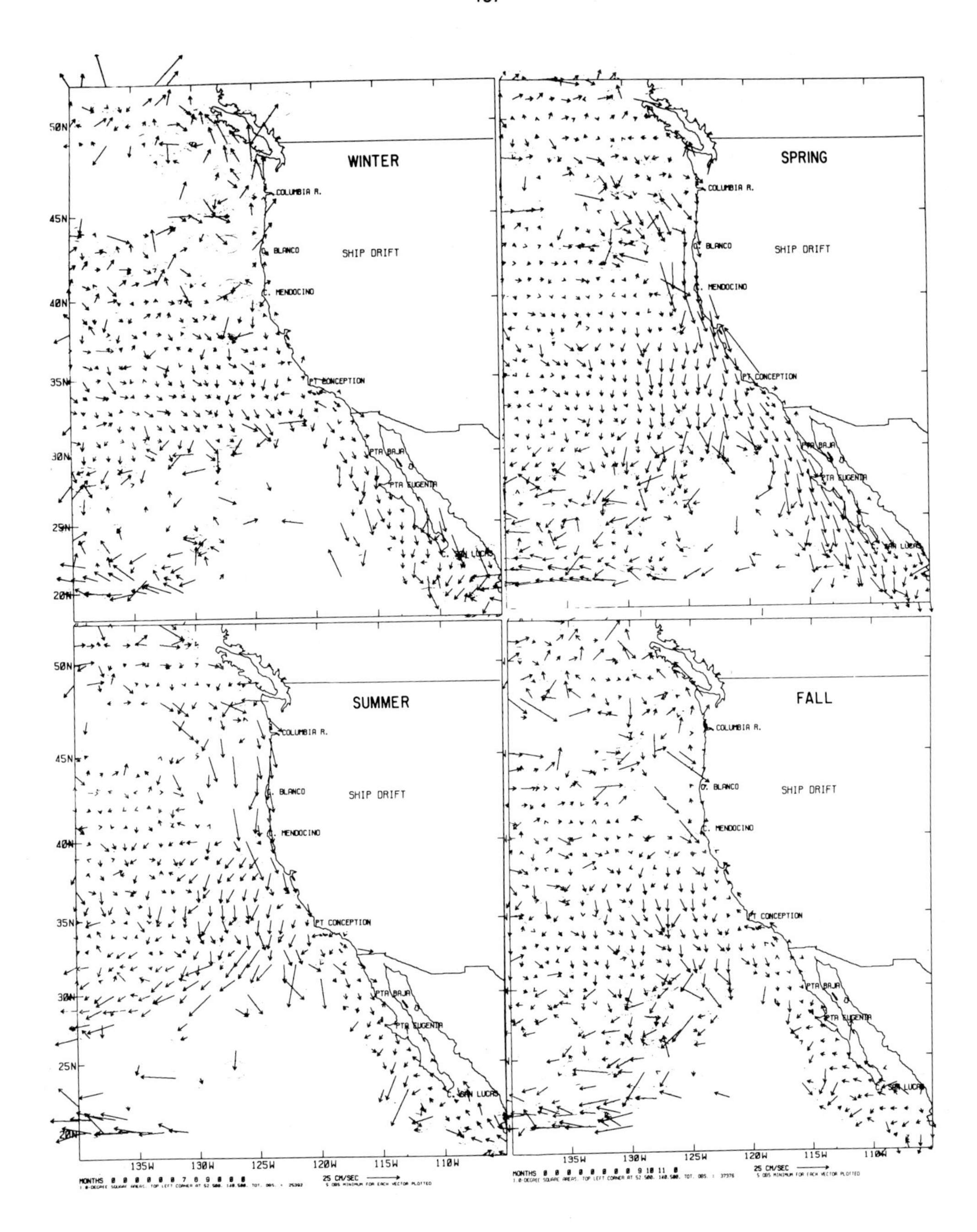

Figure 6: Historical surface ship drift in the California Current region, compiled by seasons in 1° squares. Each vector includes a minimum of five observations, the variance ellipses represents the standard error of the mean, and the entire set contains over one hundred years of data. Courtesy of A. Bakun.

resulting diffluence, which have been reported variously from geostrophic calculations (Wyllie, 1966; Reid and Mantyla, 1976, Hickey, 1979), inferred from drift bottle data (Schwartzlose, 1963; Badan-Dangon, 1972) and are supposed to cause important frontogenesis (Peláez and McGowan, 1986), are reflected neither in the ship drift data (Figure 6) nor in the mean flow computed from repeated drogue observations, but rather in the orientation of their principal axes of variance (Poulain, Illeman, and Niiler, 1987), suggesting that fluctuating flux processes should be responsible for this feature of the California Current. During the fall and early winter the shoreward side of this series of eddies appear to coalesce into a single, continuous northward flow with speeds that can attain 25 cm s^{-1}, and the Southern California Countercurrent may well flow past Point Conception and connect with the poleward flows north of that point (Reid, 1960, McLain and Thomas, 1983). Except during spring, and particularly in April, when the northwesterly winds blow persistently (Figure 6) and cause substantial upwelling, the undercurrent surfaces and constitutes a countercurrent that extends from Cabo San Lucas into the southern California Bight. The varying intensity of the winds modify the rates of advection and cause the interannual differences observed in the California Current System (Chelton, 1981). Lynn (1983) and Simpson (1983) suggest that an enhanced poleward advection results during El Niño years, in accordance with the results of McCreary (1976); Simpson (1984), however, proposes an intensified onshore advection as the principal modifying factor.

The large scale circulation off Mexico involves several water masses. The California Current consists mostly of Pacific Subarctic Water proceeding from the westwind drift; close to shore off Baja California, the surface temperature increases to the south from about 15°C to 20°C in the winter, and from 20°C to 25°C during the summer. The surface salinity ranges from about 33.5 ppt off the border with the United States, to 34.0 ppt off southern Baja California, with little seasonal changes. Farther offshore, the excess of evaporation over precipitation in the Central Pacific forms the mass of Pacific Central Water of subtropical, warm and salty, characteristics; it mixes laterally with the Subarctic Water on the edge of the California Current. The Equatorial System, and therefore the Costa Rica Coastal Current, is dominated by an excess of precipitation over evaporation; the Tropical Water carried to the north is warm and less salty. It meets the Pacific Subarctic Water of the California Current in a transition zone indicated by sharp horizontal salinity gradients (Roden, 1970, 1971). This system of fronts is usually found in the vicinity of Cabo San Lucas, at the tip of the Baja California Peninsula, where it includes some of the water that originates in the Gulf of California (Griffiths, 1965, 1968). Its position fluctuates seasonally with the strength of the currents involved and with the type of water flowing onto the North Equatorial Current. Other water masses found in the region include various proportions of Pacific Subarctic and Equatorial Waters that have combined at depth and are brought to the surface by upwelling; this explains the occasional presence nearshore of cool, high salinity water (Gómez Valdéz, 1984). No evidence has been reported of the presence of Gulf of California Water beyond a few tens of kilometers away from the mouth of the Gulf itself.

The upper structure of the California Current is now known to be dominated by intense geostrophic turbulence (Davis, 1985; Poulain, Illeman, and Niiler, 1987) and this is also probably the case with the Costa Rica Coastal Current. Gyres, whirls, and cross-shelf jets contribute most of the exchange between the nearshore coastal ocean and the offshore circulation climate (Stumpf and Legeckis, 1977). In certain locations, such as off Tehuantepec, the intense atmospheric forcing by the Tehuano winds that cross over the mountain ranges from the Gulf of Mexico (Blackburn, 1962, Alvarez, Badan-Dangon, and Valle, 1989), or from the incidence of tropical hurricanes (Enfield and Allen, 1980) provides not only peculiar localized conditions, but affect the entire coastal ocean of the Norteastern Pacific through the possible induction of a pressure gradient along the coast (Csanady, 1980) and the poleward propagation of sea-level events (Christensen, de la Paz, and Gutierrez, 1983). The mouth of the Gulf of California acts as a process of selection to the propagation of these subinertial events (Mejía, 1985); waves of certain frequencies enter the Gulf, where they dissipate and should contribute to its internal circulation. Lower frequencies appear to bypass the Gulf and proceed poleward along the west coast of the Baja California Peninsula (Enfield and Allen, 1983). The exact nature of this frequency selection has not yet been elucidated and may be accompanied by complex wave scatterings off the tip of Baja California.

DIRECT OBSERVATIONS OF POLEWARD FLOWS

The only direct observations of the coastal poleward flows off Mexico have been done on the shelf and slope off northern Baja California. They are relatively few but all provide useful measures of some of the scales involved. Amongst the earliest experiments, Reid (1963) released parachute drogues at 250 m depth, in a line extending offshore from San Quintin Bay, near 30°N (Figure 1). Only those within 50 km of the coast showed motions towards the northwest, with mean speeds of about 8 cm s^{-1}. Motions farther offshore were either to the southeast, as would be expected of the California Current drift, or close to zero. The results were compared to those made earlier near Cape Mendocino, off northern California (Reid, 1962), where drogues deployed within 70 km of the coast indicated similar net poleward motions, with speeds close to 25 cm s^{-1}. Both sets of measurements were interpreted as indicating the top of the California Counterflow, and some form of continuity between the two was inferred. However, they each lasted only a few days and could also have been a manifestation of the numerous meanders and gyres that were later described by Reid, Schwartzlose, and Brown (1963) and extensively documented by Poulain, Illeman, and Niiler (1987). The first study specific of the undercurrent off Baja California was done by Wooster and Jones (1970) off Punta Colnett, at 31°N, and included hydrographic surveys and short sets of current measurements. Their work, later complemented by Robles, Morales, García, and Flores (1981), García (1983), and Torres-Moye and Acosta (1986), all illustrate the undercurrent as an intrusion of warmer, saltier water of high nutrient and dissolved oxygen contents within 20 km of the slope, centered at about 300 m depth, and detectable to within 150 m of the surface (Figure 7). Above that level, isotherms rise to the surface near the coast, but within the undercurrent, they sink rapidly towards the slope. In those same sections, the southerly flow of Subarctic Water of the California Current is indicated by the layer of lower salinity water above 100 m. The relative currents computed from geostrophy suggested speeds close to 10 cm s^{-1} at 200 to 300 m depth. Direct measurements indicated that speeds can reach 40 cm s^{-1} in the core of the undercurrent, and at 400 m the speeds were still poleward at 20 cm s^{-1}. The hydrographic sections suggest that high levels of mixing can occur locally within the undercurrent. This configuration is remarkably like the one described later by Hickey (1979) off the northwestern United States.

Motivated by the evidence of well developed upwelling off the coast of northern Baja California, Barton and Argote (1980) developed a series of measurements near 31°N, off Punta Colonet, in summer 1976. This field effort, compiled by Morales, Argote, Amador, and Barton (1978), included detailed hydrographic measurements over the shelf and slope during times of strong upwelling, and provided direct measurements of the countercurrent. The hydrographical sections illustrate a configuration of the isopleths very similar to those reported by Wooster and Jones (1970), or those shown in Figure 7, with salinity values above 34.2 ppt typical of the Subtropical Water. The indications of this study are that at this latitude the counterflow did rarely climb onto the shelf, possibly prevented by the strong northwesterly upwelling favorable winds. Short term direct current measurements, made over the shelf with a current meter profiler, showed the water column to be flowing south most of the time, with the exceptions

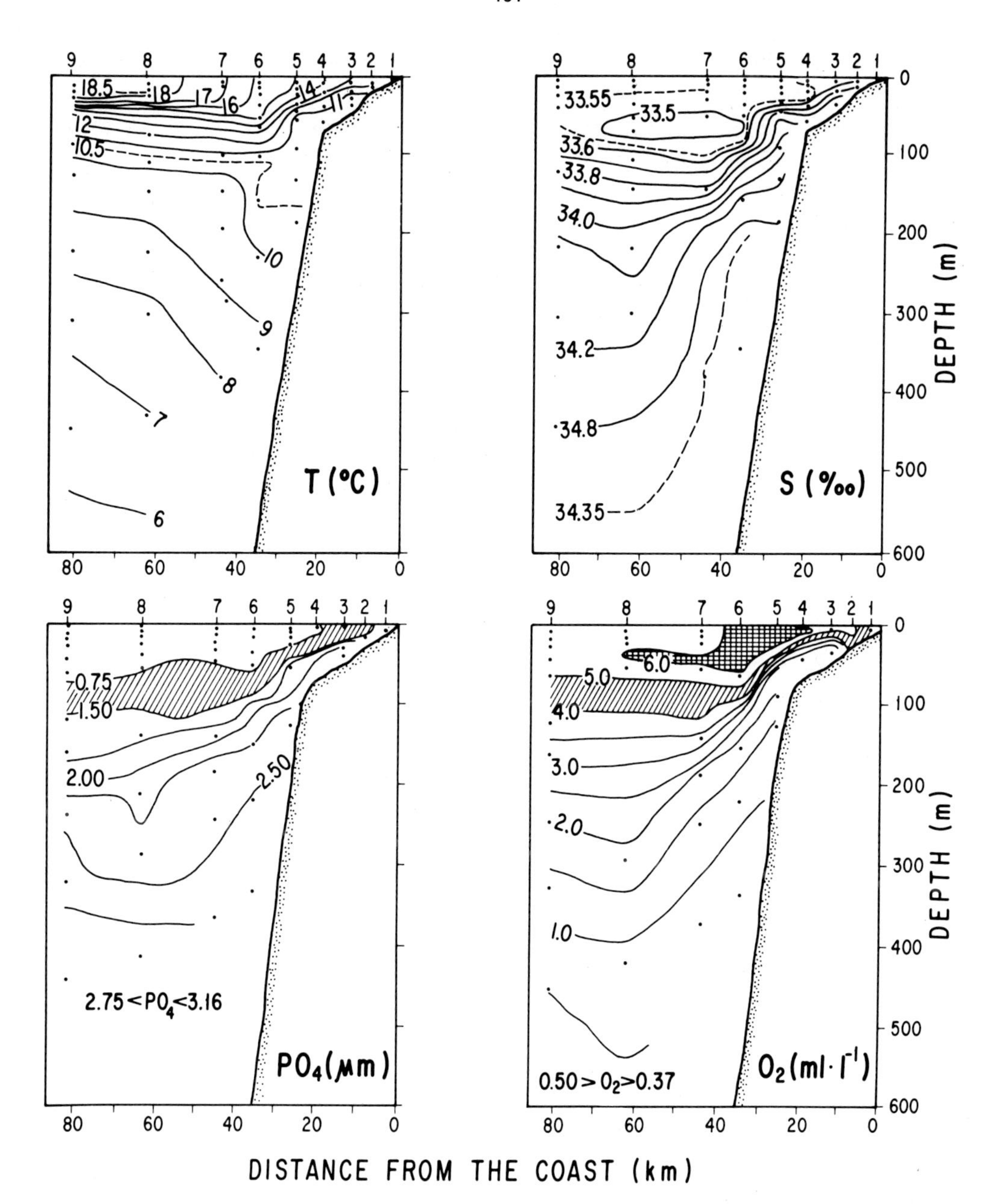

Figure 7: Hydrographic cross-sections of the California counterflow off northern Baja, California. The T and S sections are from Garciá (1983), the PO_4 and O_2 sections were redrawn from Torres Moye and Acosta (1986).

of a few small events of northward flow that suggested the countercurrent is able to climb onto the shelf. The short term variations, on the order of a few days, of the hydrography over the slope were consistent with the observed events of poleward flow over the shelf.

A later experiment was performed by Barton, Robles, Amador, and Morales (1980) off Ensenada and off Punta Camalú, straddling the locations of the earlier measurements at Punta Colnett. Direct moored current meter measurements, with a duration of one year, were made in 75 m of water over a complicated shelf off Todos Santos Bay. From October 1978 through April 1979, the mean flow was equatorward at 25 m depth, but poleward at 60 m. From April through September, the flow was poleward at all levels, with typical values of 5 cm s^{-1}. A second set of measurements was done at about 100 km to the south, from April through July, 1979. The mean flow there was more equatorward than at the northern station, but the 60 m mean flow was still directed poleward. If the mean flow suggested that a weak protrusion of the countercurrent generally exists over the shelf, there were significant events of poleward flow at all levels and at both stations, with typical amplitudes of about 40 cm s^{-1}. The current fluctuations were found to be uncorrelated with either winds or coastal sea level, in good agreement with other observations off southern California, 140 km to the north of Ensenada (Winant, 1980; Winant and Bratkovitch, 1981; Winant, 1983). The longshore current fluctuations on event scales were found to be weakly correlated with the sea level, but this relationship broke down during the summer. It appears unlikely that topographic complexities should be the source of such large fluctuations; it would rather seem that the large scale turbulent flows offshore should be responsible. Indeed, this coastal experiment showed at least one instance of anomalously large poleward events, coinciding initially with a seasonal sea level minimum at Ensenada, and continuing for a week as sea level recovered from that minimum (Figure 8). As no clear explanation existed for the event, it was tentatively attributed to the impingement of a large scale eddy on the coast (Barton, 1985). This interpretation is supported by Christensen and Rodriguez (1979), who leveled the tide gauges at Ensenada and at Guadalupe Island, distant some 300 km offshore, by considering the effects of spatially averaged subinertial currents on the sea level differences. They showed that sea level fluctuations at the offshore station, which is located in the middle of the California Current, are of the same magnitude than those at the coast, indicating that the effects of poleward currents are not necessarily confined to a narrow coastal zone. Most of the energy was found in the 15 to 30 day band, and the relative strength of the currents varied by as much as 10 cm s^{-1}. Clearly, poleward flows can occur away from the coast, but are probably intermittent and associated with eddy structures of the sort shown by

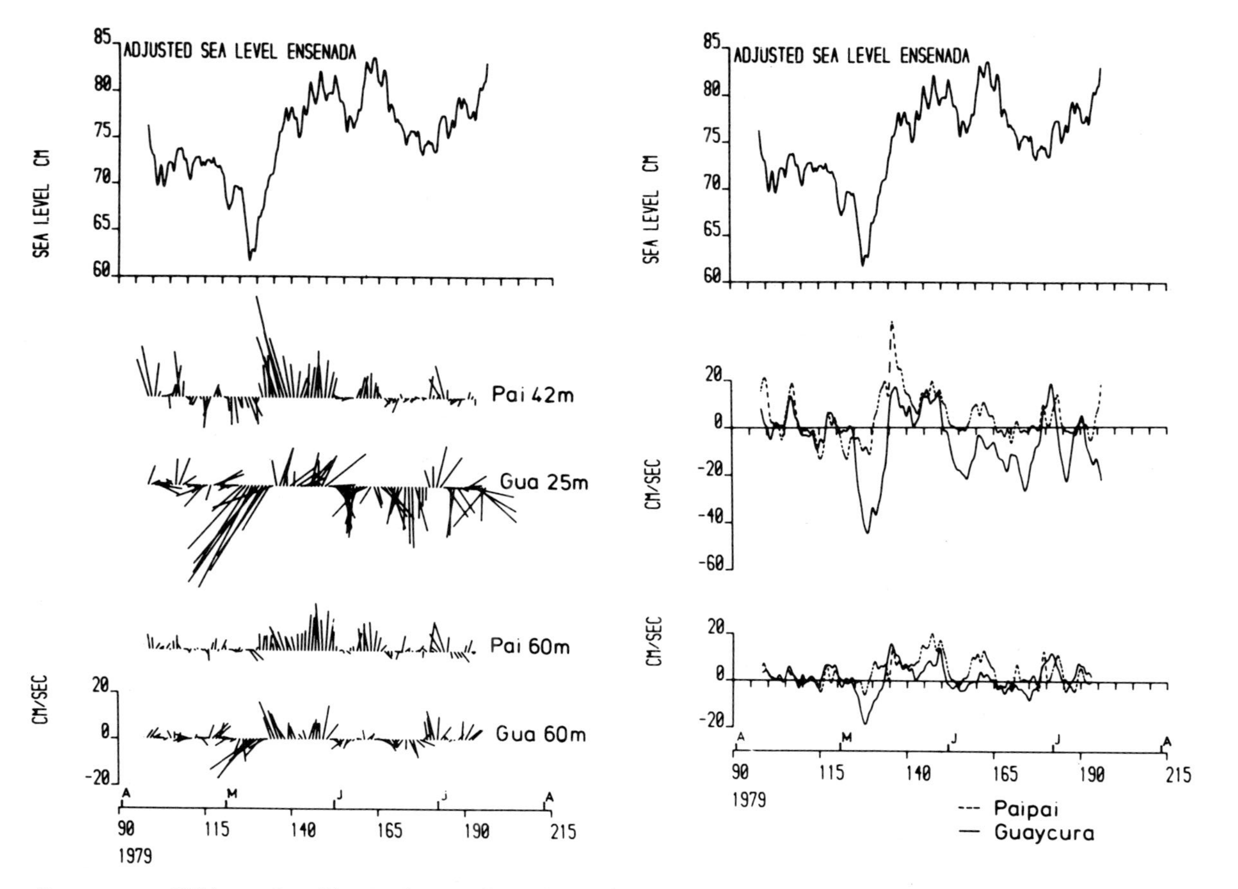

Figure 8: Low-pass filtered adjusted sea level at Ensenada and currents off Ensenada (Paipai) and off Punta Colnett (Guaycura), during the supposed passage of and eddy offshore. Left: Daily current vectors; vertically up is positive poleward along the coast. Right: Longshore component of the currents. Julian days 90 and 215 are 31 March and 3 August. From Barton (1985).

Poulain, Illeman, and Niiler (1987). Additional measurements following that period, which should confirm these results and provide more stable statistics are being analyzed.

Other studies of a more local character have been done recently off northern Baja California, consisting of short series of shallow water current measurements aimed at observing the coastal response and pollutant dispersal effects in the small scale. The first one was done off Eréndira, some 50 km south of Ensenada, in a band within 10 km of the coast (Alvarez, Durazo, Pérez, Navarro, and Hernández, 1984). The data are exemplified by the series shown in Figure 9; in the afternoon of 22 February, the wind was a steady northwesterly around 3 m s^{-1}, and the surface flow proceeded in the same direction with a speed close to 20 cm s^{-1} (Series A). During the evening of that day and the early hours of the next, the wind diminished to average close to zero; the recently redeployed floats quickly turned offshore and proceeded towards the northwest at a speed that reached 20 to 40 cm s^{-1} (Series B). Redeployed closer to shore on the afternoon of the 23rd, the floats proceeded along the coast under a well developed afternoon northwesterly breeze, but turned offshore and northwestwards during the night, as the wind died down again (Series C). These observations indicate the presence of the poleward flow close to shore, but locally obliterated at the surface by the diurnal winds, and provide a relaxation time scale from the effects of the wind in this region of only a few hours. This is congruent with the response characteristics proposed by Brink and Allen (1978) for the flow over a shallow shelf, and with the response time scales reported by Badan-Dangon, Brink, and Smith (1986). The appearance of a poleward flow soon after the subsidence of the upwelling favorable winds has been shown to happen also over the shelf and slope regions, with time scales of about one week (Kosro, 1987; Huyer and Kosro, 1987; Send, Beardsley, and Winant, 1987), as well as in broader portions of the California Current, extending over 100 km beyond the shelf break and persisting for several weeks or months (Chelton, Bratkovitch, Bernstein, and Kosro, 1988). A second series of measurements were done in July and September 1986 with a current meter deployed at 5 m off the bottom in 20 m of water, off the coast of Tijuana. The winds were light and variable northwesterlies throughout both periods of measurements. The flow, however, remained quite strong poleward at about 20 cm s^{-1} (Figure 10). The poleward flow diminished and reversed direction in late September, again without apparent relation to the wind, as had been noted earlier by Barton (1985). Accompanying Lagrangian series, not shown here, confirm the presence of poleward flow across the shallow shelf at this location (Alvarez, 1986).

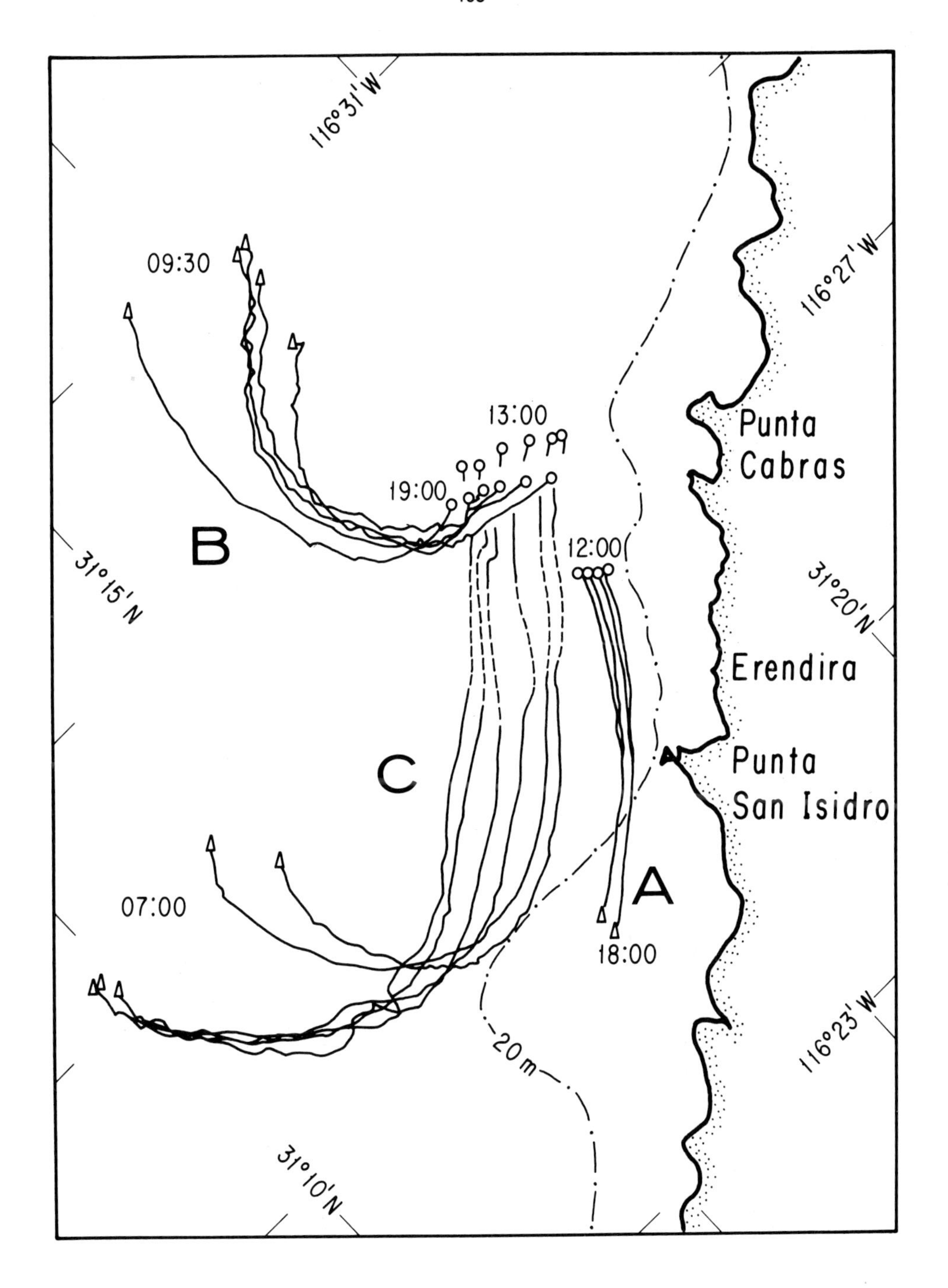

Figure 9: Lagrangian measurements of nearshore currents off Eréndira, Baja California, made in 1981. Series A begins at 1200 on 22 February 1981 and ends at 1800 on the same day. Series B starts at 1900 on 22 February and ends at 0930 on the next day. Series C begins at 1300 on 23 February and ends at 1700 on the 24 February. Redrawn from Alvarez et al (1984).

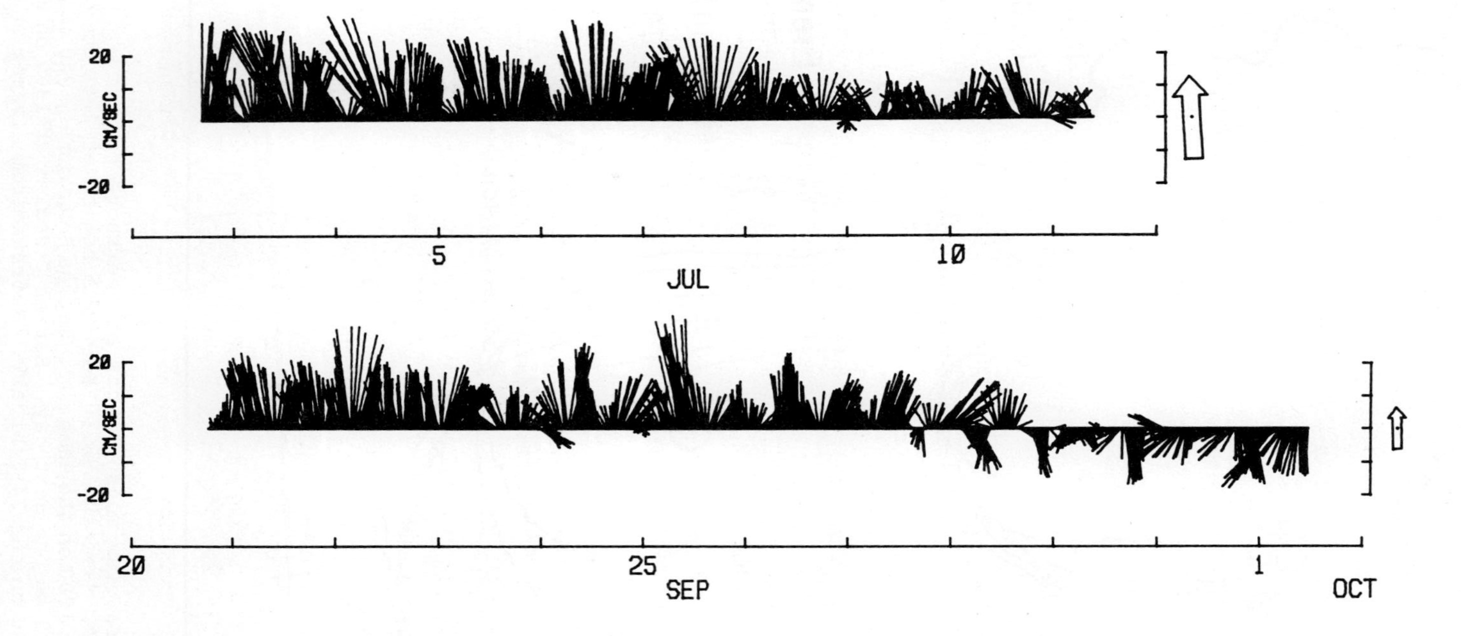

Figure 10: Vector time series of currents 5 m off the bottom, in 20 m of water off Tijuana, for the months of July and September, 1986. Vectors are drawn at 20 minute intervals and vertically on the page is positive to the north, which is also the local orientation of the coastline. The arrow on the right is the mean over the record length. Redrawn from Alvarez (1986).

In summary, direct measurements have shown that poleward flow is common over the shelf and slope off northern Baja California. The local winds obliterate the poleward flow at the surface, with a typical response time of only a few hours, but currents are otherwise weakly correlated to the wind. Barton (1985) concludes that the encroachment of the poleward undercurrent onto the shelf fluctuates seasonally, but not to the extent of being totally absent from it. Still, the distinction between undercurrent, surface countercurrent, and poleward flow over the shelf is confused, and the nature of their sources of momentum remains elusive.

THOUGHTS ON THE COUNTERFLOW

Numerous efforts have been made to understand the origins of the counterflow, undercurrent or countercurrent. Early investigations (Yoshida and Tsuchiya, 1957; Yoshida, 1967) noticed the close association of the poleward undercurrent with a rising thermocline at the coast and the presence of coastal upwelling under the influence of equatorward winds. Mass and vorticity adjustments were invoked for their nearly simultaneous development. Pedlosky (1974) indicated that equatorward winds induce equatorward currents, but that a positive wind stress curl near the coast drives a poleward undercurrent that might be coincident with upwelling, but not necessarily associated with it. Other studies have focused on the effects of the wind stress curl on scales larger than the coastal boundary, acting on the flow. Munk (1950) attempted the application of a Sverdrup balance on the California Current off Cape Mendocino, and obtained qualitative agreement with observations available then. Hickey (1979) noticed a good correspondence between the mean configuration of the poleward flow and the spatial configuration of positive wind stress curl along the coast (Nelson, 1977), but could not match their time variations on a seasonal scale. Chelton (1982), however, showed that the large scale positive wind stress curl that results from the offshore intensification of the northwesterlies is able to drive an Ekman pumping at 200 to 300 km off Baja California, and probably drives the broad poleward flow that is observed off the coast.

A longshore pressure gradient force is a well known manifestation of the inherent three-dimensionality of the coastal circulation maintained by the wind (Suginohara, 1974; Allen and Smith, 1981), and many investigators have examined it as the probable cause for the counterflow. Hickey and Pola (1983), following a formulation by Csanady (1978), provided estimates of a longshore pressure gradient off the northwestern United States, that matched the sea surface slope proposed by Sturges

(1974). Chelton (1984) indicates the poleward flow in the California Current is coherent with a large scale along-shore pressure gradient computed from differences in dynamic height, but which appears to be independent of the wind field. Models by McCreary (1981), McCreary, Shetye, and Kundu (1986), and Wang (1982), successfully developed indications of a counterflow as a result of a longshore pressure gradient force. Werner and Hickey (1983) added a longshore pressure gradient as an external force to a two-dimensional, baroclinic finite difference model of the coastal zone first developed by Hamilton and Rattray (1978) and modified by Hickey and Hamilton (1980). Their results indicate that an undercurrent can be accelerated by the longshore pressure gradient force, until bottom friction arrests its further development. In the steady state, the longshore flow is balanced by a cross-shelf pressure gradient, and the longshore pressure gradient drives a depth-dependent onshore flow, which returns offshore through a bottom frictional layer. McCreary and Chao (1985) obtained realistic coastal flows through a linear, three-dimensional, stratified model by assuming that the alongshore flow is in geostrophic balance. They show that the resulting longshore pressure gradient can drive the poleward undercurrent. However, the extension of the shelf weakens the poleward undercurrent by inducing a barotropic equatorward flow, in good agreement with the results of Suginohara (1974); it is found that a wind stress curl over the coastal region, such as results from the increase of the wind field in the offshore direction, is required to strengthen the undercurrent to realistic values. Suginohara and Kitamura (1984) speculate, further, that dissipation is required to sustain a poleward current in the steady state and that it is improbable that the California Undercurrent should be driven by the regional winds, but results rather from the more global tropical circulation interacting with the continental shelf-slope.

Early numerical formulations of the nearshore response to along-shore winds were done on an f-plane and did not produce a poleward undercurrent (O'Brien and Hurlburt, 1972). Hurlburt and Thompson (1973) and McNider and O'Brien (1973) confirmed the importance of an along-shore pressure gradient and illustrated the potential role of the variation of the Coriolis force with latitude. Indeed, McCreary (1981) and Philander and Yoon (1982) showed that for time scales long enough for β effects to develop, the westward dispersion of the along-shore flow induces a poleward flow not unlike that observed off the coast of the northwestern United States. Chelton (1982) attributes the three to four month lag he observes between the wind stress curl and the poleward undercurrent to such a β effect.

Other recent investigations have given much attention to nonlinear effects associated with offshore forcing processes and with coastal-trapped waves. Denbo and Allen (1983) consider an inviscid ocean adjoining an equally inviscid shelf interior bounded by frictional surface and bottom layers. The model is forced separately by a zero-mean periodic wind stress at the coast and by prescribed open ocean motions that impinge onto the shelf. The coastal winds induce an undercurrent of about the correct magnitude, but progressing in the same direction as the wind. The offshore forcing generates an undercurrent of the appropriate characteristics, except for its speed which is about an order of magnitude too small. Chapman and Brink (1987) nonetheless further examined the problem of offshore periodic forcing; typical periods larger than about ten days unlock the coastal system from near resonances that are otherwise found and promote a bottom-trapped along-shore jet that extends away from the forcing in the direction of propagation of coastally-trapped waves, with a velocity opposite to that within the forcing structure. That is, a poleward undercurrent should be propelled by offshore elements possessing anticyclonic vorticity. A numerical investigation by Suginohara (1974,1982), of a coastal ocean with cross-shelf topography and vertical stratification underlines the importance of coastal-trapped waves in the development of an upwelling circulation, including that of the poleward undercurrent. The onset of the wind promotes a rapid response over the shelf and slope; the passage of the first mode coastal-trapped wave contributes to confine the upwelling to the coastal area, as both the coastal jet and the poleward undercurrent grow. The arrival of the second mode trapped wave blocks further development of these currents. These results have been extended to an n-layer configuration (Yoon and Philander, 1982), and a steady state is obtained after the passage of the nth baroclinic mode. The undercurrent results from the disparity between the vertical configuration of the waves and that of the coastal jet initially produced by the wind. Simons (1983) confirms that the nonlinear effects of coastal waves can provide an explanation for the set-up of the longshore pressure gradient required to drive the poleward undercurrent; Thompson and Wilson (1983) have examined tidal rectification effects as a source of a poleward flow near the coast.

Another possible driving process for a poleward flow is the drag produced by an irregular topography (Brink, 1986). A steady current flowing in a direction opposite to that of the free shelf wave propagation generates a lee wave off each bump, whose growth constitutes the drag mechanism. When the steady current reverses, no lee waves can be formed, and thus no momentum is removed from the obstacle. Haidvogel and Brink (1986) confirm that this drag asymmetry, which is a form of flow rectification, is able to produce a mean flow in the direction of free shelf wave propagation, poleward for northern hemisphere eastern boundary currents. This principle is extended by Holloway, Brink, and Haidvogel (this volume) to the specific case of the undercurrent. The speeds that are obtained are of a few cm s^{-1} in the correct direction, but too small to justify it as an entire explanation by itself. Samelson and Allen (1987) indicate that, for certain parameter ranges, this response of the coastal flow is chaotic.

Still, the exact causes of counterflows are not yet known unequivocally, and further studies of the region we have examined should contribute useful insight into the physics of the poleward flows, in addition to complementing the geographical description of a very important segment of the world's ocean circulation. It appears fundamental to examine the composition and configuration of the Costa Rica Coastal Current, for which a specific study has not been reported, and in particular, to elucidate whether there exists a counterflow to that current. In that region, nonlinear processes of the coastal waves, topographic drag, and offshore forcings should contribute a poleward flow indistinguishable from the current itself, except for a possible secondary maximum in velocity. The longshore pressure gradient should generate an equatorward counterflow, which would therefore have a visible signature. Wind stress curl could contribute a flow in either direction, depending on the configuration of the wind field and on the relative intensity of the winds offshore; these are poorly known in the area. The presence or absence of a counterflow would suggest which of these generating mechanisms should be discarded.

The general path of the Tropical and Subtropical Water should be investigated, together with the configuration of the frontal structure near Cabo San Lucas, in order to understand how the California Countercurrent is provided for and how this influence is enhanced during ENSO events (McCreary, 1976). What water constitutes the flow to the southward of Cabo Corrientes should be known; Wyrtki (1965) suggests the flow is composed of an extension of the intensified California Current, but Hickey (1979) concludes that this current never reaches that far south, at least near the coast, in good agreement with the drogue observations of Poulain, Illeman, and Niiler (1987). The along-coast continuity and seasonal fluctuations of the California Undercurrent should be addressed; this, maybe more than any other observations, could direct to the causes of counterflows. Finally, the apparent low values of lateral mixing within the counterflow off Baja California pose a striking question. A possible explanation is that the influence of the Subarctic Water is much greater off the coast of the United States than farther to the south. Also, the flow appears to be extremely trapped dynamically to the slope and shelf, and mixing with offshore waters might depend on major disruptions of the counterflow, which are possibly a function of the offshore eddy climate, or may happen around large capes. Off Baja California, Punta Eugenia appears as the only major point where this might be possible, whereas off the United States several more capes are found.

ACKNOWLEDGEMENTS

Many useful comments were received from P.P. Niiler, M. Lavin, and E.D. Barton; unpublished data were kindly provided by A. Bakun and by L.G. Alvarez. The figures were prepared by Sergio Ramos and José María Dominguez, and Joan Semler assisted with the preparation of the manuscript. This research was funded by the Secretaría de Programación y Presupuesto of Mexico.

From: E.D. Barton, University College of North Wales, UK

On: Review and commentary to paper **POLEWARD FLOWS OFF MEXICO's WEST COAST,** by A. Badan-Dangon, J.M. Robles and J. Garcia

This paper reviews most thoroughly the knowledge of currents along the eastern boundary represented by the Pacific coast of Mexico. Because the area is one of few direct current observations, and, south of the peninsula of Baja California, of sparse hydrographic measurements, understanding of the region is at best sketchy. Even so, it has provided one of the most striking examples of the efficacy of a poleward subsurface flow in transporting waters from a remote source along the continental slope against the prevailing equatorward drift of the subtropical gyre (Wooster and Jones, 1970). The interaction between the tropical circulation on the eastern boundary and the subtropical gyre remains to be investigated in any detail, as is also the case in the Atlantic Ocean. There are a number of intriguing questions alluded to in the text. For example, what happens to poleward flow on encountering the mouth of the Gulf of California? How do the intense wind driven events which occur in the Gulf of Tehuantepec in winter affect the alongshore current regime? The lack of specific data on the nature of the poleward flow along the Mexican coast precludes useful comparison with the bulk of existing theoretical work. Nevertheless, it is to be hoped that progress in modelling such flow will lead to the formation of hypotheses which can be tested by systematic measurements in the region.

AN OVERVIEW OF THE POLEWARD UNDERCURRENT AND UPWELLING ALONG THE CHILEAN COAST

Tomas R. Fonseca
Universidad Catolica De Valparaiso
Casilla 1020, Valparaiso-Chile

ABSTRACT

INTRODUCTION

The Chilean coast, with its long meridional coastline and consistent equatorward wind stress, is well recognized as one of the most biologically productive areas of the global ocean. This highly productive region is part of the Southeastern Pacific upwelling system extending from 5°S off Peru to 40°S off southern Chile. Exporting more fish meal than any other country in the world, Chile has long relied on the national fishing industry as a significant factor in its economy. Yet, in spite of its obvious importance, the waters south of 15°S are remarkably little known. Research efforts, either by Chilean investigators or by foreign expeditions have been, for the most part, isolated and modest.

Present information indicates that there are two important factors for the high marine productivity along the Chilean coast: A poleward flowing undercurrent, which introduces nutrient-rich waters to the south, and wind driven upwelling, which brings it upward toward the surface. This paper reviews literature concerned with these two processes south of 18°S, i.e., Chilean waters. Most of the references were published locally, with a few published in international journals. With this background information and the present high level of interest in upwelling systems, it is expected that new research efforts can be triggered in Chilean waters, especially international cooperative work.

THE POLEWARD UNDERCURRENT ALONG THE CHILEAN COAST

Early Results

Two different names have commonly been used for the poleward flowing undercurrent along the eastern boundary of the South Pacific: the "Gunther Current" [e.g., Brandhorst (1963); Robles (1976)], and the "Peru-Chile Undercurrent" [e.g., (Wooster and Gilmartin (1961); Sievers and Silva (1975); Silva and Fonseca (1983)]. The latter name has gained most general acceptance and will be used in the remainder of this paper, but shortened for convenience to "PCU."

The first to report on the PCU and to recognize its importance was Gunther (1936). During the WILLIAMS SCORESBY expedition in September 1931, he explored the sea between Peru and the Golfo de Penas (47°S), sampling at depth with water bottles and inverting thermometers. From an alongshore section, Gunther clearly distinguished a layer of water of apparent subequatorial origin, with high salinity and very low oxygen content, at a core depth of 150 to 200 m. Downward bending of the isopycnals toward the coast below this core depth suggested, from geostrophic balance, a poleward flow within the layer. From above, the PCU is overlain at times, by a layer of low salinity water of subantarctic origin, modified by precipitation in the south of Chile (although Figure 1 shows distinct broaching of the surface by the PCU, particularly offshore). From below, the PCU is bounded by Antarctic Intermediate Water flowing equatorward.

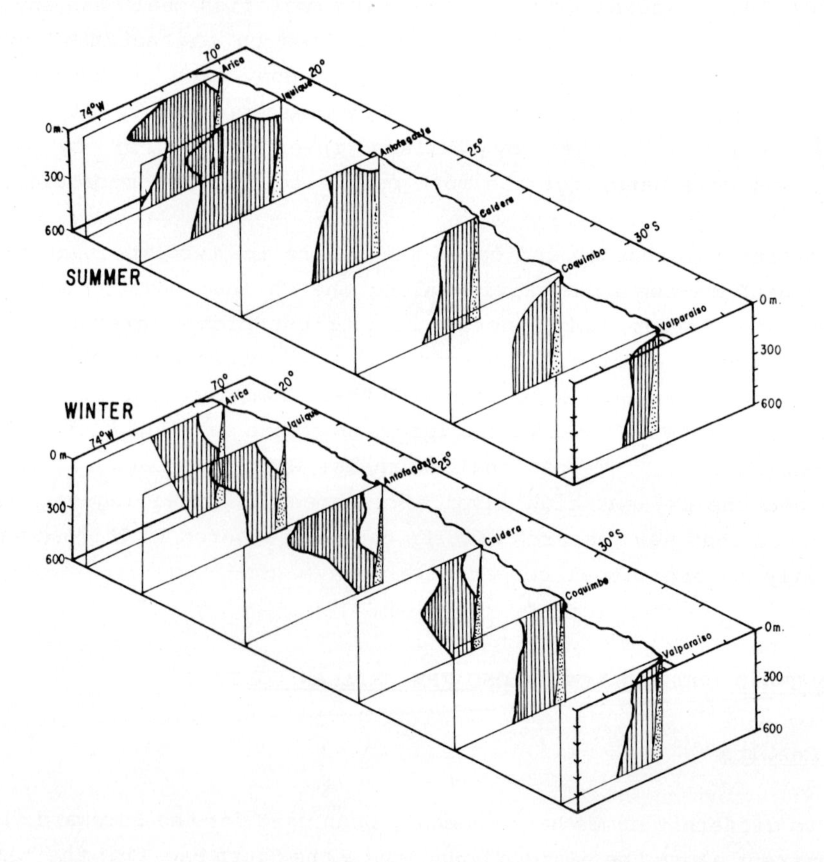

Figure 1: Simplified representation of the Peru-Chile Undercurrent (PCU) water (shaded) between the coast and 180 nm offshore in summer and winter [from Silva and Fonseca (1983)].

In the years between 1960 and 1982, twelve cruises along the Chilean coast were made with classical hydrographic stations and making indirect estimations of the flow through the dynamic method. These cruises were named MARCHILES, and were reported in Brandhorst (1971), Inostrosa (1972), Sievers and Silva (1975), Robles (1976), Silva and Fonseca (1983) and others. Silva and Fonseca (1983) processed data from several of the cruises between 1972 and 1982, looking for the most characteristic features of the currents in the region. Large variations were observed from cruise to cruise, yet there were gross features of the PCU which could be clearly distinguished: The undercurrent is present all year; it is distinguishable from near the surface to 600 m depth; its core is located around 150 to 200 m with a characteristic geostropic velocity maximum (by the dynamic method) of about 20 cm/s; and it is principally located between the coast and 200 km offshore.

A southern limit of the PCU, which can be distinguished from about 10°S off Peru, was proposed by Silva and Neshyba (1979). Based on estimations of the mass field, they suggested that 48°S (Golfo de Penas) is the southward limit, with the relative strength of the flow decreasing markedly from Valparaiso (33°S) southward (Figure 2).

Due to its characteristics and its variability, the PCU is one of the most interesting dynamic features of the Southeastern Pacific. Because of its coastal location and its association with large pelagic fish stocks along the north coast of Chile, it is a relevant and timely phenomena to study.

Direct Measurements

Direct measurements of the PCU flow are rare. Wooster and Gilmartin (1961), however, have contributed what has become a classic work. Using parachute drogues at stations off Peru and the north coast of Chile, they observed the undercurrent flowing parallel to the edge of the continental shelf at speeds of 4 to 10 cm/s. Figure 3 shows one of their stations 27 km off Antofagasta, Chile, where the poleward flow is clearly seen at drogue depths of 100 and 250 m. Coincident with the drogue measurements, a high salinity, low oxygen tongue of water was observed and reported to reach "at least to 41°S."

Further parachute drogue measurements by the Chilean oceanographer, Inostroza (1972), confirmed the presence of the PCU. Inostroza found that drogue velocities off Africa (18°S) were in agreement with computations from the dynamic method and indicated a southward flow

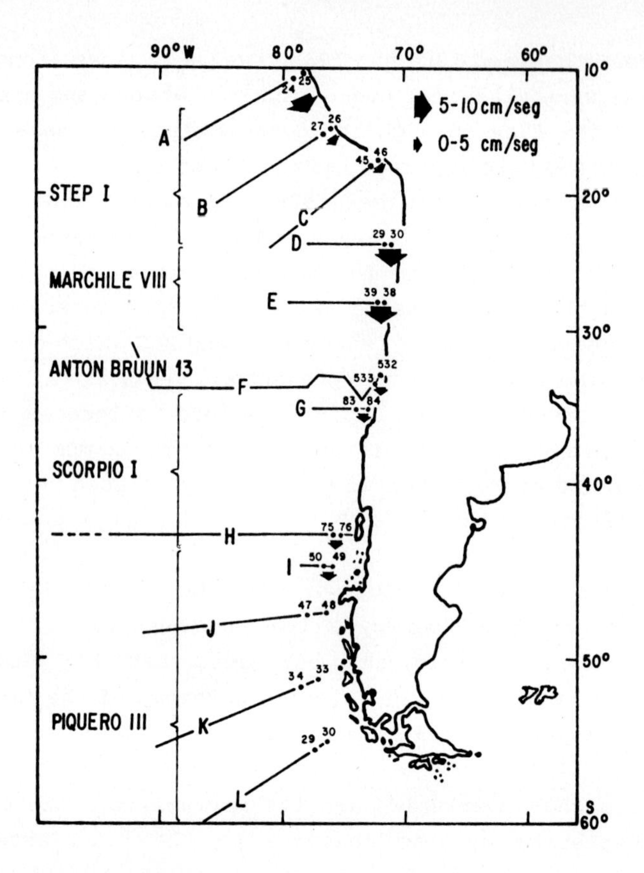

Figure 2: Composite showing the southernmost extension of the PCU [from Silva and Neshyba (1979)].

maximum of 25 cm/s at 100 m depth. In addition, a drogue at 250 m depth was recovered 13 days later 75 nm to the south, giving a minimum average drift of 12 cm/s. These measurements were taken during the winter season (July 1962).

Johnson (1980) reported direct measurements from shipboard at several stations over the shelf break off the northern coast of Chile during the MARCHILE X expedition in July 1976. Depths of poleward speed maxima were observed in clear association with the PCU water core of maximum salinity (Figure 4). Near 18°S the poleward flow was relatively weak (< 10 cm/s at 150 m), but grew stronger toward the south between 23°S and 33°S to a maximum of about 50 cm/s at 150 m.

With such sparse direct measurements, annual variations in the current are difficult to detect. However, from hydrographic data, Silva and Fonseca (1983), proposed that the undercurrent does have an annual

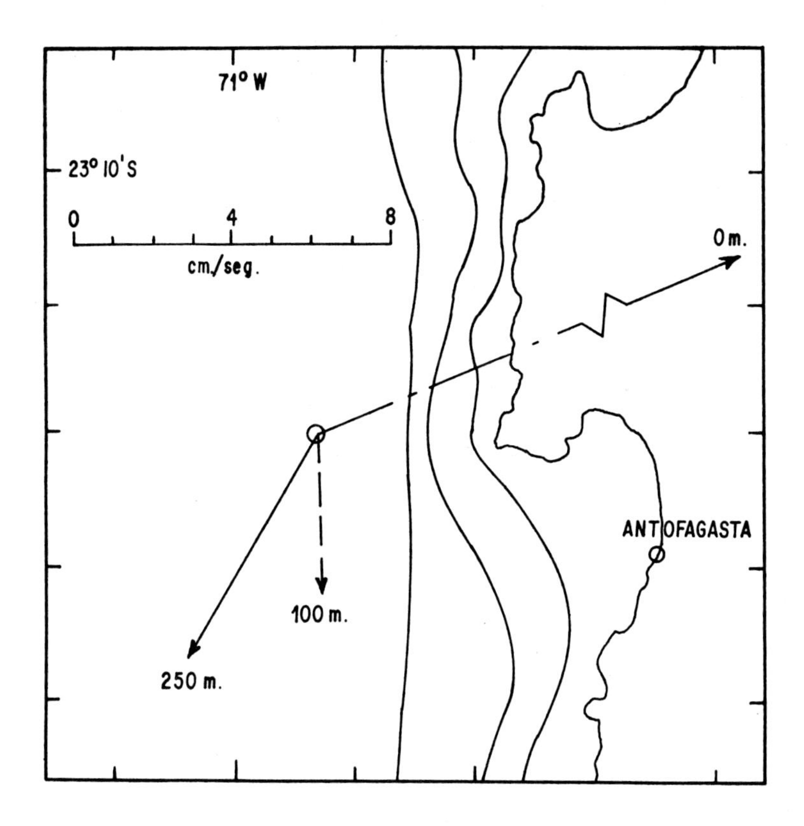

Figure 3: Mean current vectors (and bottom topography) obtained from drogues deployed at three different depths off the continental slope [from Wooster and Gilmartin (1961)].

signal. From Figure 1, it appears to be closer to the coast and broader in summer (at least off the northern coast) than in winter.

In Figure 5, salinity and temperature at 5 and 200 m depths are given for a yearly cycle over the continental shelf near Valparaiso, Chile. While the near surface signature exhibits an annual cycle, the deeper layer is relatively constant. In contrast, however, the sea level at Valparaiso (Figure 9) contains a distinctive annual signal. Furthermore, surface drifters in the nearshore region (Figure 6) appear to show seasonal reversals in direction which can be related to the variations in sea level under the geostrophic balance assumption. Sea level, then, may be a more useful indicator of flow than water mass characteristics, at least in the nearshore environment.

The predominant winds along the central and northern coast of Chile are from the SE-SW [Bakun and Parrish (1980)]. It should be expected that

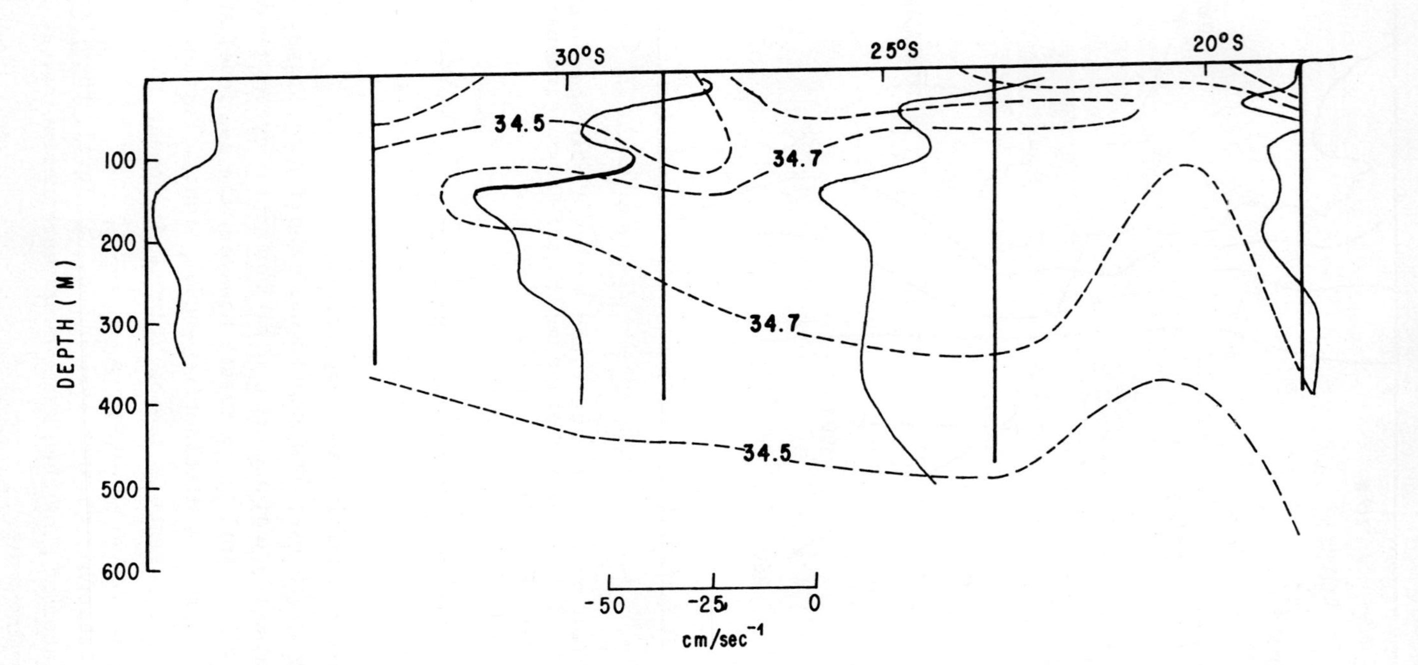

Figure 4: Along-shore salinity section (over the 1000 m isobath) with along-shore current profiles superimposed at four locations. The vertical lines representing the current axis are placed at the along-shore location of the profiles [from Johnson (1980)].

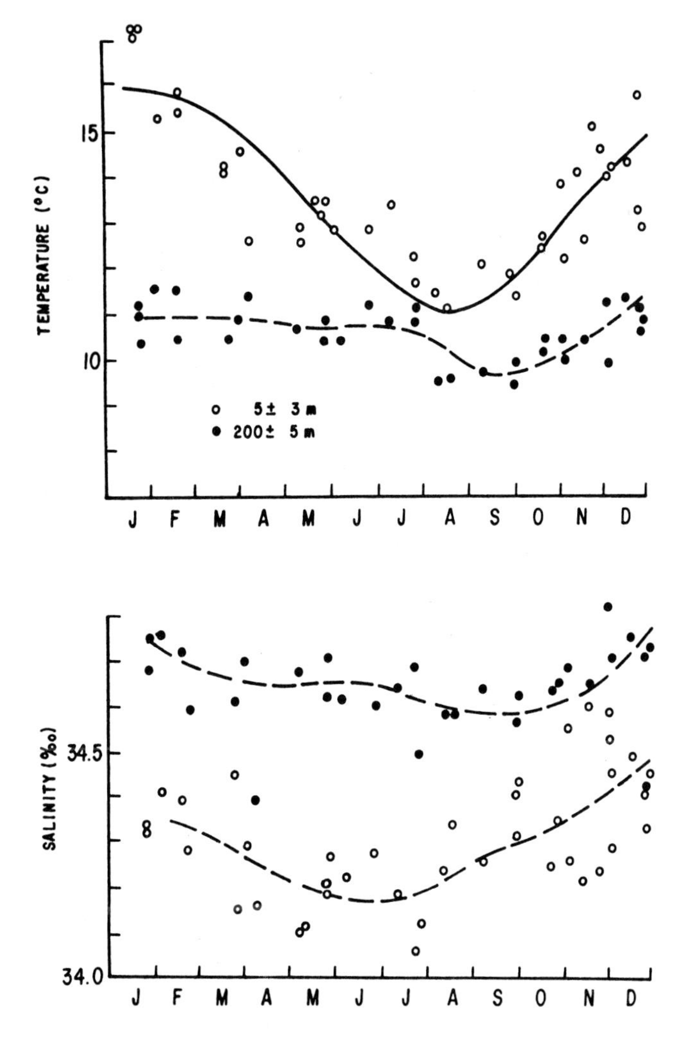

Figure 5: Annual cycle of temperature and salinity in shelf waters off Valparaiso [from Fonseca (1985)].

the northward flowing Humboldt Current would be important. However, this current has not shown a clear signal, in terms of velocity, in the first 100 miles offshore [Silva and Fonseca (1983); Espinoza, et al. (1983)]. In contrast, the PCU flowing counter to the wind stress, has been seen to be an important feature in ocean circulation off Chile, even at the surface itself (see Figure 1 and the 33°S profile in Figure 4).

In close interaction with the PCU, coastal upwelling is another feature with great influence on oceanographic conditions. Espinoza and Neshyba (1983) hypothesized that, in Chilean waters, the cross-shelf flow and local eddies associated with coastal upwelling are more important to surface conditions than the alongshore flow associated with the Humboldt Current. It is possible, and even probable, that the interaction between

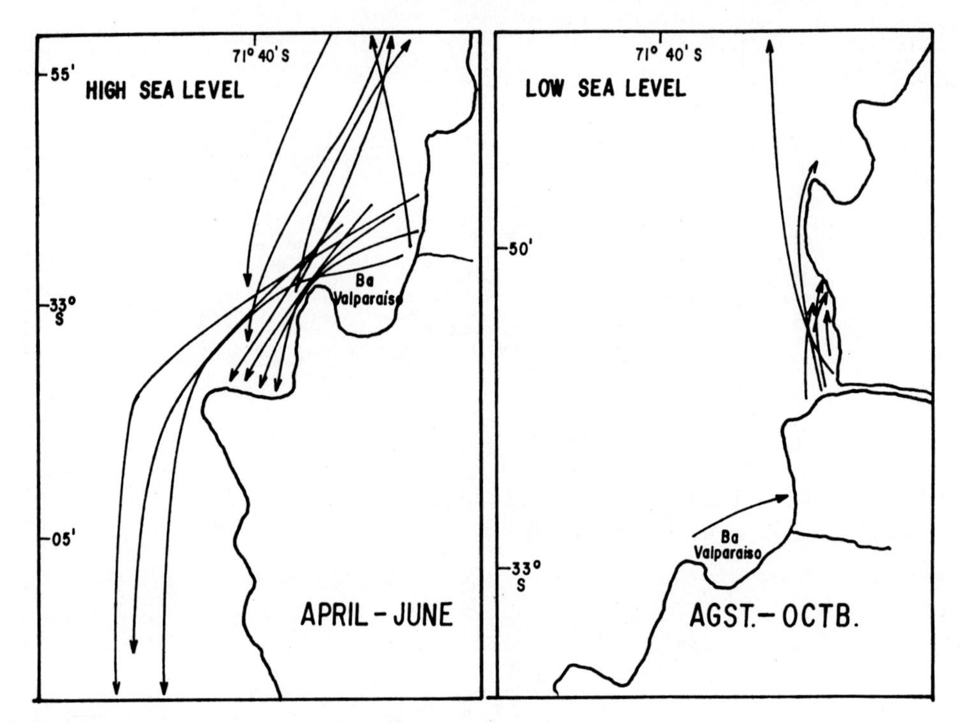

Figure 6: Coastal currents implied from drift bottles and associated with subtidal sea level measured in Valparaiso Bay [from Fonseca (1985)].

the PCU and coastal upwelling forms the basis for the intense biological growth in the area. Therefore, I feel that it is not possible to make an adequate review of the oceanography of the area without reference to coastal upwelling.

UPWELLING AND UNDERCURRENT

Centers of Upwelling

The northern and central coast of Chile, where the PCU is dominant (18°S to 36°S), is highly active in terms of fish production. Found principally in the first 100 nm off the coast, sardine, anchovy and mackerel are some of the most abundant resources. In addition, embayments close to upwelling centers support highly productive spawning areas [Rojas, et al. (1983)].

Winds are upwelling favorable at least 5 months per year. They are driven by the Southern Pacific Subtropical High, which in winter season is located off Iquique (20°S) and migrates southward to 30°S in spring-summer. South of Valparaiso (33°S), the upwelling season occurs in late spring and summer, later than on the northern coast.

In addition to its temporal displacement, upwelling is not spacially uniform, but occurs with more vigor in specific localities or centers. Satellite images are useful in locating those centers of enhanced upwelling, as can be seen in several satellite IR interpretations along the coast of Chile given in Figure 7. These centers appear to be related to the curvature of the coast, to bathymetry or simply to locally strong wind centers, and are clearly associated with areas of high productivity in the fisheries industry. In Figure 7 it can be seen that the upwelled water flows many miles offshore forming, at times, "hammer-head" or "mushroom" shaped structures.

Various authors have reported enhanced upwelling in different localities of the Chilean littoral zone. Neshyba and Mendez (1976), using surface temperature produced by the National Environmental Satellite Services (NESS), described the oceanographic conditions in relation to fisheries resources. Their maps, covering a 6-month period, show consistently colder waters at specific localities. In addition, Fonseca (1977) and Johnson, et al. (1980) refer to the upwelling center off Valparaiso (Figure 7c), and report a 2-cell type cross shelf circulation with intensive poleward undercurrent at the shelf break. Ahumada, et al. (1983) described an upwelling process off Talcahuano and showed that low oxygen waters, associated with the PCU, are raised by upwelling favorable winds toward the surface and subsequently enter Talcahuano Bay. Blanco (1984) reported on the measurements of a current meter near the surface at Talcahuano, and showed high coherence between winds and currents at 1.5 days, with an "Ekman" drift toward the west under the influence of SW winds.

The few direct measurements of the velocity field in upwelling areas [e.g., Johnson, et al. (1980) and Blanco (1984)] seem to indicate that upwelling occurs in the form of pulses or events. Those events last for five to seven days in response to intensification of SW winds. It can be expected that such events pump the nutrient rich PCU water into the surface layer and result in tongues of cold water, as seen in the satellite pictures of Figure 7, extending 30 to 40 km offshore.

In the Benguela region, Jury (1985) reported that upwelling occurs in association with capes and peninsulas similar to the Chilean experience. Also, the structure of the upwelling events observed off Valparaiso [Johnson, et al., (1980)] seems to be similar to those observed off South Africa by Nelson (1985):

Figure 7: A,B,C satellite derived sea surface temperature maps at enhanced upwelling centers on the Chilean coast. Temperature is given in degrees Centigrade.

01 NOV 85 VALPARAISO (33° S)

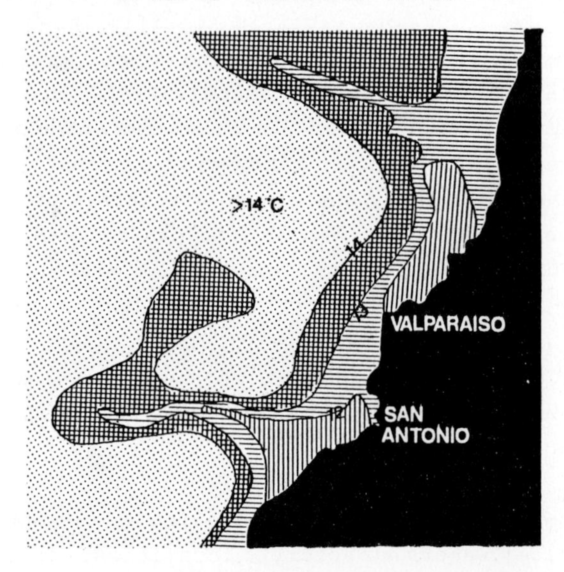

11 FEB 85 TALCAHUANO (36°40' S)

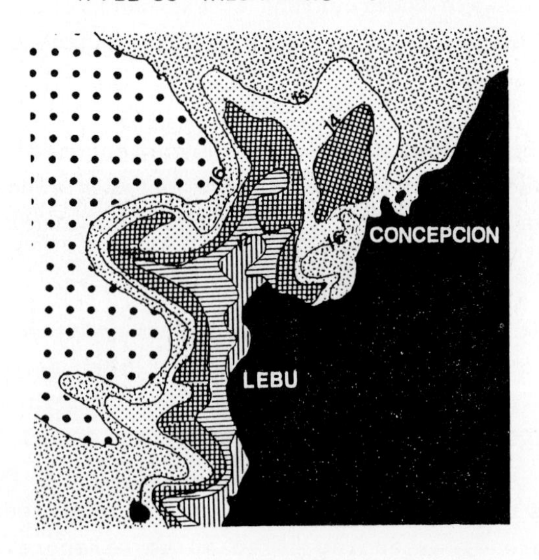

Figure 7A.

15 DIC 85 ANTOFAGASTA (23°30'S)

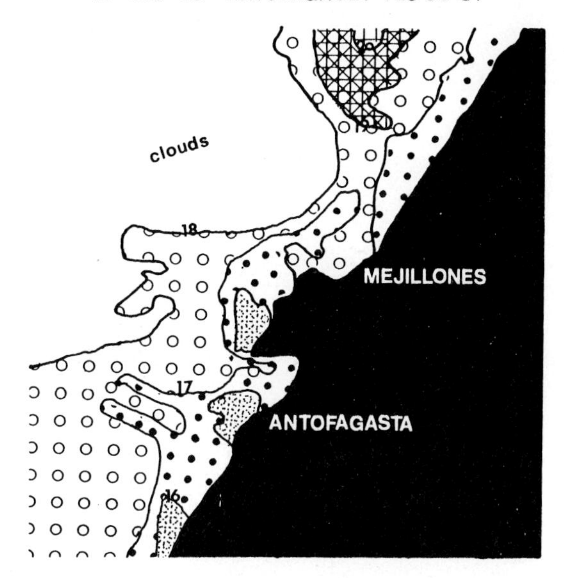

08 DIC 85 COQUIMBO (30°S)

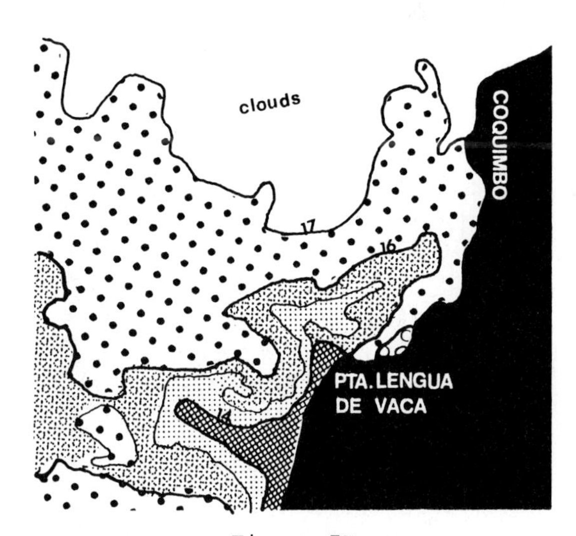

Figure 7B.

14 EN 86 IQUIQUE (20°10' S)

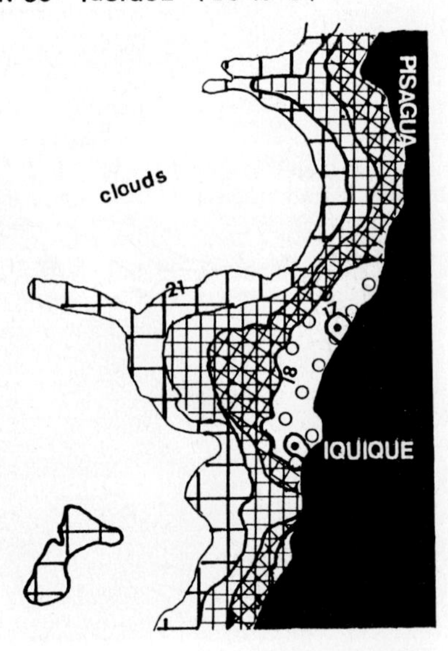

01 DIC 85 ARICA (18°10'S)

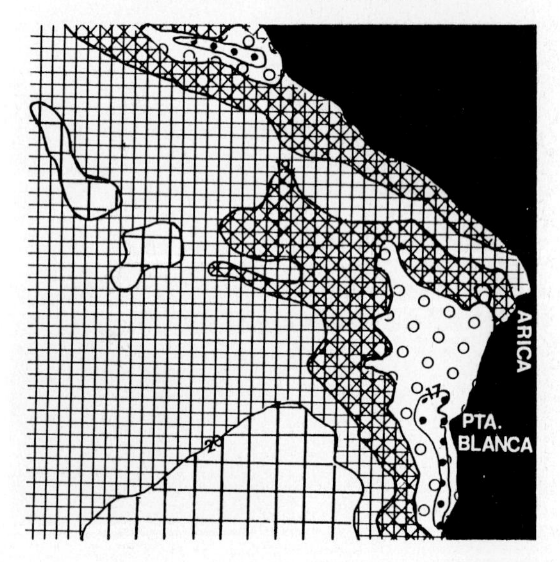

Figure 7C.

a. Curvature of the coast enhances upwelling through an adjustment of relative vorticity,

b. Zonal passage of atmospheric cyclones south of upwelling centers are associated with upwelling events of around one week in duration.

Poleward Flow, Upwelling, and Ecological Effects

Gunther (1936) and Brandhorst (1963) presented information of the characteristics of the water that is advected southward in the PCU. This water has high salinity, low oxygen, and is rich in nutrients. In Figure 8, from Brandhorst (1963), the tight association can be seen between low oxygen and high salinity of the PCU in the area between 33°S to 35°S. It would seem probable that fluctuations in the PCU must have an important effect on the demersal organisms because of the low oxygen. In fact, Brandhorst showed that Hake are concentrated near the bottom on the coast side of the 1 ml/l line. They do not occur in waters with dissolved oxygen less that 1 ml/l. Futhermore, Ahumada (1983) showed that upwelling waters with oxygen values of 0.5 ml/l penetrate inside Concepcion Bay after strong upwelling events. This penetration, clearly associated with the PCU, causes mortality and fish stranding on the beaches, apparently due to the low oxygen content.

Relationships of the physical environment to living organisms show interesting aspects of the ecological effect of the undercurrent. An interesting and provocative relationship between the depth of maximum catch of shrimp to sea level (associated with direction of surface flow) and surface temperature is presented in Figure 9. The depth of maximum catch decreases during the period of southward surface flow and concomitant decreasing sea surface temperature. The upward migration, then, appears to be associated with a rising PCU. This figure also shows the percentage of oviferous females of shrimp. The season of maximum percentage of oviferous females coincides well with low sea level and surface northward flow. It can be speculated that spawning occurs with maximum northward flow and the adults return with the PCU, thus maintaining the shrimp genetic material more or less in the same area.

McLain and Thomas (1983) presented an interesting discussion on the effect of the variability of the California Undercurrent on marine organisms inhabiting that region. Since the undercurrent systems have similarities and most of the marine organisms of each region are similar, it would be extremely valuable to compare the upwelling/undercurrent biomes in each region.

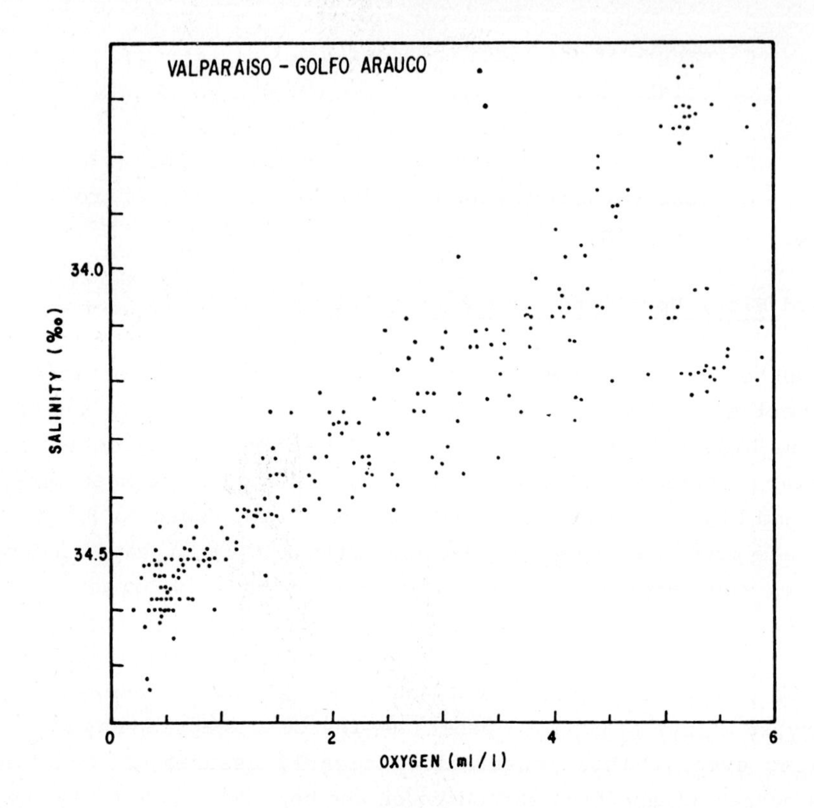

Figure 8: Oxygen/salinity diagram of shelf waters between 33°S and 35°S. The PCU has water with properties in the lower left, i.e., salinity over 34.6 parts per thousand and oxygen less than 1 ml/l [from Brandhorst, 1963].

CONCLUSIONS

The PCU or poleward undercurrent is an important feature in the current system off Chile. It dominates the flow in the first 100 nm, and it causes a great part of the oceanic variability in the area. Originating off northern Peru, it can be traced to at least 48°S. It is noticeable from the surface to at least 300 m depth. At its core, velocities range up to 50 cm/s at a depth of 100 to 150 m. There are evidences that the PCU maintains an annual cycle, with the largest transport (off the shelf) occurring during summer.

The PCU is important in the southward transport of low oxygen water, a phenomena that has serious effects on commercially important bottom organisms such as hake and shrimp. It is also an important source of upwelled water. Upwelling supplies this cold, high salinity and low oxygen water to the upper layers along the entire northern coast of Chile. However, this process is not uniform in time or space; it occurs in the

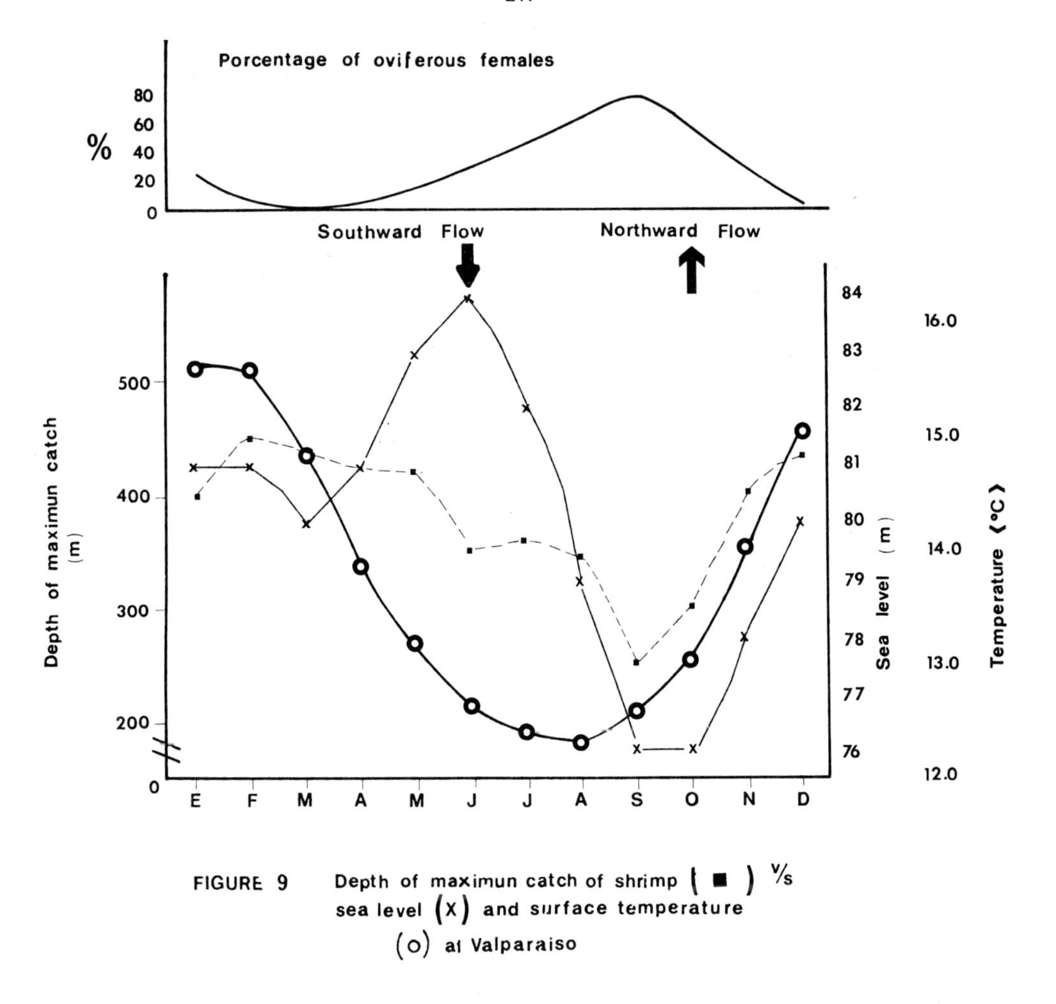

FIGURE 9 Depth of maximun catch of shrimp (■) V/s sea level (X) and surface temperature (o) at Valparaiso

Figure 9: Percentage of oviferous females and depth of maximum catch of shrimp versus surface temperature and sea level (with implied surface flow). The shrimp fishery data are from Arana and Nakanishi (1971), and the physical data from Fonseca (1985).

form of events or pulses in specific centers along the coast. From satellite images covering centers of enhanced upwellings, the surface plumes can be detected as far as 30 to 60 nm offshore.

Oceanographic conditions in this sector of the southeast Pacific coast, associated with the PCU and upwelling, seem to be analogous to those reported for the west coast of North America and the Benguela Current system, but with sufficient differences to provide important insight into the dynamics of upwelling physical/biological systems.

FINAL REMARKS

Oceanographic conditions in the Southeast Pacific, and especially the currents, are among the most poorly known of the world ocean. It would be most valuable to make a serious effort at studying this area in order to compare dynamics with similar, but better known systems. Similarities and differences could be a good base for developing theoretical understanding of this highly productive area.

Direct measurements of the currents are necessary. We need to know the limits, variability, and relative importance of the Humboldt Current and the PCU. Do branches of these currents exist as detected in studies using the dynamic method? Or, are they just eddies or other local effects? We need to have a better picture of the east-west flow and coastal eddy formation. We would like to have more information on the structure of flow conditions at enhanced upwelling centers. In-situ measurements combined with satellite imagery would be extremely helpful.

Theoretical work that allows understanding and prediction of flow changes, wind driven currents, and upwelling dynamics will help local scientists in designing efficient experimental research programs and allow more effective use of the marine resources.

ACKNOWLEDGEMENTS

I wish to express my thanks to Dr. Steve Neshyba for making this presentation possible and for his continuous cooperation with Chilean scientists. I also thank Dr. Victor Neal, Prof. Hellmuth Sievers, and Dr. Donald Johnson for their valuable comments to the manuscript. My participation in the Symposium on Poleward Undercurrents was funded by the U.S. National Science Foundation. For this support, I am grateful.

From: Donald R. Johnson Naval Ocean Research and Development Activity, Stennis Space Center, MS.

On: Review and Commentary to paper: **AN OVERVIEW OF THE POLEWARD UNDERCURRENT AND UPWELLING ALONG THE CHILEAN COAST** by T. Fonseca

INTRODUCTION

Fonseca synthesizes various pieces of broadly interdisciplinary data to form an interesting and credible picture of the Chilean coastal ocean. With a country 4270 km long, running in a nearly north/south direction across latitudinal changes from 8°S to 56°S, clearly the Chilean ocean provides an enormous challenge for that country's oceanographers. A number of Chilean universities, plus several government agencies and a small, but economically important oil industry are all engaged in a range of ocean-related science and engineering activities. If I have one negative comment about Fonseca's paper, it is that it does not include enough of the results of Chilean research activities, and consequently, does not provide an adequate view of the Chilean efforts in marine sciences.

Fonseca's paper presents three topics for consideration: the Peru-Chile Undercurrent (PCU), coastal circulation, and some concepts of ecological interactions with the PCU, upwelling and the Chilean fish stocks. My comments will take the same route.

PERU-CHILE UNDERCURRENT

Fonseca presents an overview of the geographic extent of the PCU along the Chilean coast (Figure 1) using a three-dimensional water mass representation for summer and winter seasons. (In this discussion we will conform to Fonseca and use PCU in reference both to Equatorial Sub-Surface Water and to the undercurrent.) Regardless of interpretations concerning seasonal variations or flow conditions, the most striking feature is the dominant presence of the PCU along a major portion of the Chilean coast. Second, and equally striking, is the surface broaching of the "undercurrent" (offshore in the northern sections and inshore in the southern sections). Because of the broaching, nutrient-rich water is in contact with the euphotic zone, and the stage is set for dynamic growth in the biota.

A third, and somewhat obvious, point that should be made with Figure 1, is that the strength, direction, and dynamics of the flow system cannot be easily determined from "classical" measurements. Not only does the water mass carry a complex history with it, which is difficult to unravel from its characteristic tags, but implied dynamics are least certain over the slope and shelf regions where the flow is normally most vigorous. Unfortunately, with limited resources, the Chilean experience with the PCU has mostly had to concentrate on surveys using the classical method.

In Figures 2 to 4, Fonseca provides a rough view of PCU flow characteristics from three different, but limited, methods: dynamic calculations, lagrangian drogues, and shipboard current meter profiles. Obviously missing, and much needed, are long-term moored current meter measurements. From dynamic calculations and drogue measurements, Fonseca indicates that the PCU seems to flow between 5 and 25 cm/sec. These measurements were made in deeper water just off the slope, and at discrete depths (drogues) which were poorly resolved in the vertical. The shipboard current profiles were measured over the slope itself, and are intriguing in that they suggest a vigorous current of 30 to 50 cm/sec. Although these latter measurements were "snapshots" of the current system, some rational speculations can be made after examining them in more detail.

The current profiles presented in Fonseca's Figure 4 were obtained on MARCHILE X, a hydrographic survey during July, 1976, which covered the eastern boundary regime from Valparaiso to the border with Peru, and offshore to about 200 nm. At four stations along the coast (Arica, Antofagasta, Huasco, and Valparaiso), current profiles were made over a period of 12 hours, and averaged in order to eliminate tidal variations. In addition, a similar station was made over the O'Higgins Seamount (approximately 120 nm offshore of Valparaiso). The three northern coastal stations were made near the 1000 m isobath along the continental slope, and the Valparaiso station was obtained near the 400 m isobath, very near the shelf break.

Figure A-1 shows the current profiles from Antofagasta, Huasco, and Valparaiso plotted together. In addition, a profile using the dynamic method at the central station (Huasco) is presented since this method tends to filter out non-geostrophic "noise." The similarity in shape and amplitude of the measured profiles both in the along-shore and in the cross-shelf components is striking and suggests that the snapshot-type measurements may be providing a characteristic look at the currents over the slope region. Furthermore, although the calculated profile does not reflect smaller scale structure in the upper layer, its general adherence to the measured profiles gives additional assurance that the measurements were not spurious. The Arica current profile was relatively weak in comparison and will be discussed separately.

Referring to Figure A-1, it can be seen that the maximum southward flow occurs at about 150 m in depth, in direct association with the salinity maximum and dissolved oxygen minimum. Current speeds in the core of the undercurrent are in the 30 to 50 cm/sec range. At Valparaiso, the southward flow reached the surface although high salinity water at that location did not do so (the strong southward surface flow was confirmed by measured ship drift toward the south against the windstress). It should be recalled that the Valparaiso profile was taken in shallower water and may be as much indicative of a cross-shelf variation as an along-shore variation.

In order to obtain a better feeling for areal distribution of the PCU, we can examine some of its characteristics through isentropic analysis of the MARCHILE X data. In Figure A-2, the 26.5 potential density surface has been chosen as characteristic of the level of maximum salinity and minimum oxygen indicative of the PCU. Since this surface nominally lies at approximately 175 m in deep water, the 175 m level was also chosen for comparative current analysis. Apparent Oxygen Utilization (AOU) was used as a water mass tag in Figure A-2 instead of dissolved oxygen since it provides a parameter which is primarily independent of both temperature and salinity tags. Dynamic topography at 175 m referred to 1000 m is represented in Figure A-2 as maxima (ridges) and minima (valleys), with superimposed direct current measurements interpolated to 175 m.

From Figure A-2, PCU water appears to cover a broad offshore area from Antogasta northward, but decays southward into "fingers" of water turning away from the coast. The alignment of maxima and minima in dynamic topography and, consequently, the calculated flow alignment tend to conform to the water mass tags.

A close examination of the dynamic topography in Figure A-2 together with measured currents (note the inclusion here of the current measurement over O'Higgins Seamount) reveals at least one interesting feature: the presence of the maximum just offshore in the northern section suggests a northward flow between this maximum and the continental slope. Whether it is a semi-permanent or a temporary feature, the point is that there appears to be (not too surprisingly) a high speed core of the PCU, locked to the continental slope (as indicated in the direct current measurements of 30 to 50 cm/sec), and a weaker southward drift in deeper water (as indicated in dynamic calculations and parachute drogue measurements of 5 to 10 cm/sec).

A second noteworthy feature in Figure A-2 involves the splitting of the PCU south of Chanaral. Is this an indication of gradual slowdown and decay of the PCU, or is it being dynamically wedged apart by a northward

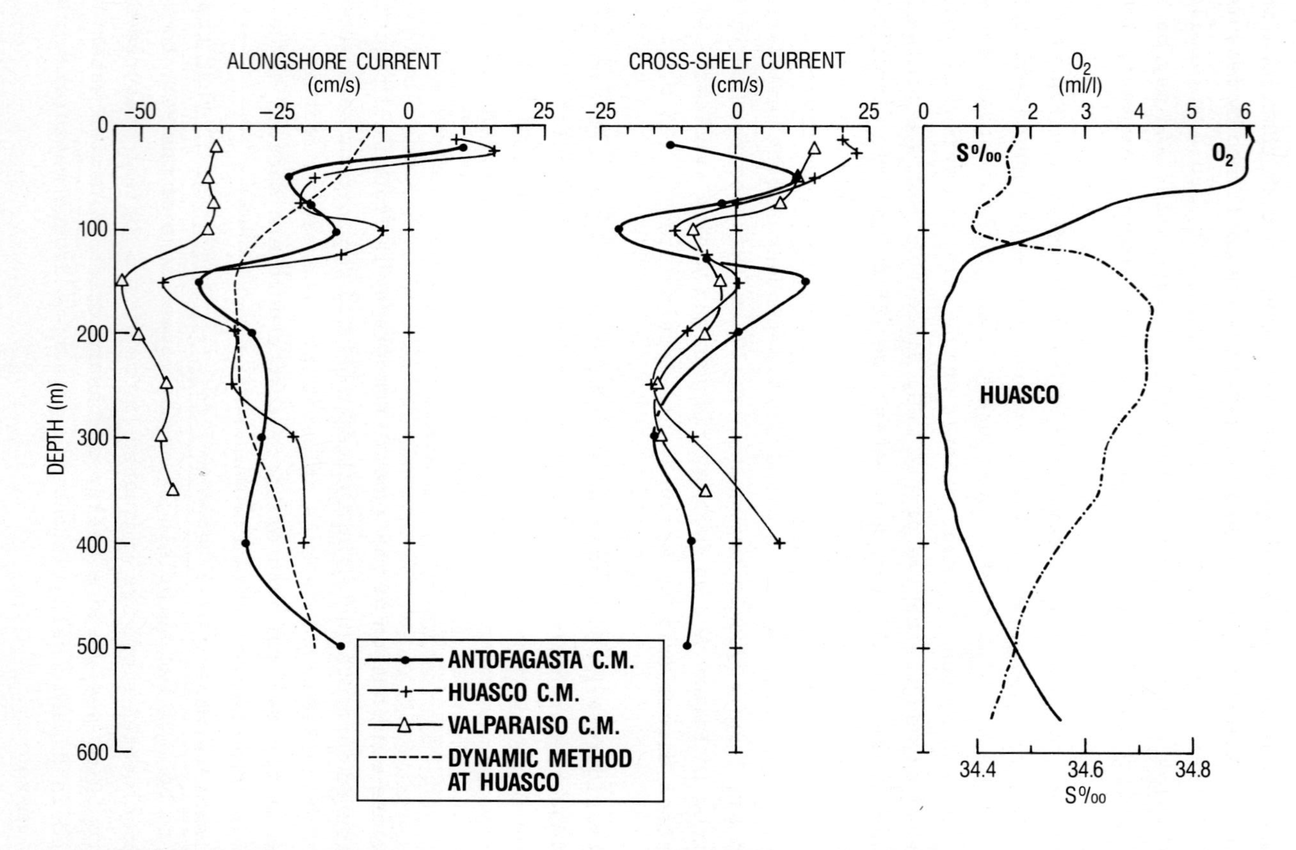

Figure A-1: Current profiles at continental slope/shelf break stations (see Figure A-2 for locations), along with a salinity/oxygen profile and a profile of currents using the dynamic method at Huasco. Equatorward and shoreward are positive directions.

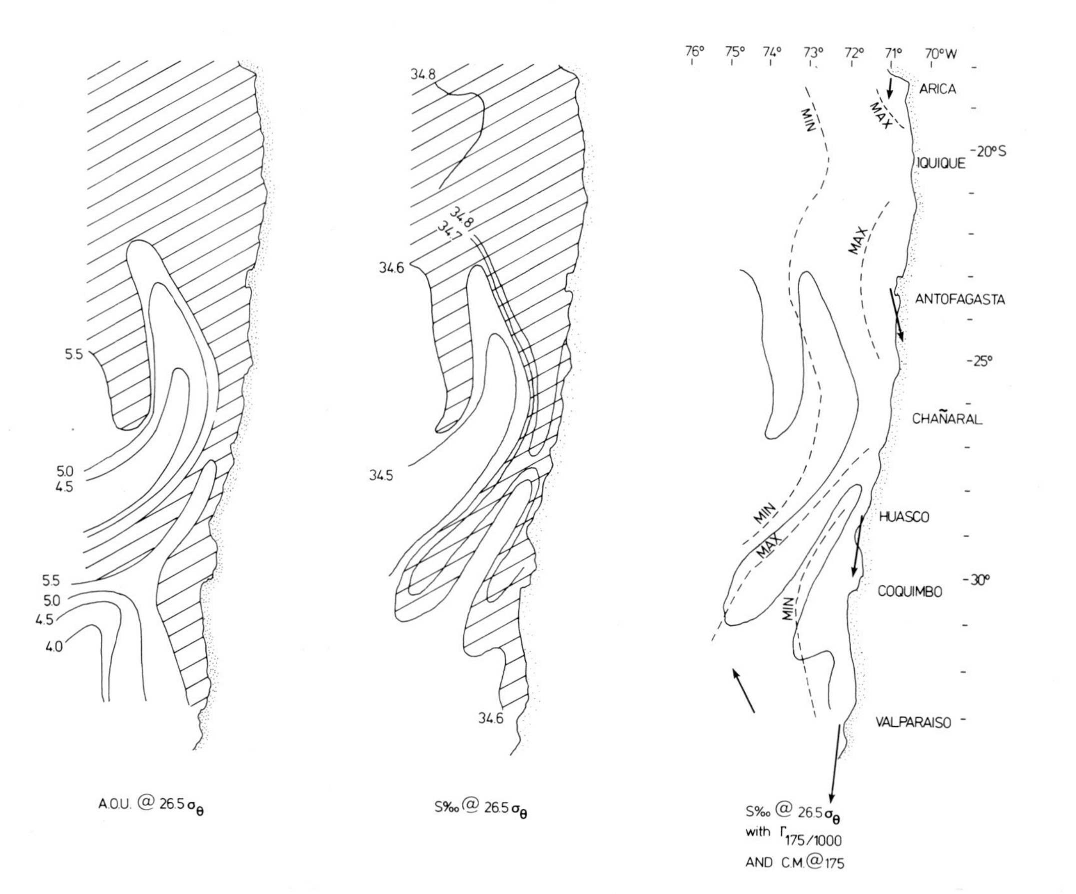

Figure A-2: Isentropic analysis on the 26.5 potential density surface. Left-Apparent oxygen utilization. Center-Salinity. Right-Salinity and dynamic topography sketch with measured current vectors at five stations.

intrusion? Note that the current measurement over O'Higgins Seamount is in alignment with dynamic topography and, at approximately 30 cm/sec toward the north, is relatively strong. Does the PCU gradually decay, or does it end in more vigorous interactions? In Figure A-3, three sections of salinity are presented along 33°S (Valparaiso), from data obtained over a 49-day period in early 1960. Changes in the PCU during this time were dramatic. Although a variety of interesting speculations can be made on the geometry and dynamics of the PCU from these sections, they do serve to illustrate the rapidity and scale of changes which occur over a relatively short time. Since hydrographic cruises, such as MARCHILE X, cover the northern and central coast area in approximately 30 days, it is apparent that such smeared snapshots may not provide more than a generalized picture of the area, and may be somewhat misleading in detailed interpretation. It should be noted that Fonseca has been quite cautious in his presentation of large scale hydrography.

COASTAL CIRCULATION AND UPWELLING

From Hastenrath and Lamb (1977), the climatological monthly wind direction along the north and central Chilean coast is clearly upwelling favorable over the entire year. In addition, a relatively strong negative windstress curl (upwelling favorable) covers the coastal area out to 200 km offshore. From the persistence of the winds, then, we should expect consistent upwelling conditions. Fonseca, however, makes the point that upwelling is both spatially and temporally varying along the Chilean coast, with a propagating seasonal signal in evidence. Fonseca also suggests that spatial/temporal changes in windstress, along with local curvature of the coastline and mid-scale along-shore changes in bathymetry tend to produce localized areas of enhanced upwelling. In spite of the first impression of a relatively straight coastline, Fonseca's satellite derived sea surface temperature maps demonstrate graphically that upwelling on the Chilean coast is indeed spatially dependent and geographically highly complex. In fact, as far back as Gunther (1936), it has been suggested that permanent "centers" of upwelling occur along the Chilean/Peruvian coast.

In a further examination of Hastenrath and Lamb's (1977) atlas, the wind, windstress curl, and sea surface temperature fields along the South American coast (to 30°S) show minima in wind and windstress curl, and a maximum in SST occurring around 18° to 22°S. This area is located at the "bend" in the coastline, separating Peru and Chile. Wooster (1970) makes an interesting speculation of this area: "...that most of the Chile Current goes off to the west before reaching 20°S, that the Peru Current

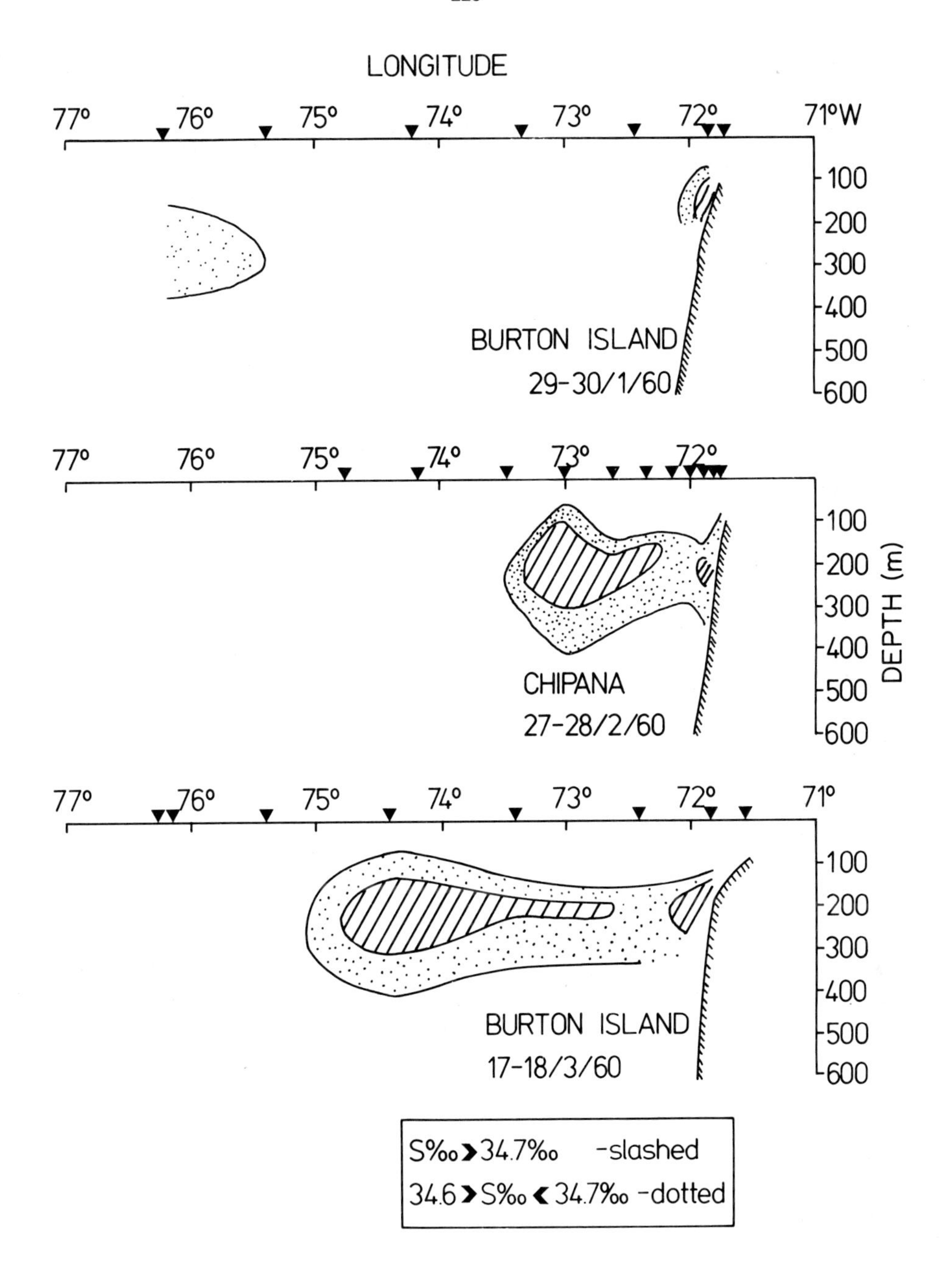

Figure A-3: Cross-sections of salinity over 49-day period off Valparaiso showing rapid change in PCU core water.

arises along the southern coast of Peru, that this current is fed largely from the west and from below, and that the 18° to 22°S zone off northern Chile represents a quiet backwater between the two circulations." Although the Arica current profile in Fonseca's Figure 4 is only a brief snapshot of the currents in this zone, it does show substantially weaker currents than those profiles taken at locations further south and, at least, does not contradict Wooster's speculation.

In spite of the persistent winds, it seems clear that large scale spatial variations, as well as propagating seasonal changes, occur in upwelling along the South Pacific's eastern boundary. Embedded in these variations, local areas of enhanced upwelling appear to be evident. It is somewhat surprising, then, to find that the cross-shelf current profiles in Figure A-1 are so similar to each other, considering the nearly 600 km separation and 8-day intervals between coastal stations. Although these stations were obtained over the shelf break where high vertical modes can be expected, it should be pointed out that they suggest the presence of multi-cellular-type cross-shelf flows, such as have been observed off Oregon, off NW Africa, and previously, off Chile [Johnson and Mooers (1980)].

ECOLOGICAL EFFECTS

Fonseca's relationship between sea level at the coast, with implied flow direction, and the depth of maximum catch of shrimp is intriguing. His suggestion that the shrimp may be using the PCU and coastal surface currents in their survival strategy is logical. If this relationship does exist, and coastal sea level does indeed provide a means of discerning flow conditions, then one would hope to find a predictive relationship between interannual sea level variability during the critical larval dispersion and recruitment stages and future shrimp harvest. In addition, if the amount of low oxygen PCU in the coastal zone can be predicted based upon variability of upwelling favorable winds, then detrimental effects on certain local fish stocks may be estimated and considered in fishery management plans.

Although it should be clearly noted that fisheries develop according to very complex biological and environmental interactions, some simple relationships as noted above can be quite useful to fishery managers in estimating the relative effects of environmental change and fishing pressure on the fish stocks. Coastal tide gauges, wind, and sometimes, coastal temperature records are relatively accessible sources of data for comparison with fishery records. Although many flaws exist in both types of data sets, statistics can be decently forgiving if applied toward long-term trends.

RECOMMENDATIONS

Chile's remote location makes collaboration with northern hemisphere colleagues a difficult problem. Although such collaboration is earnestly desired on both sides, the sharp reduction in funding of International Programs in the National Science Foundation and in NOAA Sea-Grant has considerably hampered development of collaborative investigations. International cooperative programs, such as that maintained by Oregon State University with Chilean counterparts, should be commended for persistence in spite of bleak funding, and should be encouraged. These kinds of programs are essential in order to set the stage for meaningful large scale cooperative field programs as requested by Fonseca.

But Chile's remote location has one advantage that should be exploited. During the first half of the next decade, several oceanographic satellites with all-weather sensors will be launched. The European Space Agency will launch its ERS-1 satellite with altimeter, scatterometer, and synthetic aperature radars on board. Japan will launch its J-ERS-1 SAR/optical satellite. The U.S. Navy will continue its GEOSAT series (altimeter); TOPEX/POSEIDON (altimeter) will be launched in a cooperative U.S./French effort; and Canada will launch its RADARSAT. Not only will the all-weather sensors aboard these satellites provide a means of sampling through the cloudy conditions along the Chilean coast, but satellite oceanography may be the best method of extracting information to match the spatial and temporal scales.

All of these satellites will need in-situ validation, particularly during wintertime extreme conditions. If the launches are made during northern hemisphere summer, the validations will either be delayed or moved to southern hemisphere locations. Participation in the validation effort can lead to access to satellite data, as well as to the entire validation data base. I recommend that Chilean oceanographers begin vigorous efforts to contact the program directors for these satellites and to offer a Chilean national response to the announcements of opportunity. The Union of Marine Sciences of Chile would be an excellent forum to begin developing such a plan.

Probably more than their northern hemisphere counterparts, Chilean scientists depend on applicability of their research to attract internal funding. With this in mind, the need for management and effective exploitation of fish stocks depend, to a large extent, on understanding the environmental effects during critical stages of biological development. As mentioned in the previous section, long term measurements of relatively accessible parameters such as coastal sea-level, wind, and temperature can be related to fisheries and, at the same time, can provide insight and direction for more focused field studies. But establishing quality long-term measurements will require a national cooperative effort.

Fonseca's paper lays a good foundation for investigation of this interesting, but relatively neglected (by northern hemisphere oceanographers) area of the world's ocean. Like Fonseca, I hope that new international cooperative work can be triggered in Chilean waters.

THE INDIAN OCEAN

THE LEEUWIN CURRENT

John A. Church
George R. Cresswell
and
J. Stuart Godfrey

CSIRO Division of Oceanography
GPO Box 1538
Hobart Tasmania 7001 Australia

ABSTRACT

The Leeuwin Current is a surface stream of warm, low-salinity tropical water that flows poleward (southward), against the climatological mean equatorward winds, from northwestern Australia to Cape Leeuwin and then eastward towards the Great Australian Bight. It flows principally, but not exclusively, in autumn and winter. In the north, it is broad and shallow (200 km by 50 m), tapering and deepening in the south to a relatively narrow current (less than 100 km wide), with a top speed of 1.8 ms^{-1} and strong vertical and horizontal shears. There is an equatorward undercurrent below. The warm, low-salinity Leeuwin Current water intrudes between the continent and the southgoing high-salinity flow described by Andrews (1977). As the current approaches Cape Leeuwin, it extends down to about 200 m and is most commonly centered at the shelf edge. After rounding Cape Leeuwin, it spreads onto the shelf and develops seaward offshoots that result in cyclone/anticyclone eddy pairs. A strong poleward geopotential gradient is thought to drive the Leeuwin Current.

INTRODUCTION

Equatorward winds characterize the ocean off Western Australia and the Benguela, Canary, Peru, and Californian current systems of the Atlantic and Pacific Oceans. However, the behavior of the ocean off western Australia seems to be different: there is no evidence of a steady, continuous equatorward flow near western Australia (Andrews, 1977) or of coastal upwelling, despite the strong equatorward wind stress. Instead, historic oceanographic observations have consistently shown a poleward (southward) surface flow in winter, against the equatorward wind stress, near the shelf edge from at least 22°S to 34.5°S (the southwest tip of Australia). The strong southward flow of warm, low-salinity water, called the Leeuwin Current by Cresswell and Golding (1980), is readily apparent in satellite infrared images (Figure 1). Further offshore, the flow in the upper 1000 m is clearly equatorward (Warren, 1981).

Figure 1: A satellite infrared image of western Australia showing the Leeuwin Current starting as a broad fan-shaped form off northwestern Australia and then tapering and hugging the continental shelf edge as it heads southward to Cape Leeuwin and then eastward towards the Great Australian Bight. The shallow water of Shark Bay is cool in response to mid-winter cooling; an eddy is located off Perth; and two offshoots of the Leeuwin Current can be seen off the south coast. Note that black shading indicates warm water and grey indicates cool water; there are some clouds (white) in the bottom left of the image.

The earliest evidence for the Leeuwin Current came from the observations by Saville-Kent (1897)—later confirmed by Dakin (1919)—of tropical marine fauna and warm waters around the Abrolhos Islands (28.5°S to 29.0°S). More recent evidence comes from ship drift observations (Nederlandsch Meteorologisch Institut, 1949), time series water property data (Rochford, 1969a), historical bathythermograph data (Gentilli, 1972), research vessel surveys (Kitani, 1977), and biological data sets (Wood, 1954; Colborn, 1975; Krey and Babenerd, 1976; Markina, 1976). The Leeuwin Current acts as a conduit to bring tropical marine fauna and flora to southern Australia, and appears to play an important role in the life cycles of a number of subtropical marine creatures (Maxwell and Cresswell, 1981; Wells, 1985; Cresswell, 1986). It may also influence the range and breeding seasons of seabirds (Dunlop and Wooller, 1986).

In this review, we will discuss the water masses of the Leeuwin Current and then consider various aspects of it in three regions: the region northwest of Australia, the region directly west of Australia, and its eastward extension south of Australia. We will also discuss the effect of the Leeuwin Current on the continental shelf off western Australia and the various attempts to model it. In the next section, we will briefly describe the LeeUwin Current Interdisciplinary Experiment (LUCIE) and, then present some preliminary results from LUCIE. The nature of the geography of the region and place names to which we will refer are shown in Figure 2.

THE LEEUWIN CURRENT—GENERAL FEATURES

The Leeuwin Current was defined by Cresswell and Golding (1980) as a surface stream of warm, low-salinity tropical water that flows southward from northwestern Australia to Cape Leeuwin and then eastward to and across the Great Australian Bight. The Leeuwin Current flows principally, but not exclusively, in autumn and winter and is most commonly encountered near the continental shelf edge. It is a relatively narrow current (less than 100 km wide) with top speeds exceeding 1.5 m s^{-1} . The depth of the current varies with time and position, and there is a tendency for the current to deepen as it moves southward. Despite this temporal and spatial variability, it is clear that the current is shallow and that below it there is an equatorward undercurrent. Godfrey and Ridgway (1985) indicate that the intermediate level of minimum motion is between 200 m and 300 m.

Andrews (1977) described a northeastward flow that turns at 29° to 31°S and flows poleward near the continental shelf. This flow dominates

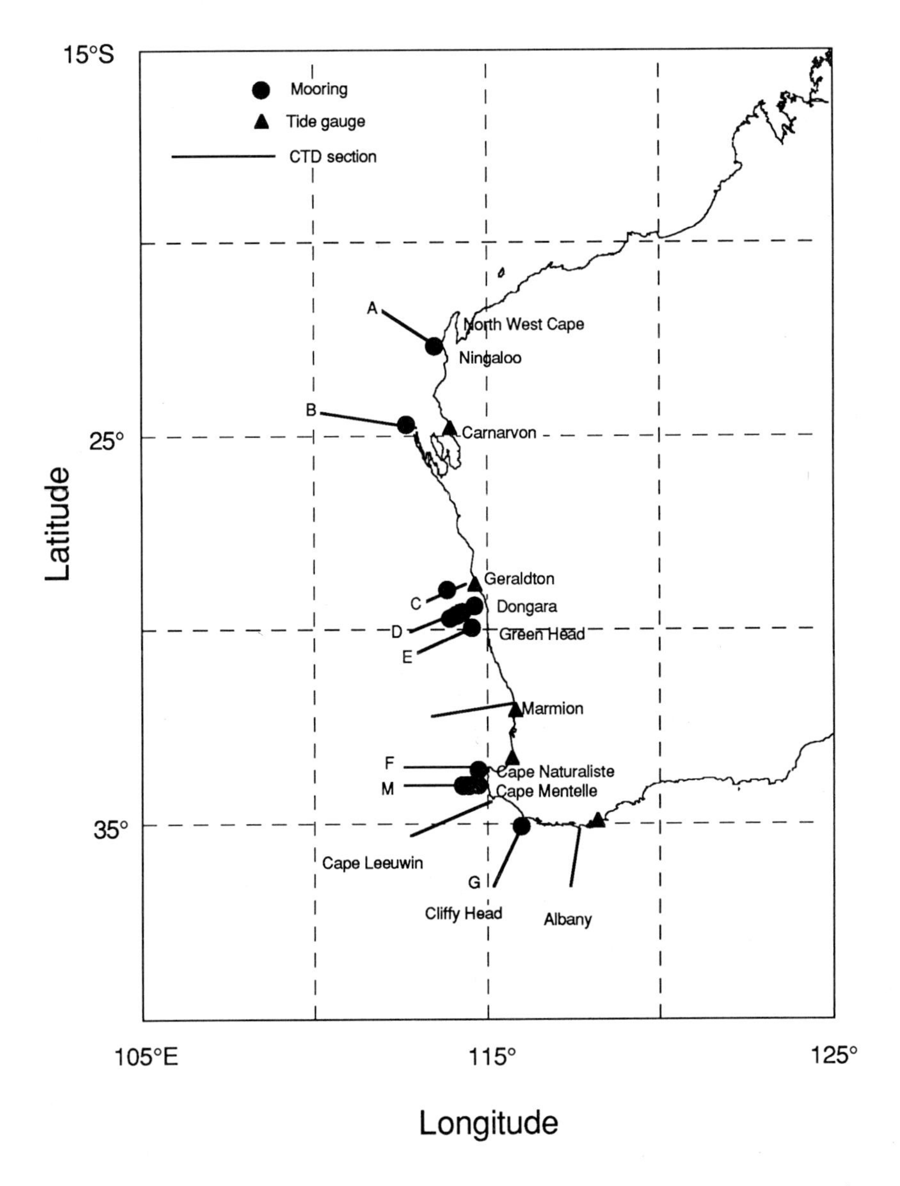

Figure 2: The location of the CTD sections and moorings used in LUCIE.

during the summer period, carries high-salinity water, and was called the West Australian Current by Andrews. It can be seen in Godfrey and Ridgway's (1985) steric sea-level maps.

WATER MASSES OF THE LEEUWIN CURRENT

The hydrological regime in which the Leeuwin Current flows was first described from a series of seasonal biological cruises conducted along the 100°E meridian between 9° and 32°S with HMAS "Diamantina" and HMAS "Gascoyne" in 1962/63. These enabled various water types to be identified in the upper 500 m off western Australia (Rochford, 1969a). Confining our attention mainly to the subtropics, the water types were:

1. Surface (0-50 m), low-salinity (less than 35.00 ppt), high temperature (greater than 25°C), low-nutrient, tropical waters that spread south of 20°S in autumn and winter—the Leeuwin Current. At 32°S the corresponding winter salinity was 35.20 to 35.40 ppt, with the increase from 20°S being due to mixing; the Leeuwin Current holds the sea surface temperature at 32°S above 22°C until after June (Rochford, 1969b; Cresswell and Golding, 1980).

2. Surface (0-50 m), high-salinity (greater than 35.90 ppt), lower-temperature (20-22°C), low-nutrient, subtropical waters at 30°S carried northward to 25°S and down to 100 m depth in summer by the West Australian Current. Webster, Golding, and Dyson (1979) identified these waters during late summer surveys between 29° and 32°S in 1974; the waters gave rise to a secondary salinity maximum between the surface and the South Indian Central Water (see below).

3. Tropical oxygen minimum—subsurface (100 to 150 m), low-salinity (less than 35.00 ppt), low-oxygen (less than 3.5 ml/l), tropical water spreading southward to about 26°S on the 25.00 sigma-t surface in late summer and autumn.

4. Subsurface high-salinity (greater than 35.80 ppt), subtropical water of 17° to 19°C of the South Indian Central region, spreading northward on about the 26.00 sigma-t surface to about 12°S in summer, and to about 16°S in winter. At 33°S this water is at less than 100 m while at its northern limit it has sunk to 300 m. Webster, Golding, and Dyson (1979) found that the Leeuwin Current spread southward over, and mixed with, the South Indian Central Water, thereby becoming progressively more saline and cooler with increasing latitude. Seasonal variations in solar heating, evaporation-precipitation effects, dynamical uplifting of the South Indian Central Water, mixing, and the supply of Leeuwin Current Water were found to be

controlling factors for the temperature and salinity of the surface layer.

5. Subtropical oxygen maximum—subsurface (400 to 500 m), low-salinity (less than 35.00 ppt) waters of the subtropical oxygen maximum (greater than 4.50 ml/l) drifting northward on about the 26.80 sigma-t surface to about 14°S in winter.

Ridgway and Loch (1987) determined mean TS relationships and confirmed a number Rochford's results. For the offshore region, they found that the high-salinity South East Indian Central Water penetrated as far north as 15°S. North of 15°S, this penetration is inhibited by the westward flow of relatively low-salinity water within the South Equatorial Current. Closer to the coast between 5°S and 20°S, the TS properties are more continuous and indicate the presence of a southward-flowing current of low-salinity water which prevents the high-salinity water from approaching the shelf.

THE NORTHWEST REGION

The waters northwest of Australia appear to contain the source region for the Leeuwin Current. Historical bathythermograph data were interpreted by Gentilli (1972) to show that the throughflow from the Pacific to the Indian Ocean in autumn and winter was isolated by a reversal of the flow in spring. This water then achieved thermal homogeneity over the summer to become a "raft" of warm water, which spread southward during the following autumn and winter. A qualitatively similar seasonal progression of the Leeuwin Current was deduced from objective maps of steric sea level through the year (Godfrey and Ridgway, 1985). A near-synoptic picture of this kind can be seen in unpublished research vessel sections ("Diamantina" cruise 4/71) taken at 2° steps of latitude from 20°S to 34°S in August 1971. It was broad and shallow (200 km by 50 m) and mainly off the continental shelf at 20°S, and then tapered, deepened, and spread onto the shelf as it moved southward as the Leeuwin Current. Satellite infrared images such as Figure 1 in this paper and also Figure 1 of Legeckis and Cresswell (1981) support this general picture.

The nature of the Leeuwin Current at the southern part of the North West Shelf was determined by Holloway and Nye (1985) from current meters moored at a number of sites from January 1982 to July 1983. The current ran parallel to the bottom topography and was strongest at the shelf break, reaching a maximum speed of 0.25 m s^{-1}. It was strongest between February and June. The southeast trades, which blow from March to August, did not generate alongshore currents that strengthened the Leeuwin Current.

Satellite drifters with 20 m tethers to their sea anchors showed weak random currents in January and February, 1983 (Cresswell, unpublished data). In March and April the situation changed rapidly with the onset of the Leeuwin Current: on the shelf it had speeds up to 0.5 m s^{-1}; after it rounded North West Cape the maximum speed reached was 1.0 m s^{-1}. At 20°S, 113°E one drifter was trapped for a month in an anticyclonic eddy 80 km in diameter and completed three revolutions.

THE LEEUWIN CURRENT OFF WESTERN AUSTRALIA

Southward flow above the continental slope from 29°S to Cape Leeuwin appears to be a common feature, although the source can vary. Surveys by Andrews (1977) in the summers of 1972 and 1973 revealed that high-salinity surface water from the West Australian Current was carried by a large-scale zonal flow (cf. Hamon, 1965) northeastwards towards the coast between 30° and 33°S, where it turned to flow southward, meandering and breaking into eddies. When it reached Cape Leeuwin, it turned to run to the east. Research vessel and satellite-tracked drifter data suggested to Cresswell and Golding (1980) that the warm, low-salinity Leeuwin Current water intruded between the continent and the south-going high-salinity flow described by Andrews (1977). The drifters showed that the waters of the Leeuwin Current and the cyclonic eddies on its seaward side were mixing. These drifters registered accelerations and water temperature changes as they interchanged between the current and the eddies. While the eddies that trapped the drifters were most frequently cyclonic, two drifters were carried into an energetic anticyclonic eddy and held in it for two months (Cresswell, 1977). The maximum speeds travelled by drifters were 1.7 m s^{-1} along the south coast (Cresswell and Golding, 1980) and 1.8 m s^{-1} near Cape Leeuwin (Godfrey, Vaudrey, and Hahn, 1986).

Eddies were a feature of Kitani's (1977) observations from "Kaiyo Maru" in November 1975. They occurred at 30°S where south-going, high-temperature (21°C), low-salinity (35.2 ppt), northern water encountered north-going low-temperature (18° to 19°C), high-salinity (35.5 to 35.6 ppt) southern water. The south going flow split so that part of it was on the continental shelf and part was farther out to sea. Kitani reasoned that if the south-going flow was a stable feature of the region, then it could be defined as a countercurrent to the north-going West Australian Current farther offshore.

Thompson (1984) took observations of the Leeuwin Current between 22°S and 28°S and suggested that the strong along-coast geopotential gradient could drive the current. He found the Leeuwin Current to be low in dissolved oxygen and high in nutrients (where it touched bottom at the shelf edge) and that only a small portion of its $4 \times 10^6\ m^3 s^{-1}$ flux came from the North West Shelf. Several hundred meters below the Leeuwin Current was an equatorward flow of salty, high-oxygen, low-nutrient water—or what we would identify as Rochford's (1969a) South Indian Central Water. The undercurrent was also identified in mean steric height patterns from historic Nansen bottle and bathythermograph data (Godfrey and Ridgway, 1985).

EFFECTS ON SHELF WATERS

The effects of the various currents on the waters of the continental shelf were first monitored at the Rottnest Island 50 m hydrology station by Rochford (1969b). From the drift cards he released over a year, he concluded there was a southward flow of low-salinity water in autumn/winter and northward flow of high-salinity water in summer. The northward flow is opposite to the southward open ocean summer flow reported by Andrews (1977) and is probably the effect of a predominantly southerly wind stress on the shallow waters of the shelf. A region of shear at the outer part of the shelf was shown by the drifters of Cresswell and Golding (1980). The arrival each autumn of the low-salinity waters of the Leeuwin Current at the Rottnest Island 50 m hydrology station produces an annual salinity variability of roughly 0.5 (Figure 3). An exception, for which we have no explanation, occurred in 1954 when there was no low-salinity winter signal.

Changes in the strength of the Leeuwin Current—or at least of the net poleward geostrophic flow—are reflected in mean sea level changes at coastal stations. These typically show an annual amplitude of 15 to 20 cm, with highest values in March (North West Shelf) to May or June (southwestern Australia); this seasonal movement of the sea-level maximum reflects the southward passage of the Leeuwin Current pulse.

THE SOUTHERN AUSTRALIAN COASTLINE

Satellite images of the Leeuwin Current along the south coast of western Australia showed that it could have five or six jet-like offshoots to seaward in various stages of development (Legeckis and Cresswell, 1981). These were separated by about 200 km and were probably associated

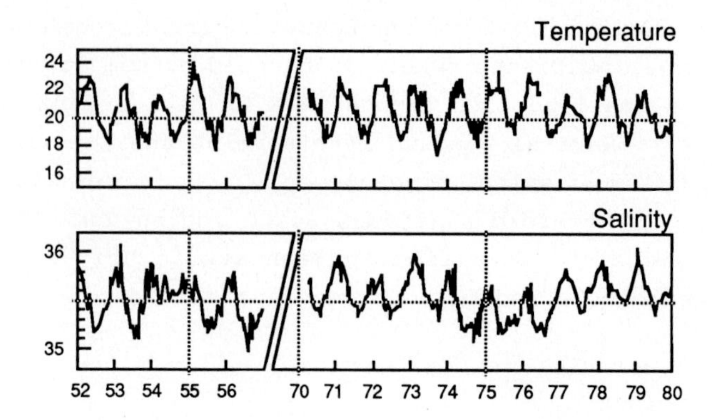

Figure 3: Surface salinities and temperatures from the Rottnest Island 50 m hydrology station.

with the eddies detected earlier by the drifters. Griffiths and Pearce (1985) found that eruptions of the Leeuwin Current to seaward along the south coast of western Australia temporarily diverted the flow into unstable baroclinic waves and cyclone-anticyclone pairs, similar to a phenomenon observed in the laboratory.

The Leeuwin Current was found from research vessel and satellite drifter data to extend along the south coast of Australia from Cape Leeuwin (115°E) to Portland (142°E) (Godfrey, Vaudrey, and Hahn, 1986). Cyclonic eddies breaking away from the shelf west of 124°E were thought to have probably removed most of the warmest water from the continental shelf. The research vessel data revealed fronts having very low Richardson numbers and, therefore, turbulent interfacial stresses.

INTERANNUAL VARIABILITY

The extent of interannual variations in the strength of the Leeuwin Current is not known. Allan and Pariwono (1987) showed that anti-ENSO (ENSO: El Niño---Southern Oscillation) events are accompanied by higher than normal sea level at Darwin and Port Hedland. Pearce and Phillips (1988) have suggested that the enhanced sea level along western Australia is associated with a stronger than normal Leeuwin Current. This suggestion finds support in the Rottnest Island data (Figure 3), which, for the 1973-76 anti-ENSO event, show low salinities suggestive of an enhanced flow of low-salinity water from the tropics.

MODELING THE LEEUWIN CURRENT

Thompson and Veronis (1983) were the first to attempt to model the Leeuwin Current. They suggested that the winter winds on the North West Shelf and slope generated the Leeuwin Current, and they also predicted a southward flow off the west coast of Tasmania. However, this model is inconsistent with Holloway's observations that the southeast trades, which blow from March to August, did not strengthen the Leeuwin Current. Thompson (1984) rejected this model and suggested instead that the strong along-coast geopotential gradient could drive the current. He felt that winter deepening of the mixed layer might allow the geopotential gradient to overcome the equatorward wind stress.

Godfrey and Ridgway (1985) show that seasonal variations of both the alongshore pressure gradient and the wind stress favor a late autumn maximum in the speed of the Leeuwin Current. In particular, they conclude that a 27 cm rise of steric sea level off North West Cape from January to May is the primary driving agency of the current. While each of the eastern boundary current regions have poleward surface pressure gradients, Godfrey and Ridgway (1985) show that this gradient is largest in the eastern Indian Ocean region.

The idea that the southward geopotential gradient drives the Leeuwin Current is a central element of the models of McCreary et al. (1986) and Thompson (1987). These models invoke this poleward pressure gradient as the principal force driving the Leeuwin Current against the equatorward wind stress. In the McCreary et al. model, vertical mixing is important, and all currents vanish as the vertical viscosity and diffusivity go to zero. This model also produces an equatorward undercurrent. In both models, the Leeuwin Current is fed by a geostrophic inflow from the west rather than an inflow from the north.

The Leeuwin Current is different from other eastern boundary currents in a number of ways, which appear to be linked in a feedback loop. The first unusual feature of the Leeuwin Current is that the current flows directly into the wind, thereby advecting warm water polewards. This is probably the reason for a second unusual feature: there is a strong heat flux out of the ocean from the waters of the Leeuwin Current, rather than into the ocean as in all other eastern boundary currents (e.g., see Figure 1 of Hsiung, 1985). This heat loss favors convection, and hence, the formation of deep mixed layers, which are a third unusual feature of the Leeuwin Current (Thompson, 1984). Thompson (1984) further argues that these deep mixed layers are essential to the maintenance of the observed large alongshore pressure gradient (a fourth unusual feature of the

Leeuwin Current). But—as already noted—all authors agree that it is the alongshore pressure gradient that drives the Leeuwin Current into the prevailing wind, while the deep mixed layers allow the alongshore pressure gradient to overcome the equatorward wind; qualitatively at least, this closes the loop.

So far, no modelers have successfully explained why all these linked features occur off western Australia and not in other eastern boundary currents. Godfrey and Ridgway (1985) suggested that it might relate to the existence of an opening to the warm waters of the equatorial Pacific, north of New Guinea; this might maintain the unusually high steric heights (warm waters) off the North West Shelf, and hence, the gradient driving the current. In Kundu and McCreary's (1987) model, an Indonesian throughflow of 7.3 Sverdrups produced a secondary flow with Leeuwin Current-like properties, but was about a factor of five too weak. This question is still very much open.

THE LEEUWIN CURRENT INTERDISCIPLINARY EXPERIMENT

The Leeuwin Current Interdisciplinary Experiment (LUCIE) was a field experiment coordinated by the CSIRO Division of Oceanography during 1986/87. LUCIE was aimed at improving the description of the Leeuwin Current and understanding the processes that drive it, along with understanding its role in the combined ocean/atmosphere system and its influence on the life cycles of marine species. The work is expected to result in realistic models of the current.

LUCIE begin in September 1986 when an array of self recording instruments were deployed. Some of the instruments were recovered and a large array was deployed in January/February 1987. The moorings included an alongshore array of pressure gauges and current meters (Figure 2) to measure the alongshore pressure gradient and to examine the continuity of the current along the shelf edge. There were also arrays of current meters at 29.5°S (Dongara) and 34°S to examine the cross-shelf structure of the current (Figure 4). Buoys recording meteorological data were also deployed at these two latitudes. The moorings were finally recovered in August 1987, and the data are presently (December 1987) being processed. Additional time series data are being gathered from coastal meteorological and tidegauge stations.

On the mooring recovery and deployment cruises, a series of CTD and acoustic Doppler current profiler (ADCP) sections were completed. CTD and ADCP data were also collected on two other cruises. Data was also

Figure 4a, 4b, 4c: Locations of current meters on (a) the Dongara section for the first deployment, and (b) the Dongara section for the second deployment, and (c) the 34°S section. A solid circle indicates useful current data was obtained and an open circle implies no useful current data was obtained. The record from the top of the D3 mooring has not yet been processed.

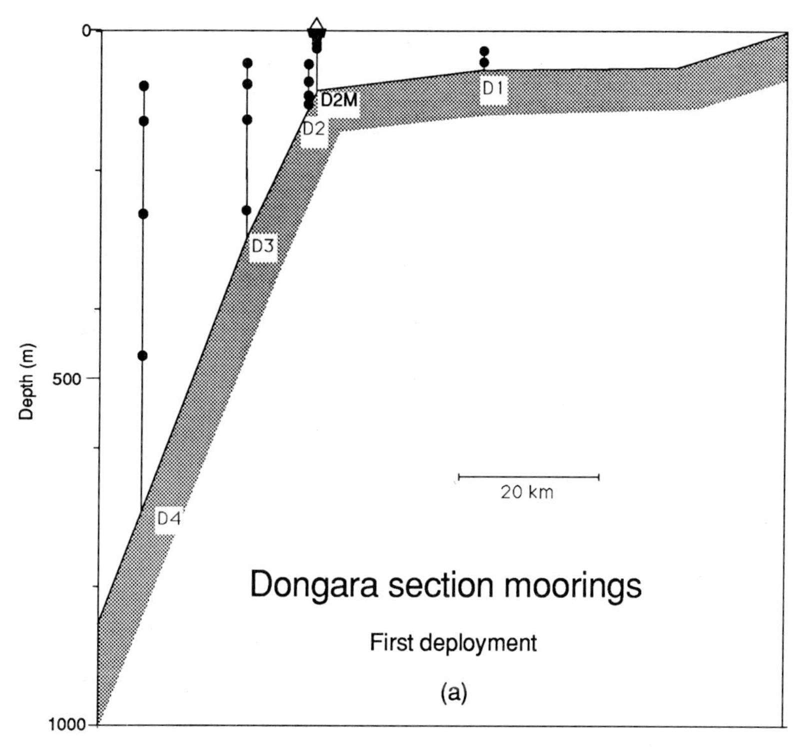

Figure 4a.

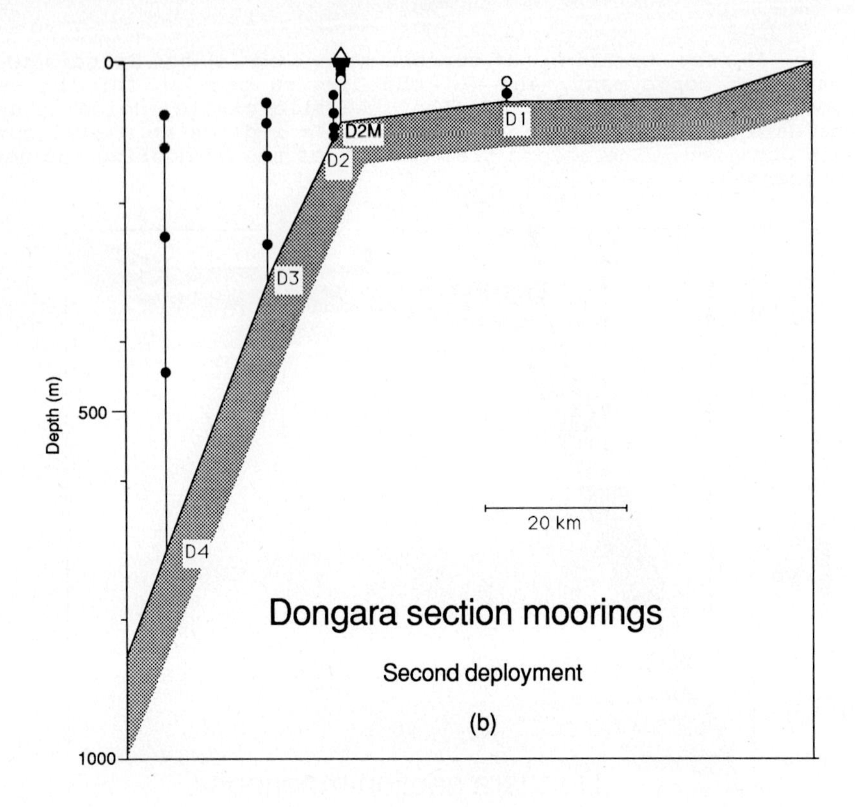
0
500
1000
Depth (m)
D1
D2M
D2
D3
D4
20 km
Dongara section moorings
Second deployment
(b)

Figure 4b.

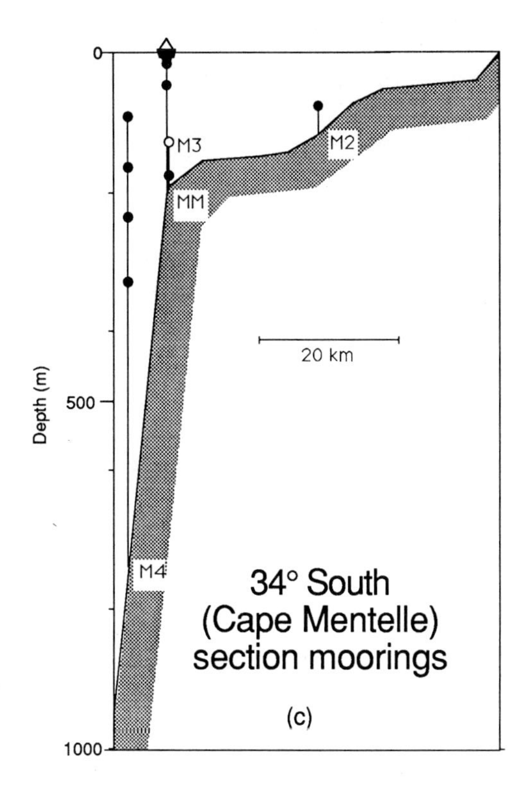

Figure 4c.

collected from satellite-tracked drifters and from BUNYIP (a towed vehicle measuring temperature, conductivity, and microstructure data as it follows an undulating path).

RESULTS

The observational phase of the experiment is now complete and the data are being processed. Some preliminary results are presented here.

CURRENT METERS OBSERVATIONS

The current meters were first deployed (September 1986 to January 1987) at a time when the Leeuwin Current was expected to be weak or non-existent. However, the mean currents at 29.5°S over this period (Figure 5a) do show a southward flow. The southward flow seems to be confined to a narrow stream at the shelf edge and over the upper slope. The maximum mean southward flow recorded was also 20 cms-1 at a depth of 75 m over the 300 m isobath. On the slope below this southward stream, there was a northward flow of 15 cms^{-1} and the flow of the shelf was weakly northward.

The second deployment period (February 1987 to August 1987) included the period when the Leeuwin Current should be flowing strongly. The mean currents (Figure 5b) are southward in the upper 400 m of the water column and indicate a southward jet of almost 40 cms^{-1} on the upper slope. Only the deepest current meter, at 450 m on the 700 m isobath, shows a weak northward undercurrent.

Figure 5a and 5b: The mean along-shore currents at Dongara for (a) the first period and for (b) the second period. Only records from the current meters indicated have been used.

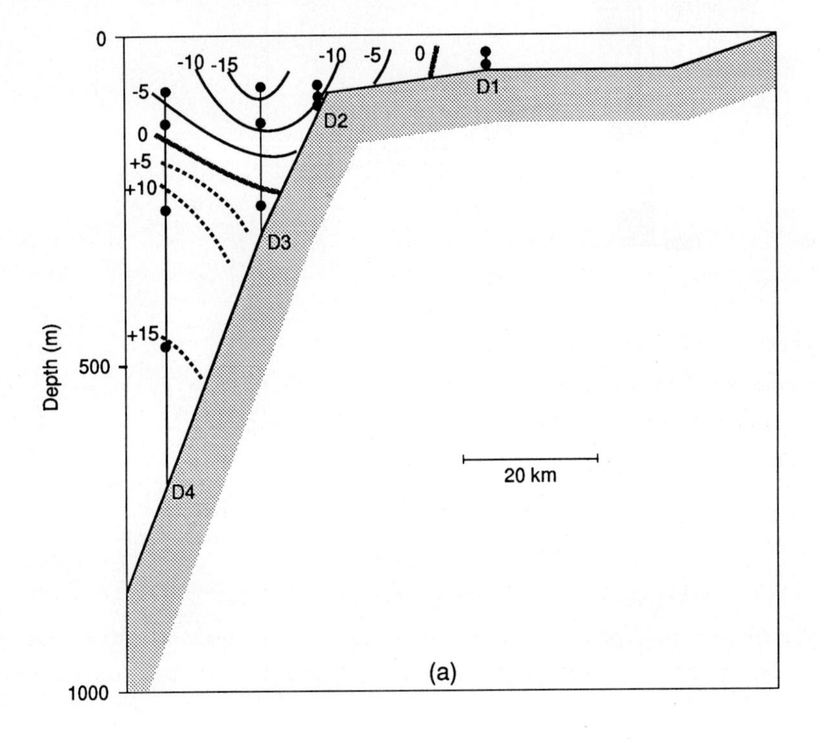

Figure 5a.

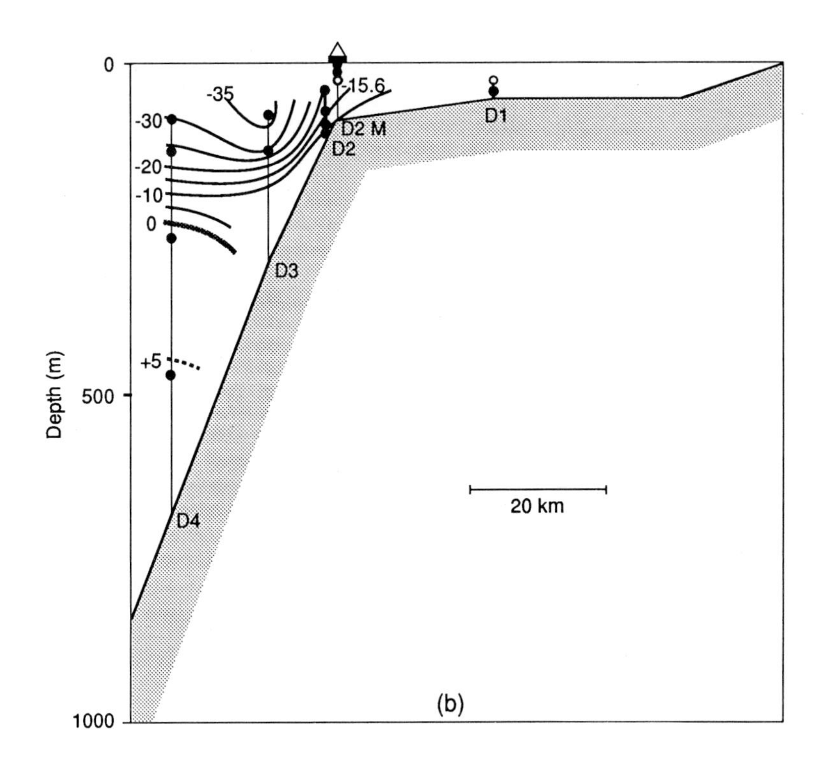

Figure 5b.

The record from the top current meter on the 300 m mooring (Figure 6) indicates a difference between spring/summer and autumn/ winter. During spring/summer, the currents were very variable with periods of southward and northward flow. However, during autumn/winter, the first four months of the record show an almost continuous southward flow, and only in the last two months of the record was there a significant weakening of the flow. The currents from the slope and shelf break mooring appear to be correlated, but this correlation has not yet been quantified.

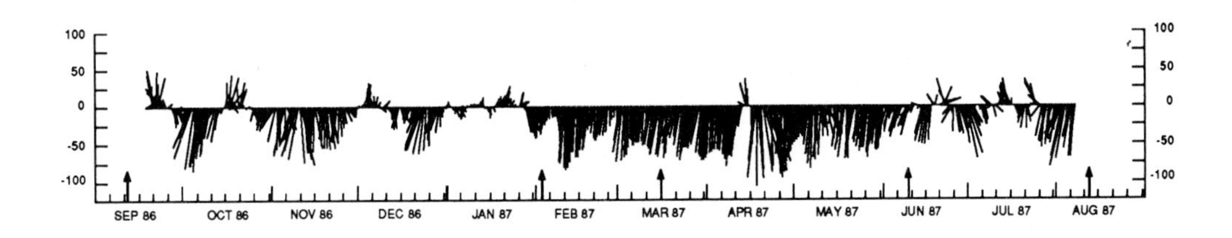

Figure 6: Currents recorded by the meter at a depth of 79 m over the 300 m isobath. A short gap between the two periods has been artificially filled. The arrows indicate the times when CTD sections were occupied.

CTD SECTIONS

CTD sections completed adjacent to the current meter array indicate the changing currents and water properties over the period of the experiment. The approximate times when the CTD sections were completed are indicated on the current meter time series (Figure 6).

Two of the sections (March and June 1987) are sufficient to show the main water types in the study area (Figure 7):

The Leeuwin Current, which has low salinity and was confined to the upper waters at the shelf-edge. Using 35.5 ppt salinity as an arbitrary upper limit for its presence, we found that it was absent in the summer months of February and March. It was widest (> 70 km) and deepest (~200 m) in June.

The South Indian Central Water, with salinities as high as 35.9 ppt. This was a subsurface feature at 100 to 200 m depth; it sank beneath the Leeuwin Current near the continental slope.

Figure 7: CTD sections of salinity and temperature from the Dongara line: (a) March and (b) June 1987, taken as part of LUCIE.

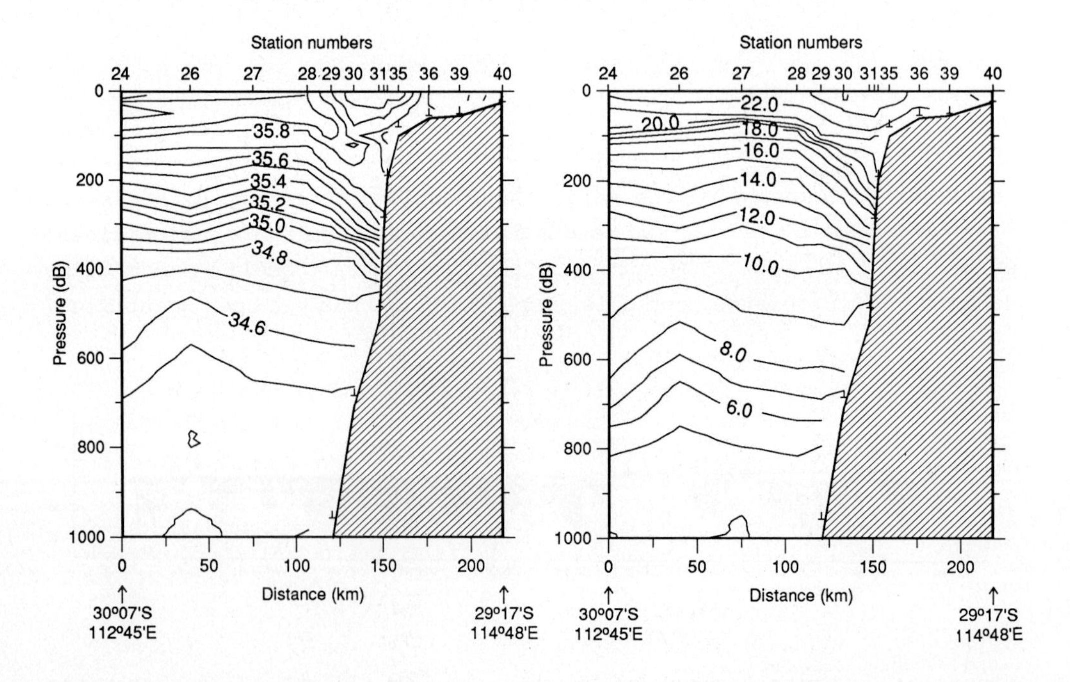

Figure 7a: Dongara section
Salintiy
14-16 March 1987.

Figure 7a: Dongara section
Potential temperature
14-16 March 1987.

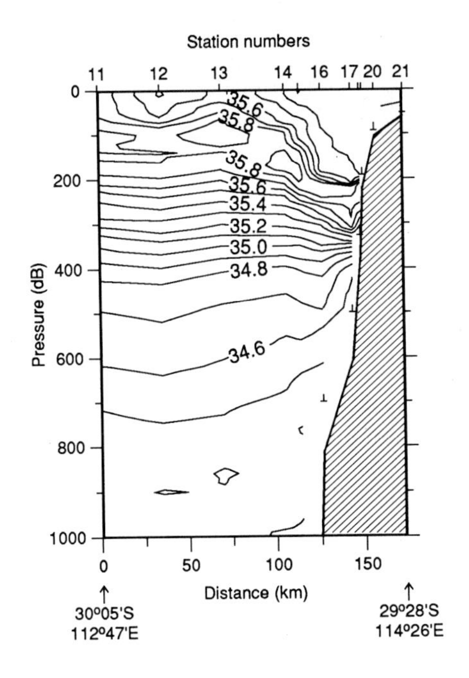

Figure 7b: Dongara section
Salinity
7-9 June 1987

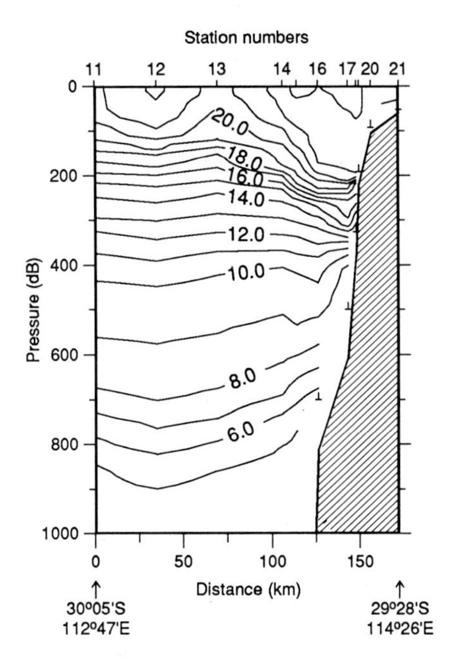

Figure 7b: Dongara section
Potential temperature
7-9 June 1987

The waters of the West Australian Current, which were found at the surface in February and March. These overrode the South Indian Central Water, as Webster et al. (1979) have described.

The σ_θ sections (not shown) are qualitatively similar to the temperature and salinity sections: the geostrophic current shears above the continental slope inferred from them are southward down to about 400 m in September 1986; southward down to about 400 m in February 1987; to about 600 m in March; to about 350 m in June; and to about 400 m in August. In February and June there are strong equatorward shears below 400 m.

The June 1987 survey included sections out from Cape Mentelle, between Capes Naturaliste and Leeuwin, and Cliffy Head on the south coast. At Cape Mentelle (Figure 8), the Leeuwin Current could be best defined as having salinity less than 35.8 and temperature greater than 20°C; in other words, cooler and more saline than farther north at Dongara. The Leeuwin Current extended from very near the coast to some 130 km offshore

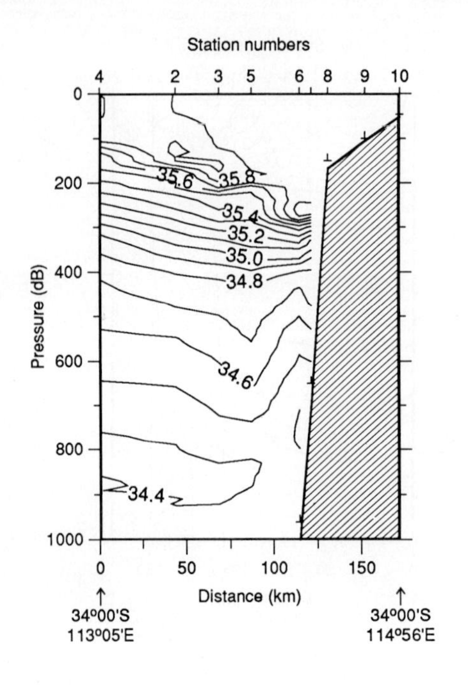

Figure 8a: Cape Mentelle section
Salinity
3-5 June 1987

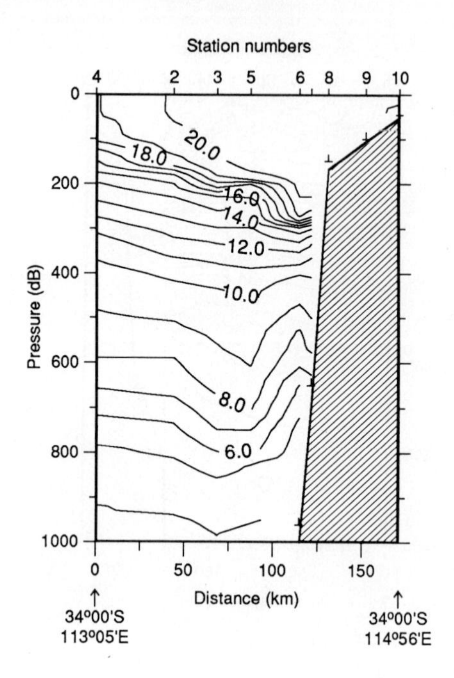

Figure 8b: Cape Mentelle section
Potential temperature
3-5 June 1987

Figure 8: CTD sections of salinity and temperature for the Cape Mentelle line in June 1987.

and was mixed down to 220 m. At Cliffy Head (Figure 9), the situation was interesting because the Leeuwin Current had flowed from a regime where the offshore waters were more saline to one where they were less saline. The Leeuwin Current was narrower and slightly cooler and the 18° to 19°C surface isotherms that marked its offshore edge descended to the shelf break. Preliminary results from an acoustic Doppler current profiler obtained on the same day and roughly along the Cliffy Head section are shown in Figure 10. The Leeuwin Current can be seen to have considerable vertical and horizontal shear.

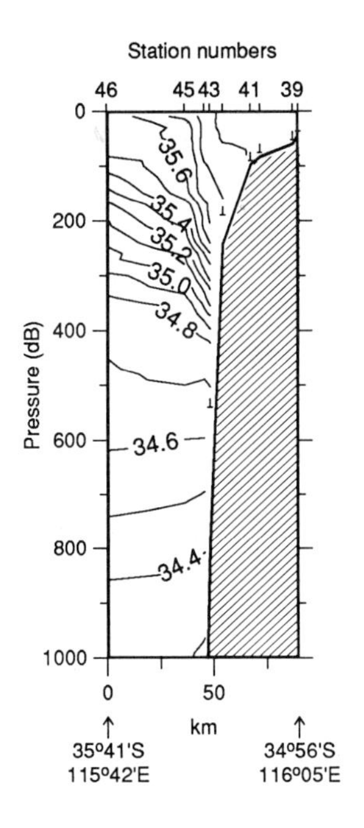

Figure 9a: Cliffy Head section
Salinity
15-17 June 1987

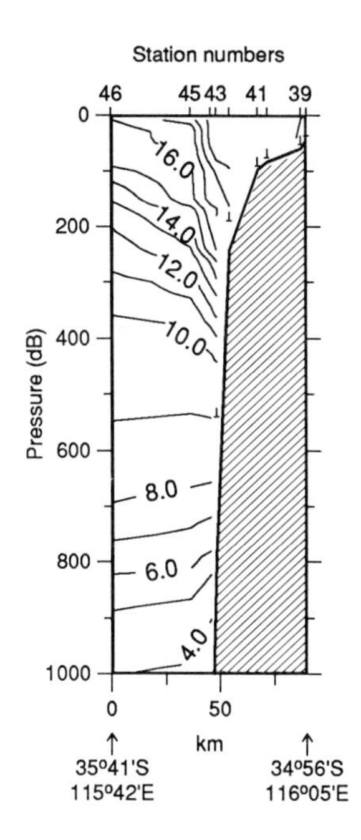

Figure 9a: Cliffy Head section
Potential temperature
15-17 June 1987

Figure 9: CTD sections of salinity and temperature for the Cliffy Head line in June 1987.

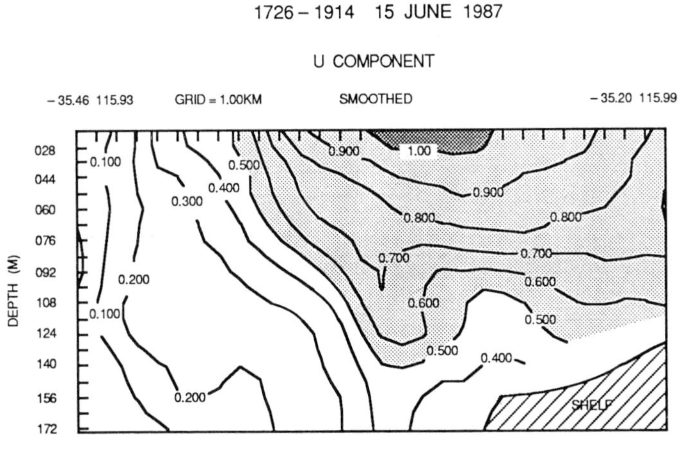

1726 – 1914 15 JUNE 1987

U COMPONENT

–35.46 115.93 GRID = 1.00KM SMOOTHED –35.20 115.99

DEPTH (M)

Figure 10: A section of the east component of the current measured by an acoustic Doppler current profiler roughly along the Cliffy Head CTD section on the same day as for Figure 9.

DISCUSSION

The historic observations clearly show the intrusion of low-salinity tropical water that flows poleward (southward) down to Cape Leeuwin and then eastwards towards the Great Australian Bight. This low-salinity flow (defined as the Leeuwin Current by Cresswell and Golding, 1980) was found to occur chiefly in autumn and winter, although there was some poleward low-salinity flow in November/December (Cresswell and Golding, 1980). In contrast, a northeastward inflow of high-salinity water, which turns to flow poleward (southward) on approaching the coast, dominates the summer situation. Andrews (1977) referred to this flow as the West Australian Current. Low-salinity water is also drawn down from the north in summer (Andrews, 1977 and 1983) and the combined flow subsequently turns east towards the Great Australian Bight. Summer satellite images sometimes indicate a narrow poleward flow starting from near North West Cape. Presumably this is a stream of low-salinity tropical water. The summer LUCIE cruise occurred after two months of weak or zero poleward flow (Figure 6), and no low-salinity water was found. In contrast to Andrews (1977), Kitani (1977) and Rochford (1969a) referred to the northward flowing West Australian Current. There is clearly a problem of terminology of the flow in the eastern Indian Ocean; unfortunately, we have been unable to find a clear resolution to this difficulty.

The source of the low-salinity water seems to be a fan-shaped area off northwestern Australia, as depicted in Figure 1. Gentilli (1972) suggested that autumn and winter throughflow from the Pacific Ocean to the Indian Ocean was isolated off northwestern Australia over summer and then became the source for southward flow in the next autumn and winter. Godfrey and Ridgway's (1985) steric height maps are consistent with this interpretation. Thompson (1984) suggested that only a small portion of the Leeuwin Current flux came from the North West Shelf. Recent theoretical work (McCreary et al., 1986; Thompson, 1988) suggests that much of the inflow to the Leeuwin Current is the result of geostrophic near-surface inflow from west of Australia. However, this inflow is of high-salinity water, and there must be a northern source region as well.

The differences between the Leeuwin Current and other eastern boundary currents are illustrated by Figure 11 (Godfrey and Ridgway, 1985). This shows annual mean temperature sections at similar latitudes in the Leeuwin Current, and off California and South Africa (representative of the other four major eastern boundary currents). Even at 500 km offshore, there are substantial differences: inspection of Figure 11 shows that throughout the top 300 m, 500 km offshore, temperatures off western Australia are higher by about 2° to 4°C than those off California,

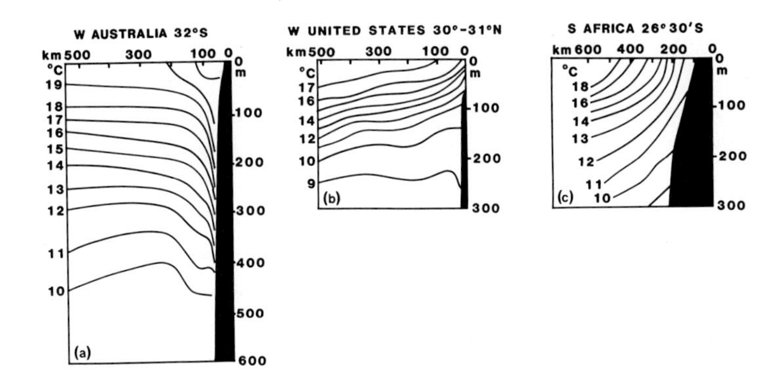

Figure 11: Annual average temperatures, along (a) a section at 32°S off western Australia, (b) a section at 30° to 31°N out from California, and (c) a section at 26°30'S off the west coast of South Africa (from Godfrey and Ridgway, 1985).

and by 1° to 2°C than those off South Africa. At the shelf edge these differences increase greatly because of the southward flow and downwelling associated with the Leeuwin Current, and the upwelling off the other coasts. Thus, at 200 m at the shelf edge, annual mean temperature off western Australia is some 8°C warmer than in the other two current systems!

Some of these differences are compensated by higher salinity, but specific volume anomaly (SVA) is generally higher all along the western Australian coast than along any of the other ocean eastern boundaries. Godfrey (1988) has developed a simple extension of the familiar Sverdrup models, that appears to explain (in a depth integrated sense) these differences between different ocean basins. Specifically, he has shown using Levitus' (1982) annual mean hydrographic data that depth-integrated steric height (or double depth-integrated SVA) in the top 1000 to 2000 m, relative to appropriate "depths of no motion," have roughly characteristic values for each of seven eastern boundaries: eastern Pacific, western New Zealand, western Australia, Indonesia-Malayasia-Burma, western India, western Malagasy, and eastern Atlantic. Furthermore, the quite large differences between these seven characteristic values are quite well modeled by certain integrals of the annual mean wind stress. In particular, western Australian values of double depth-integrated SVA are higher than those in the eastern Pacific because of the action of zonal winds along the equatorial Pacific, which pile up warm, light water near Indonesia. This effect may propagate via baroclinic waves along the western Australian coast.

The result provides a partial explanation for the generally high level of temperature in Figure 11a relative to Figure 11b (there is a similar explanation for the low levels in Figure 11c), but it does not explain the existence of the Leeuwin Current. Godfrey (1988) suggests the Leeuwin Current might be indirectly driven by heat loss to the atmosphere associated with these high surface temperatures. The heat losses at high latitudes create a strong along-shore pressure gradient which in turn drives the Leeuwin Current. These heat losses are unique in eastern boundary regions (see Hsiung, 1986). LUCIE was partly designed in the hope that it might throw some light on such speculations.

Through the year-long use of the CTD and acoustic Doppler current profiler on Franklin, along with the current measurements across the continental shelf and slope and satellite imagery, the Leeuwin Current, during LUCIE, has been observed more precisely than in the past. Data analysis in progress is addressing the relationship between the currents and the pressure and wind stress forcing functions. By combining the various data sets, it is anticipated that a much more accurate and detailed description will become available, leading to a greater understanding of the Leeuwin Current dynamics and the development of useful models.

ACKNOWLEDGEMENTS

We would like to thank everybody who participated in the planning and execution of LUCIE.

From: Joseph L. Reid, Scripps Institution of Oceanography

On: Review and Commentary to paper **THE LEEUWIN CURRENT**
by John A. Church, George R. Cresswell, and J. Stuart Godfrey

This is a clear and straightforward description of the present information about the Leeuwin Current. I have only three comments or questions that the authors may wish to consider.

First, the poleward flow in winter, with an equatorward flow offshore, is rather like the California Current. It is the equatorward subsurface flow that is so remarkably different.

The equatorward surface flow in summer seems to carry subtropical water, however, while the equatorward flow off North and South America carries higher-latitude, colder, and less saline waters. Is it the difference in extent of latitude (35°S) of the end of Australia from that of the eastern Pacific continents that accounts for this? South Africa ends at about that latitude. How do the flow and characteristics of the Benguela differ from western Australian waters?

Second, the Leeuwin Current is discussed alone, without reference to the general circulation. For example, is the equatorward undercurrent part of a larger-scale anticyclonic flow in the central South Indian Ocean, with the Leeuwin Current a very interesting aberration from most eastern boundary currents? Is the underlying salinity minimum near 600 m a part of the Intermediate Water, with the oxygen maximum above it, corresponding to the structure seen off South America in the Scorpio data?

Third, is it the narrowness of the Leeuwin Current that has made it so little known until recently? With speeds of 1.5 m s^{-1}, seamen must have been aware of it.

Reply to comments by Professor Reid, sent by J. Stuart Godfrey:

> We have added text and a figure to the original manuscript that address [Prof. Reid's] first and second points. With respect to the third point, I believe the explanation is as follows:
>
> Ships generally crossed the narrow Leeuwin Current en route to South Africa or the Red Sea, and reported ship's drifts from about a day's run. Hence, the Leeuwin Current was missed with nearshore (equatorward) flows and offshore eddies. The

Current does <u>not</u> show up very clearly in current atlases, e.g., the very comprehensive Dutch atlas "Sea Areas Around Australia," published in 1947. Also, early hydrographic studies usually took no stations until they were in water over 1500 m deep, to get a good "depth of no motion"—this was offshore from most of the Leeuwin Current!

PART III:

SPECIAL TOPICS

AN APPLICATION OF TURBULENCE DATA TO THE CALIFORNIA UNDERCURRENT

Rolf Lueck
Hidekatsu Yamazaki

Chesapeake Bay Institute
The Johns Hopkins University
315-711 W 40th Street
Baltimore, MD 21211

ABSTRACT

Persistent and large rates of the dissipation of kinetic energy were observed in the intrusive region of the California Undercurrent. Viscous dissipation can extract the total mechanical energy (potential plus kinetic) of the undercurrent in 6 months and its kinetic energy in only 11 days, suggesting the occurrence of geostrophic adjustments on similar time scales. A simple mechanical energy budget does not show a balance between dissipation and the work done by the pressure field contrary to findings for the Equatorial Undercurrent.

INTRODUCTION

The California Undercurrent is a poleward flow of relatively warm and saline water along the west coast of North America originating from the Equatorial Undercurrent and the subsurface South Equatorial Countercurrent (Tsuchiya, 1981). Wooster and Jones (1970) show that the California Undercurrent "jet" reaches average speeds of 0.3 m s^{-1} off northern Baja California with speeds as high as 0.4 m s^{-1} in August, 1966. They estimated a width of 20 km and a vertical scale of 300 m. Reid (1962) observed a maximum speed of 0.2 m s^{-1} at a depth of 250 m in December, 1961 on a zonal section at 36°. Cannon, Laird, and Ryan (1975) estimate a geostrophic speed of 0.05 m s^{-1} for the undercurrent off Washington State. The undercurrent appears to be a permanent feature throughout the year as far north as 50° (Reed and Halpern, 1976).

Our turbulence measurements, made in April, 1982 over the San Diego trough near the 1000 m isobath (Yamazaki and Lueck, 1987), show persistent and high levels of kinetic energy dissipation in an interleaving region between 150 and 250 meters depth. This region had the T-S signature of the undercurrent, but the presence of the undercurrent over the San Diego trough during our observations is ambiguous. Monthly means of geopotential anomaly (Tsuchiya, 1980) show an undercurrent of 0.05 m s^{-1} at our observation site. Current meter observations by Denbo, Polzin, Allen, Huyer, and Smith (1984) show a mean northward flow of 0.35 m s^{-1} at 139 m off Purisuma Point (34.7°N) between February and May, 1982. Observations by Tsuchiya (1980) show that the California Undercurrent in the San Diego trough is generally northward at 0.05 m s^{-1}. The surface flow is southward at 0.1 m s^{-1} in March and April and weakly to the north during the rest of the year. Osborn and Cox (1972) repeatedly observed enhanced levels of temperature gradient microstructure over the San Diego trough near 90, 180, 200, and 400 meters depth. Gregg and Cox (1972) also observed persistent temperature and conductivity microstructure over the trough and suggested that it is driven by either salt fingering or shear produced by intrusions.

TURBULENT DISSIPATION

The results from 17 microstructure velocity profiles made over the San Diego trough using Camel II (Lueck, Crawford, and Osborn, 1983) have recently been reported by Yamazaki and Lueck (1987). A profile (Figure 1) of two orthogonal components of shear (1-100 cpm) temperature and its gradient shows the typical vertical distribution of turbulence near the 1000 m isobath. Turbulent layers, 1 to 20 m thick, were found between 50 and 70 meters and near 155, 200, and 240 m. The most dissipative layers are associated with the interleaving of water types. A CTD profile (Figure 2), made 35 minutes after the microstructure profile, shows fine structure very similar to that observed by Camel II. However, many of the profiles that were made approximately 30 minutes apart show changes in the vertical scale of temperature inversions, suggesting that the intrusions between 150 and 250 m are evolving rapidly in space and time. The temperature inversion near 215 m in the microstructure profile (220 in the CTD profile) is salinity compensated and occurs at S = 34.12, T = 8.7°C and σ_t = 26.48. This isopycnal is only 0.1 lighter than the one used by Wooster and Jones (1970) and Tsuchiya (1980) to trace the core of the undercurrent. The average dissipation between 150 and 250 meters, where intrusions with water of southern origin were observed, is 1.4×10^{-6} W m^{-3}. For comparison, Lueck et al. (1983) reported 0.53×10^{-6} W m^{-3} as the average dissipation between 200 and 500 m over the continental slope off Vancouver Island.

For the California Undercurrent to maintain its identity as a current system, it cannot dissipate its energy over a period shorter than the time scale of its longitudinal flow. The time that a water parcel is in the current system is L/U = 4×10^{7} seconds (15 months), where we have used

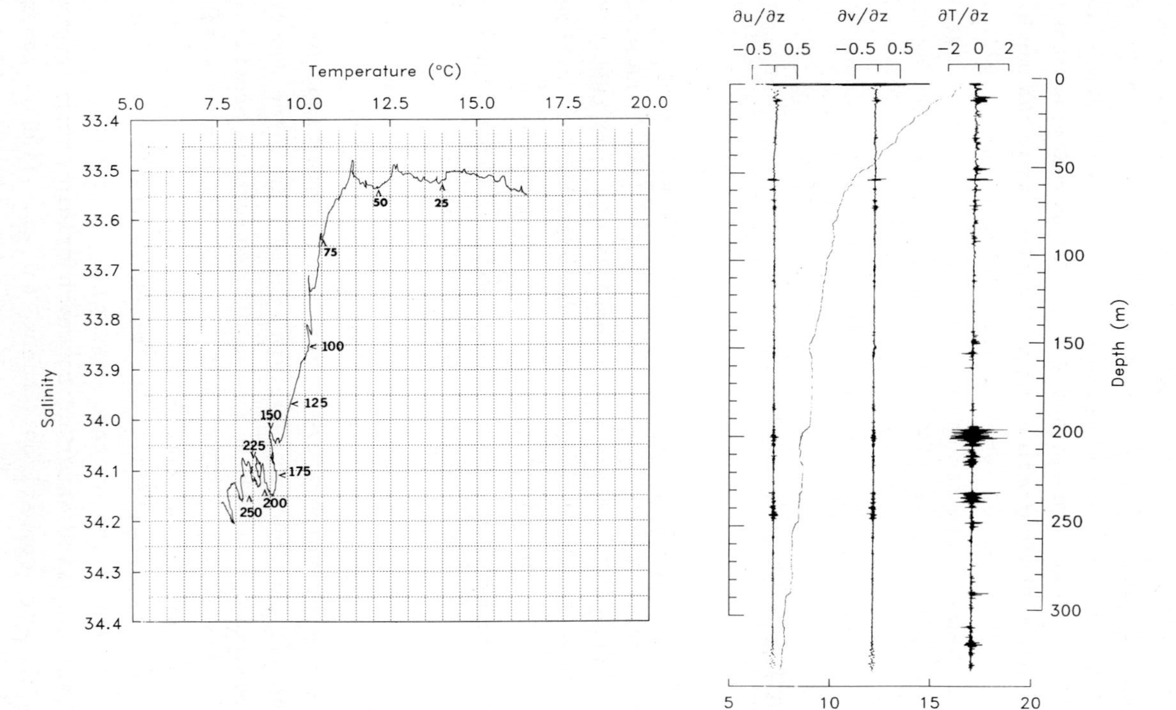

Figure 1: Vertical profile of the microstructure of temperature, its vertical gradient, and two components of the vertical shear of horizontal velocity (right half). The dissipation of kinetic energy is proportional to the variance of the shear. The variance of the temperature gradient is proportional to the dissipation of temperature variance. Bipolar gradients of temperature at microscales indicate overturning motions at such scales even if the fine scale density is salinity compensated. Temperature-salinity diagram from a CTD cast made 35 minutes after the microstructure profile (left half).

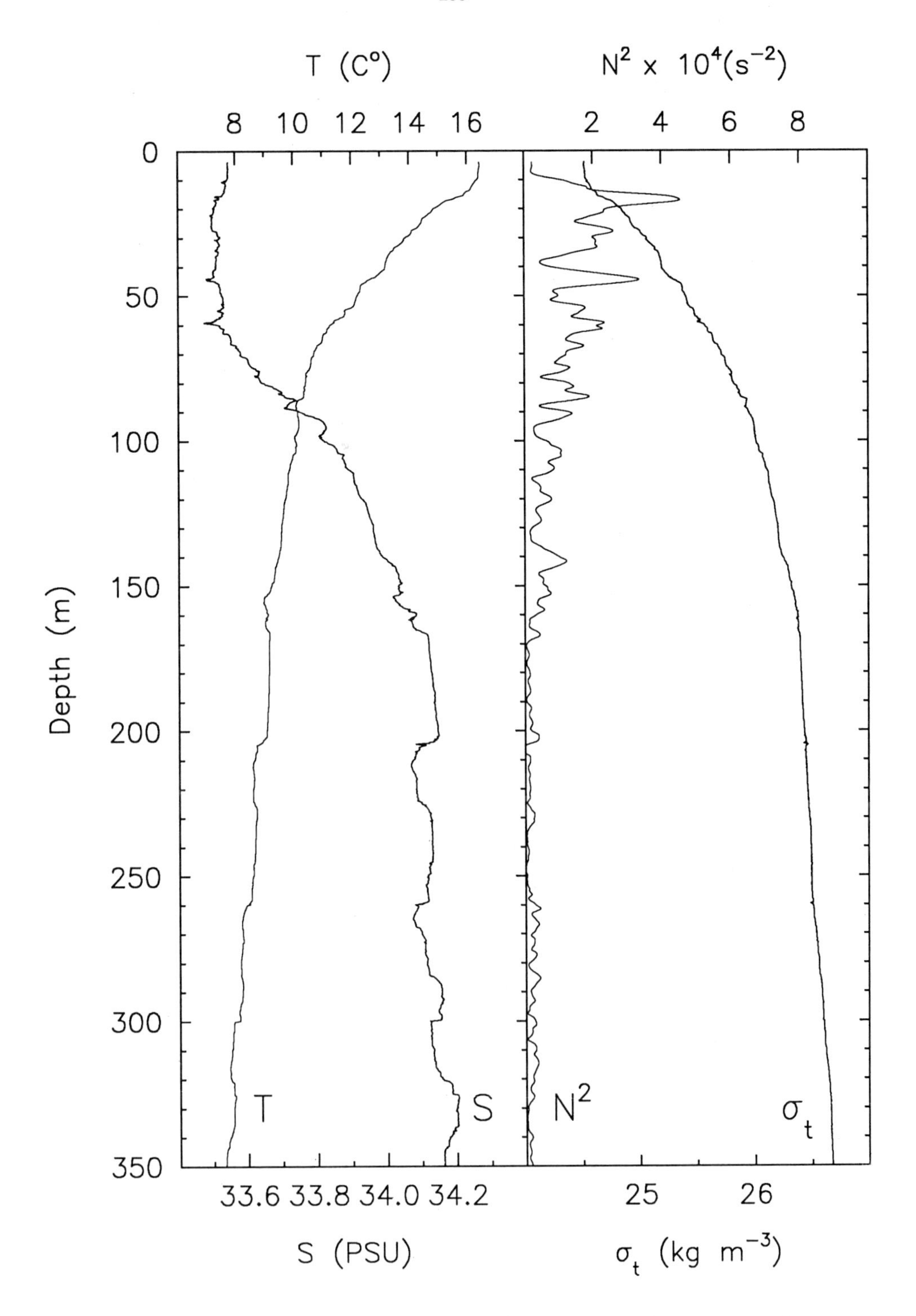

Figure 2: Profile of temperature, salnity, σ_t, and buoyancy frequency squared from CTD cast made 35 minutes after microstructure profile in Figure 1.

the distance separating southern California from mid-Vancouver Island ($L \approx 2 \times 10^6$ m), and a speed scale $U \approx 0.05$ m s^{-1}. The time scale for the dissipation of the current's kinetic energy is only $0.5\rho U^2/\langle\varepsilon\rangle \approx 1 \times 10^6$ s (11 days), where $\langle\varepsilon\rangle = 1.4 \times 10^{-6}$ W m^{-3} is the average measured dissipation. Unless we have underestimated the speed of the undercurrent by a factor of 3.5, the rate of dissipation of kinetic energy will extract the kinetic energy of a fluid parcel off San Diego long before it reaches Vancouver Island. The current speed that would yield a dissipation time scale equal to the longitudinal time scale is 0.18 m s^{-1}, but the undercurrent is weakest in April (Tsuchiya, 1980 and Hickey, 1979). Our work, of course, represents only one position in a current that extends over 2000 km and so does not necessarily give a mean dissipation value for the entire current system. Local effects such as internal tides and basin oscillations may bias our dissipation values upwards. However, a bias, if present, has to selectively favor the intrusion region because that is the only depth range where dissipation rates are anomalously large (Yamazaki and Lueck, 1987).

A dissipation time scale much shorter than the longevity of a current system is, however, not inconsistent with the balance of total energy. Lueck and Osborn (1986) noted that the time scale for the dissipation of the kinetic energy of a Gulf Stream ring was also much shorter than the life time of a typical ring. However, the time scale for the dissipation of total energy (kinetic plus potential) was 2 to 3 years. The ratio of potential to kinetic energy is (Gill, 1982) $\left(k_H N/mf\right)^{-2}$ where k_H is the horizontal wavenumber, N/mf is the Rossby radius of deformation, m is the vertical wavenumber, N is the buoyancy frequency, and f the local Coriolis parameter. To estimate the ratio of potential to kinetic energy we will take $k_H \approx (40000)^{-1}$, $m \approx (200)^{-1}$, $N = 0.004$ s^{-1}, and $f = 8.0 \times 10^{-5}$ s^{-1}. This corresponds closely to Tsuchiya (1980) and gives a ratio of 16. Therefore, the time scale for the dissipation of the total energy of the undercurrent is 16×10^6 s (6 months), and it is not inconsistent with the times scale of the longitudinal flow. Thus, a fluid parcel in the undercurrent dissipates about e^{-1} of its total energy traveling from California to Vancouver Island. To maintain a geostrophic balance, energy must be transferred from the potential into the kinetic field with a time scale comparable to the dissipation time.

The axis of the California Undercurrent is approximately parallel to the coast of North America because of the confines imposed by Coriolis acceleration and the eastern boundary. The axis of the Equatorial Undercurrent is restrained zonally by the reversal of the sign of the Coriolis acceleration. This motivates the notion that the two current systems may be similar. A mechanical energy budget for the Atlantic Equatorial Undercurrent (Crawford and Osborn, 1980) shows a close balance between the work done by the pressure field and the rate of dissipation of kinetic energy with minor contributions from the advection and the local time rate of change of kinetic energy. A similar budget is made here, but the historical mean pressure field in the California Undercurrent is too poorly resolved to permit a definitive mechanical energy budget. Following Gill (1982, equation 4.6.3), the mechanical energy equation is

$$\partial/\partial t \left(1/2\rho\mathbf{u}^2\right) + \nabla \cdot \left\{\left(p + 1/2\ \rho\mathbf{u}^2\right)\mathbf{u} - \mu\nabla\left(1/2\mathbf{u}^2\right)\right\} = -gw\rho' - \varepsilon + p\nabla\cdot\mathbf{u}$$

where p is the perturbation pressure, ρ' the perturbation density, and $\mathbf{u}^2 = \mathbf{u}\cdot\mathbf{u}$. We will assume that the fluid is incompressible, $\nabla\cdot\mathbf{u} = 0$; the viscous transports are negligible, $\mu\nabla\cdot u^2 = 0$; and that the buoyancy flux is small compared to viscous dissipation, $gw\rho' << \varepsilon$. Further, we assume that the undercurrent is contained in an imaginary box so that flow normal to its surface vanishes everywhere except at the zonal boundaries. The zonal boundaries are taken orthogonal to the eastern boundary formed by the continental slope. The upper boundary is the interface between the surface current and the undercurrent, where the horizontal velocity vanishes; the bottom boundary is where the along-shore velocity is very small compared to the mean speed of the undercurrent. We assume that the western boundary can be defined by a zero-flow streamline. The cross-sectional area of the southern zonal boundary is A_1, which may be different from the area of the northern boundary, A_2. These assumptions exclude zonal eddy fluxes and internal wave radiation from further consideration. Finally, we assume that the current system is contiguous, a fact not yet established. The volume integral over the box, divided by the contained volume, then yields

$$d/dt\left(1/2\rho\langle\mathbf{u}^2\rangle\right) + v_1A_1\left(AL\right)^{-1}\left\{\left(p_2 - p_1\right) + 1/2\rho\left(\mathbf{u}_2^2 - \mathbf{u}_1^2\right)\right\} = -\langle\varepsilon\rangle \quad (1)$$

where $A = 1/2\left(A_1 + A_2\right)$, angled braces denote a volume average. We have used $v_1A_1 = v_2A_2$, and v represents the along-shore (meridional) mean velocity. The intensity of the undercurrent varies seasonally and so a lower bound for the time scale for the rate of change of the kinetic energy is one month. The magnitude of the current is approximately $|\mathbf{u}| = 0.05\ m\ s^{-1}$ and can change up or down by about the same amount in a month or longer. The first term in equation (1) then is $\pm 5\times10^{-7} W\ m^{-3}$. The historic mean pressure field for January is estimated from the 200/500 dBar dynamics height field of Schartzlose and Reid (1972), which indicated a change of 0.00 to 0.02 dynamic meters over 12 degrees of latitude $\left(L \approx 1.3\times10^6\ m\right)$. However, the contour interval is only 0.02

dynamic meters making this estimate quite unreliable. The pressure difference over 12 degrees is $p_2 - p_1 = 10\rho\Delta D = 0$ to -200 J m^{-3} where ΔD is the dynamic height difference in dynamic meters. The kinetic energy difference is maximized when $\mathbf{u}_2^2 \ll \mathbf{u}_1^2$ and is at most -1 J m^{-3} for a 0.05 m s^{-1} flow making it negligible compared to the pressure difference. The work by the pressure field then is less than $-A_1A^{-1} \times 8 \times 10^{-6}$ W m^{-3}. The last term in equation (1), the rate of dissipation of kinetic energy, is 1.4×10^{-6} W m^{-3} based on our observations at one site only. So the magnitude of the terms in equation (1) is

$$\pm 5 \times 10^{-7} + 4 \times 10^{-8} A_1 A^{-1} \{(0 \text{ to } -200) + (-1)\} = -1.4 \times 10^{-6}$$

It is not clear by how much the undercurrent broadens towards the north, but a balance between dissipation and pressure work requires a northern boundary 10 times larger than the area of the southern boundary, which probably is unrealistic. Zonal eddy fluxes may be important, and further measurements of both the dissipation and the pressure field are needed before a reliable energy budget can be made. The importance of zonal eddy fluxes is also implied by low values of the vertical eddy diffusivity of density (Yamazaki and Lueck, 1987).

CONCLUSION

The kinetic energy of the California Undercurrent is dissipated on a time scale much shorter than its transit time from San Diego to Vancouver Island. The time scale for the dissipation of the current's total energy is comparable to its transit time. Turbulent dissipation is sufficiently large to cause a geostrophic imbalance if there is not transfer of energy from the potential to the kinetic field. In contrast to the Equatorial Undercurrent, it is not possible to show that the rate of dissipation is balanced by the work done by the pressure field, largely due to uncertainties in the dynamic height anomaly of the California Undercurrent. The available data suggests that the two are not in balance and that lateral fluxes may be important in the mechanical energy budget.

MODEL SIMULATIONS OF A COASTAL JET AND UNDERCURRENT IN THE PRESENCE OF EDDIES AND JETS IN THE CALIFORNIA CURRENT SYSTEM

Mary L. Batteen
Department of Oceanography
Naval Postgraduate School
Monterey, CA 93943

ABSTRACT

A high-resolution, multi-level, primitive equation ocean model is used to examine the response of an initial baroclinic jet in an idealized, flat-bottomed, oceanic regime along an eastern ocean boundary. The initial system, intending to represent the mean California Current System during the upwelling season, consists of an equatorward coastal jet overlying a poleward undercurrent. After ~30 days, eddies and jets are generated, due to the baroclinic instability associated with the vertical shear between the Coastal Upwelling Jet (CUJ) and the California Undercurrent (CUC). An examination of both the near-surface and subsurface flow fields in the presence of the eddies shows a complex structure. In the presence of cyclonic (anticyclonic) eddies close to shore, the near-surface (subsurface) current meanders cyclonically (anticyclonically) around the eddies, thereby extending its offshore domain. In the presence of anticyclonic (cyclonic) eddies, the near-surface (subsurface) flow continues equatorward (poleward) in a narrowly confined nearshore region of ~40 km for the surface flows (~8-20 km for the subsurface flows).

INTRODUCTION

Recent observations have shown that, superimposed on the broad, slow (~5 cm s^{-1}), climatological mean flow in the California Current System (CCS), are highly energetic, mesoscale eddies and meandering jets (Bernstein et al., 1977; Mooers and Robinson, 1984; Rienecker et al., 1985). The eddies have wavelengths of several hundred kilometers, which can intensify over several months and be "cutoff," creating isolated eddies (Bernstein et al., 1977). Strong baroclinic jets with peak velocities of ~50 cm s^{-1} are embedded in this field of cyclonic and anticyclonic eddies. These jets are ~40 km wide, extend from the surface to at least 100 m depth and have offshore extensions of several hundred kilometers (Mooers and Robinson, 1984; Rienecker et al., 1985; Flament et al., 1985).

The dynamical processes responsible for the generation and evolution of these intense and complex eddy and jet patterns in the CCS have yet to be fully identified. One possible generation mechanism currently under investigation is that the eddy and jet fields are the result of either a baroclinic or barotropic instability of the mean California Current System (CCS) which, during the upwelling season (from ~April to October), consists of an equatorward coastal upwelling jet overlying a poleward undercurrent (Ikeda and Emery, 1984).

Both baroclinic and barotropic instabilities have been suggested to generate eddies and jets in the Northeast Pacific (Wright, 1980). Evidence has been shown for baroclinic instability being an important mechanism for eddy generation off Vancouver Island (Emery and Mysak, 1980; Thomson, 1984), and off the coast of Oregon and northern California (Ikeda and Emery, 1984). In particular, Ikeda and Emery (1984) have hypothesized that current meanders are triggered by along-shore variations in the coastline (capes) close to the California coast and grow as a result of the baroclinic instability of the coastal, equatorward, near-surface jet (~40 km wide and associated with coastal upwelling) and the poleward California Undercurrent. As these unstable meanders intensify, they carry the cool, upwelled water offshore and are often cutoff, creating pairs of isolated eddies or "vortex pairs" consisting of a cyclonic and an anticyclonic eddy (Bernstein et al., 1977). Evidence for barotropic instability as a dominant mechanism is still forthcoming, although one case of eddy-eddy-jet interaction observed by OPTOMA in the summer of 1983 showed characteristics of barotropic instability (Robinson et al., 1984).

In this study a high-resolution, multi-level, primitive equation ocean model is used to examine the response of an idealized, flat-bottomed oceanic regime along an eastern boundary to an imposed baroclinic jet. Baroclinic instability, associated with the vertical shear between the Coastal Upwelling Jet (CUJ) and the California Undercurrent (CUC), occurs after ~30 days, resulting in the generation of both cyclonic and anticyclonic eddies and jets. Significant changes from the imposed initial conditions in the CUJ and CUC fields are due to the presence of the eddies and jets.

MODEL DESCRIPTION AND INITIAL CONDITIONS

Model Equations

The numerical model used in this study was developed by Haney (1985) and is a multi-level, primitive equation (PE) model of a baroclinic ocean on an f-plane. The model is based on the hydrostatic and Boussinesq approximations with salinity neglected and the rigid lid approximation made. Depth of the model is variable; however, this study will only consider a flat bottom. In addition, the depth-averaged current is neglected in these experiments. The governing equations, written in sigma coordinates, are as follows:

$$\frac{du}{dt} = \frac{-1}{\rho_o}\frac{\partial p'}{\partial x} + \frac{\sigma}{\rho_o D}\frac{\partial p'}{\partial \sigma}\frac{\partial D}{\partial x} + fv - A_m \nabla^4 u + \frac{K_m}{D^2}\frac{\partial^2 u}{\partial \sigma^2} + \delta_d(u) \quad (2\text{-}1)$$

$$\frac{dv}{dt} = \frac{-1}{\rho_o}\frac{\partial p'}{\partial y} + \frac{\sigma}{\rho_o D}\frac{\partial p'}{\partial \sigma}\frac{\partial D}{\partial y} - fu - A_m \nabla^4 v + \frac{K_m}{D^2}\frac{\partial^2 v}{\partial \sigma^2} + \delta_d(v) \quad (2\text{-}2)$$

$$\frac{\partial w}{\partial \sigma} + \frac{\partial u}{\partial x} + \frac{\partial v}{\partial y} = 0 \quad (2\text{-}3)$$

$$p' = D\int_{\sigma}^{o} B\,d\xi - \int_{-1}^{o}\left[D\int_{\sigma}^{o} B\,d\xi\right]d\sigma \quad (2\text{-}4)$$

$$B = \alpha g\left(T - T_o\right) \quad (2\text{-}5)$$

$$\frac{dT}{dt} = -A_H \nabla^4 T + \frac{K_H}{D^2}\frac{\partial^2 T}{\partial \sigma^2} + \delta_d(T) \quad (2\text{-}6)$$

In the above equations, sigma, denoted by σ, is equal to z/D. All horizontal partial derivatives are on constant sigma surfaces. In addition, the variables used are defined in Table 1, while Table 2 gives other symbols in the model equations along with the corresponding values of constants used throughout this study.

Domain Size and Resolution

The study region extends approximately 500 km offshore from the west coast of North America, and it spans the California coastline from Point Sur in the south to Cape Blanco in the north (Figure 1). The model domain extends 6° in longitude and 6° in latitude (512 km by 640 km) and has 8 km by 10 km horizontal resolution with ten levels in the vertical.

Finite Difference Scheme

In the horizontal, a space-staggered B-scheme (Arakawa and Lamb, 1977) is used while a sigma coordinate system controls the vertical. The noise-free version of the hydrostatic equation, advocated by Arakawa and Suarez (1983), and Batteen (1988), is used.

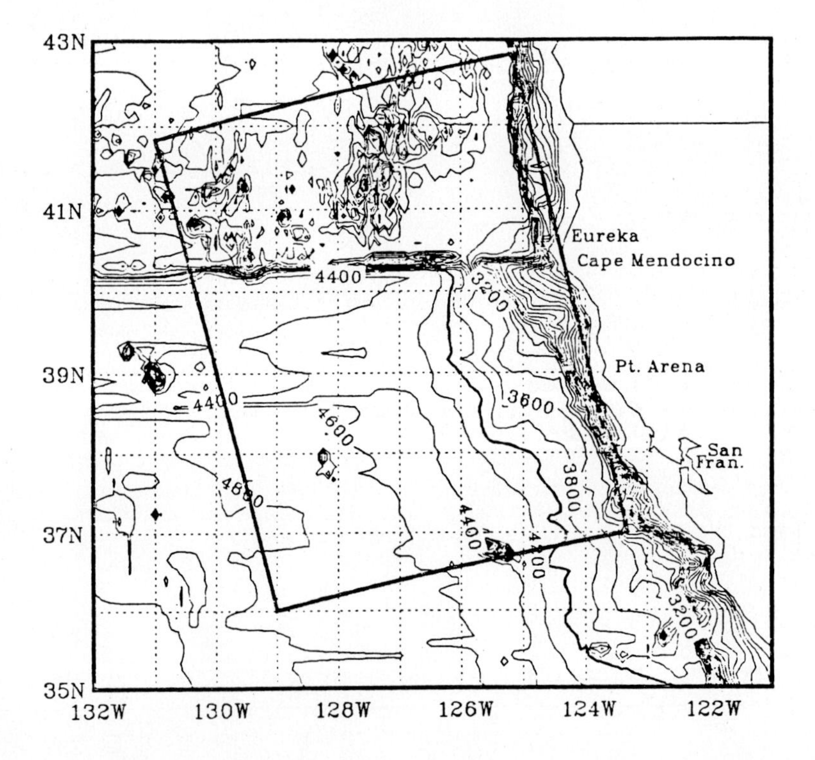

Figure 1: Study domain.

TABLE 1

DEFINITIONS OF VARIABLES USED IN THE MODEL

SYMBOL	DEFINITION
z	height (increasing upwards)
t	time
p'	pressure perturbation from a vertical average
T	temperature
u, v, w	eastward, northward and vertical (sigma) velocity components, respectively
B	buoyancy
δ_d	dynamic adjustment term

TABLE 2

VALUES OF CONSTANTS USED IN THE MODEL

	VALUE	NAME
Ω	$2\pi \text{day}^{-1}$	earth rotation rate
T_o	278.2°K	constant reference temperature
ρ_o	1.0276 gm cm^{-3}	density of sea water at T_o
α	2.01×10^{-4} $(°K)^{-1}$	thermal expansion coefficient
K	10	number of levels in the vertical
ΔX	8.0×10^{5} cm	cross-shore grid spacing
ΔY	1.0×10^{6} cm	along-shore grid spacing
D	4.5×10^{5} cm	total ocean depth
$\emptyset_o$	36.5°N	latitude of southern boundary
$\emptyset_m$	42.5°N	latitude of northern boundary
λ_o	124°W	longitude of eastern boundary
λ_m	130°W	longitude of western boundary
Δt	800.0 s	time step
f_o	0.93×10^{-4} s^{-1}	mean Coriolis parameter
g	980.0 cm^{-2}	acceleration of gravity
A_m	2.0×10^{17} cm^4 s^{-1}	biharmonic momentum diffusion coefficient
A_H	2.0×10^{17} cm^4 s^{-1}	biharmonic heat diffusion coefficient
K_m	0.5 cm^2 s^{-1}	vertical eddy viscosity
K_H	0.5 cm^2 s^{-1}	vertical eddy conductivity

Boundary Conditions

The northern, western and southern boundaries are open using a modified version of the radiation boundary conditions of Camerlengo and O'Brien (1980). The eastern boundary, representing the west coast of North America, is modeled as a straight, vertical wall. In this study, a symmetric, free-slip condition on the tangential velocity is invoked at the coastline.

Initial Conditions

The model can either be spun up from rest by a surface wind stress or heat flux, or it can be initialized with a specified current field. In this study, the model is unforced and initialized, following Ikeda and Emery (1984), with representative vertical (Figure 2a) and horizontal (Figure 2b) profiles of mean flow during the upwelling season. The initial flow pattern is based on the type of along-shore baroclinic coastal jet which Hickey (1979) observed in summer after about a week of strong equatorward winds favorable for upwelling. An estimation of the magnitude and sign of terms in the quasigeostrophic potential vorticity equation for the imposed initial conditions was used to determine that the necessary condition for baroclinic instability would be satisfied, i.e., since the potential vorticity gradient changes sign in the domain, the coastal jet is expected to become baroclinically unstable.

The initial mean stratification used in this study is a pure exponential temperature profile with a vertical length-scale of H=450 m. The exact form is

$$T(z) = T_B + \Delta T \, e^{z/450} \quad , \qquad (2\text{-}7)$$

where $T_B = 2°C$ is the temperature at great depth, and $\Delta T = 13°C$ is the increase in temperature between the bottom of the ocean and the surface.

This temperature profile was derived from observations used to support the Princeton Dynalysis model (Blumberg and Mellor, 1987) and is representative of the long-term, mean climatological temperature stratification for the California coastal region. Additionally, the

Brunt-Vaisala frequency profile, $N^2(z)$, was calculated analytically using the temperature function in Eq. (2-7) from

$$N^2 = \alpha\, g\, \partial T/\partial z \; . \qquad (2\text{-}8)$$

The model temperature profile and resultant N^2 profile are shown in Figures 2c and 2d, respectively. Using the method of Feliks (1985), the model baroclinic radius of deformation was calculated to be ~30 km, which compares favorably with observed values (Emery et al., 1984).

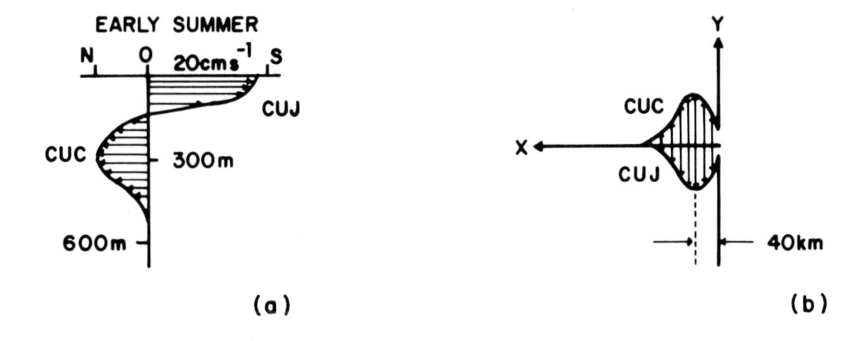

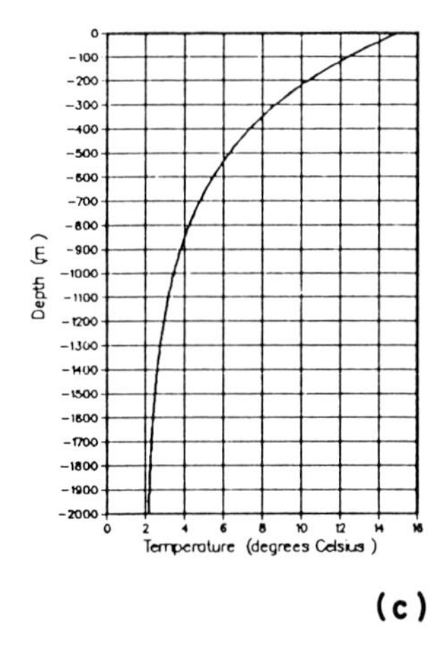

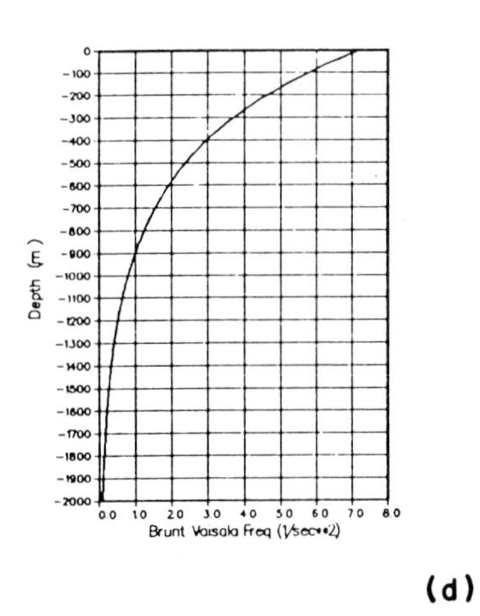

Figure 2: Model initial conditions. (a) Vertical and (b) horizontal profiles of mean flow for the early summer (after Ikeda and Emery, 1984). (c) Model initial temperature profile and (d) initial profile of square Brunt-Vaisala frequency N^2.

RESULTS

Baroclinic instability, associated with the vertical shear between the CUJ and CUC, occurs after ~30 days, resulting in the generation of eddies and jets. Figures 3 and 4 show the near-surface (~50 km depth), instantaneous perturbation pressure (Figure 3) and temperature fields (Figure 4) at five day intervals from days 35 to 50. The CUJ with equatorward flow is discernible in the pressure fields (Figure 3) as is the establishment of anticyclonic vortices, and pinched off cyclonic eddies on the seaward side of the CUJ. Also formed are meanders and dipole pairs of eddies with strong jets between them. An important feature in the temperature fields (Figure 4) is the presence of squirts (regions of intense seaward flow) of cold water with offshore extensions of ~100-200 kilometers. A comparison of the temperature and pressure fields (Figures 3 and 4) shows that the squirts occur in the vicinity of the (geostrophic) flow between dipole pairs of eddies.

A significant evolution of the flow from the imposed initial conditions is that the CUJ, instead of being strictly a southward flow, now is also a meandering flow around the nearshore cyclonic eddies, resulting in offshore extensions of ~100-200 kilometers. In the presence of anticyclonic vortices, the CUJ flows equatorward in a narrowly confined region of ~40 km inshore of the eddies.

The subsurface (~350 m depth), instantaneous perturbation pressure (Figure 5) and temperature fields (Figure 6) are shown at five day sequences from days 35 to 50. The CUC, instead of being strictly a poleward flow, as prescribed in the initial conditions, now is also a meandering flow around the nearshore anticyclonic eddies, resulting in offshore extensions of ~100 kilometers. In the presence of cyclonic eddies, the CUC flows poleward in a very narrowly confined region of ~8-20 km inshore of the eddies. As in the near-surface temperature fields, an important feature in the subsurface temperature field is the presence of squirts of cold water offshore, which also occur in the vicinity of the flow between dipole pairs of eddies. A comparison of the near-surface and subsurface pressure and temperature fields (Figures 3-6) shows that these are the same squirts and dipole pairs of eddies shown at 50 m and at 350 m, and that there is a slight tilt (toward the warm water) with depth of these features, indicative of baroclinic instability.

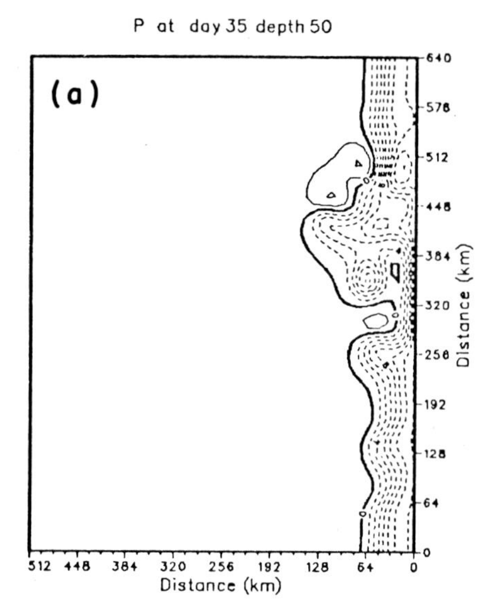

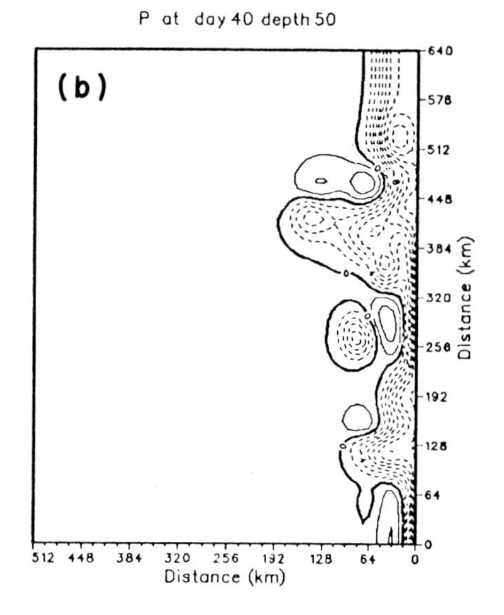

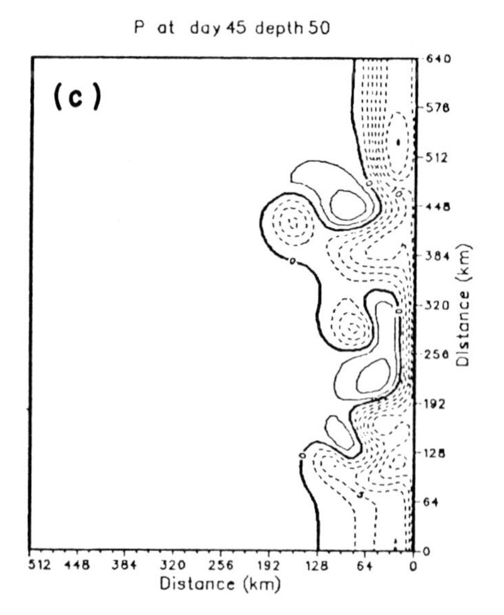

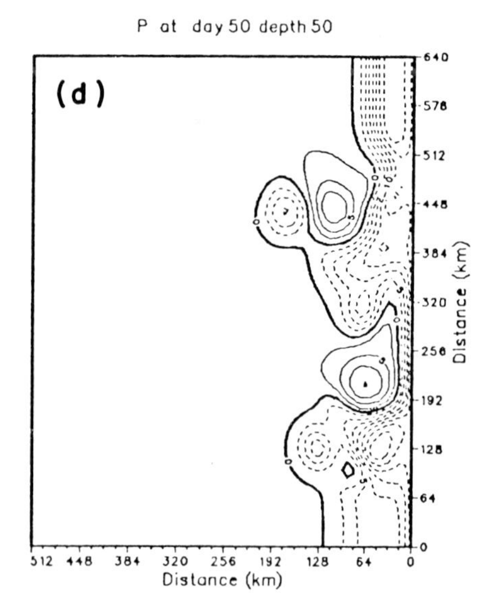

Figure 3: Instantaneous perturbation pressure fields at 50 m depth for days a) 35, b) 40, c) 45, and d) 50. Solid closed contours denote anticyclonic eddies.

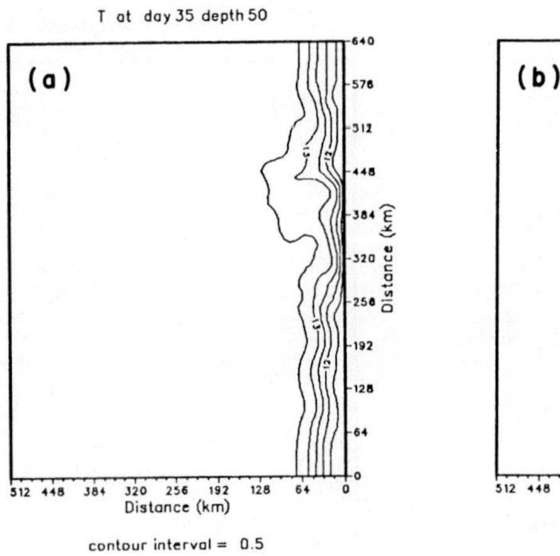

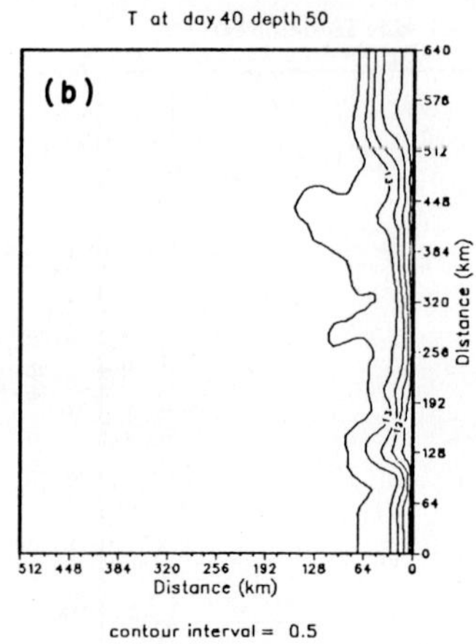

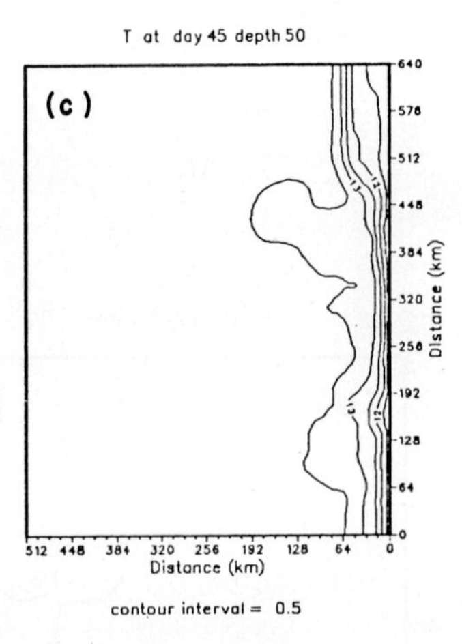

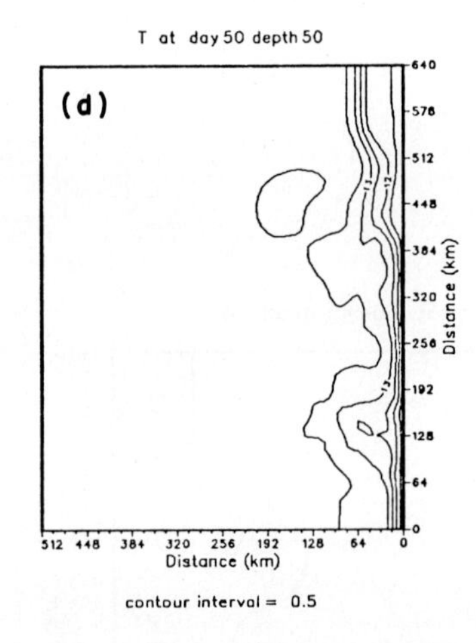

Figure 4: Temperature fields at 50 m depth for days a) 35, b) 40, c) 45, and d) 50. Contour interval is 0.5 degrees Celsius. The temperature is coldest at the coast.

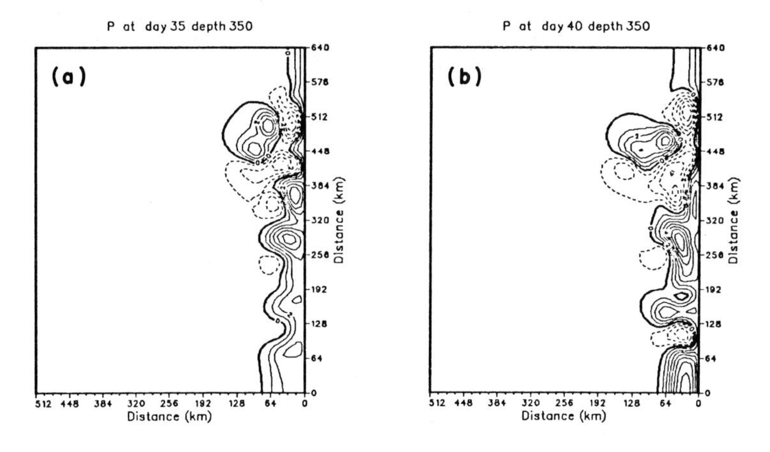

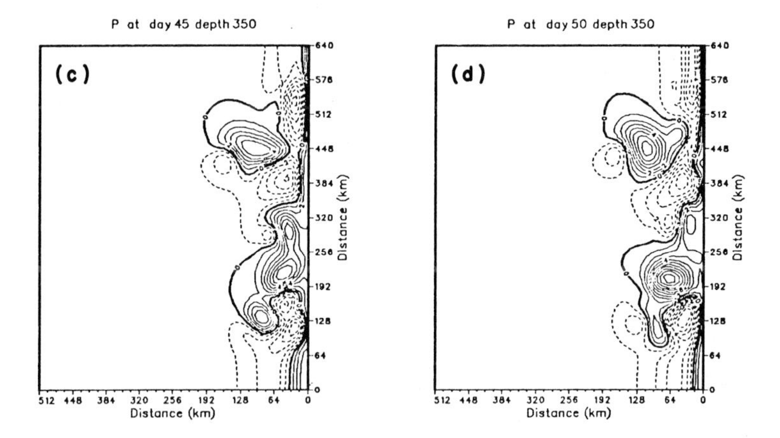

Figure 5: Same as Figure 3 but for 350 m depth.

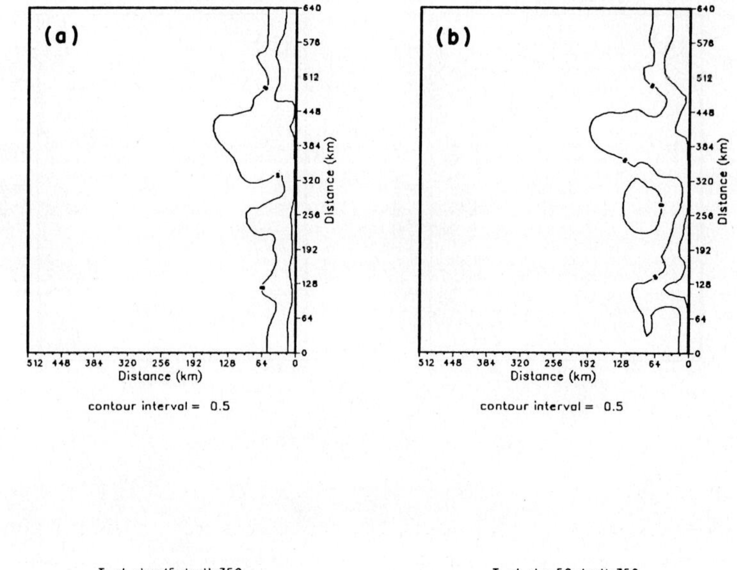

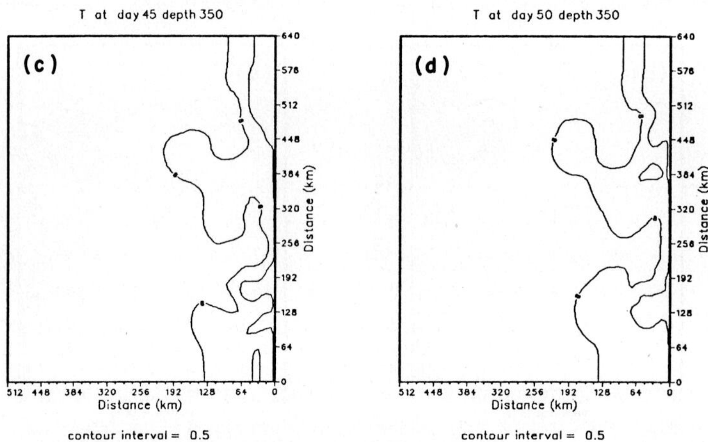

Figure 6: Same as Figure 4 but for 350 m depth.

Figure 7 shows instantaneous horizontal (Figures 7a, b) and vertical cross-sectional (Figure 7c) maps of the northward velocity (v) at day 50. Consistent with the horizontal pressure maps at depths 50 and 350 m (Figures 3d and 5d, respectively), horizontal maps of the v-fields show a cyclone with a meridional velocity of ~20 cm/s at 50 m depth adjacent to an anticyclone with a meridional velocity of ~35 cm/s at 50 m depth. This vortex pair is offshore of the coastal jet with peak velocities greater than 50 cm/s at 50 m depth.

The vertical cross-shore section of v at day 50 at y = 290 km (Figure 7c) shows that the vertical extent of the cyclone is ~1000 m (between ~25-50 km width) with maximum values of ~40 cm/s at the surface (~120 km from the coast). The vertical extent of the anticyclone is ~1200 m depth (~50 km width) with maximum values of ~70 cm/s at ~100 m depth (~50 km from the coast). The vertical extent of the coastal jet is ~400 m (~20 to 40 km width) with maximum values greater than 50 cm/s at the surface. A poleward undercurrent of ~15-20 km width and velocities of ~10 cm/s (maximum velocity at ~450 m depth) is discernible beneath the coastal jet nearshore.

Figure 8 shows vertical cross-sections of the along-shore and time-averaged (ten-day) v-fields before the onset of eddies (Figure 8a) and after the onset of eddies (Figure 8b). A comparison of the cross-sections shows significant changes in the vertical structure of both the coastal jet and the undercurrent. The depth of penetration of the coastal jet has shallowed nearshore, and the core of the equatorward jet has moved 20 km farther offshore. Instead of near-zero surface velocities at the coast, maximum surface CUJ velocities of ~50 cm/s are discernible. The undercurrent, instead of lying between ~15 to 50 km offshore, adjoins the coast.

A comparison of the instantaneous (Figure 7c) and along-shore and time-averaged (Figure 8b) vertical cross-sections of the v-field shows that, in the presence of eddies, the vertical and offshore structures of the coastal jet are readily evident in the time-averaged but not in the instantaneous cross-section of the v-field. This is consistent with recent observations of Kosro (1987) and Huyer and Kosro (1987), who suggest that a classical two-dimensional upwelling-induced coastal jet may not be very representative of actual conditions in the CCS. Their observations show that there is great variability in time and space in the CCS and that the classical coastal jet has not been frequently observed off California. Only one of their six surveys off Point Arena resembled a simple two-dimensional, equatorward, baroclinic coastal jet associated with coastal upwelling, and this was during a period of wind relaxation instead of winds

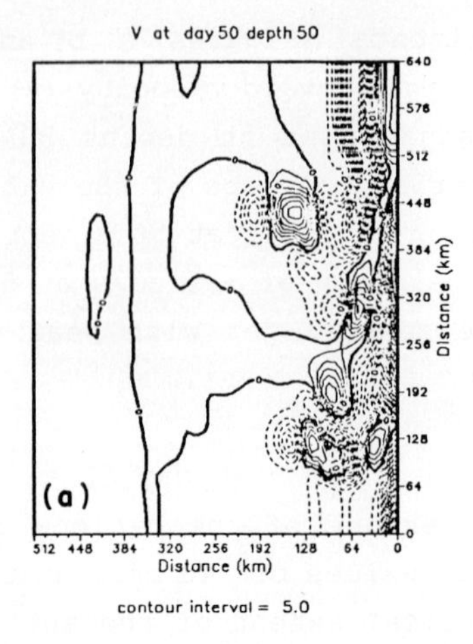

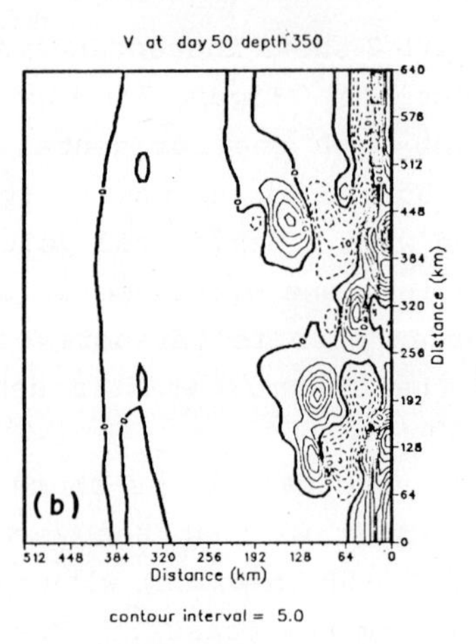

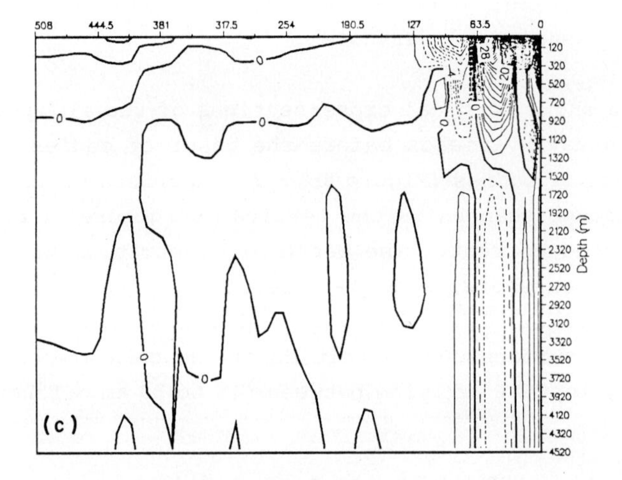

Figure 7: Instantaneous meridional velocity (v) fields at day 50. Solid closed contours denote anticyclonic eddies, while solid contours nearshore denote the undercurrent or poleward flow. The horizontal maps are shown at depths 50 m (a) and 350 m (b). The vertical cross-section (c) of v has been taken at y = 290 km. Contour interval is 5 cm/s.

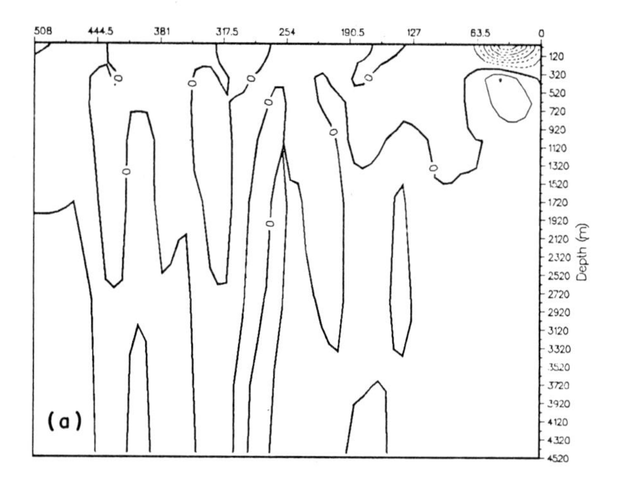

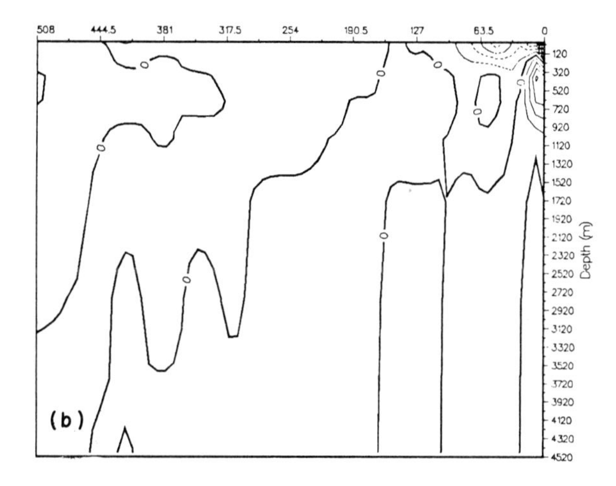

Figure 8: Vertical cross-sections of the v-field which are along-shore and time-averaged (a) over the first 10 days and (b) over days 41-50 of model simulation time. Contour intervals is 5 cm/s. The solid contours nearshore denote the undercurrent.

favorable for upwelling. The other five surveys showed complex circulation patterns which included meandering equatorward flow, cyclonic and anticyclonic eddies, and inshore countercurrents. The overall averaged fields did, however, show a relatively simple pattern: a broad equatorward surface flow and a poleward undercurrent. The present results show that the model coastal jet, after the onset of eddies, is more readily discernible on a time-averaged basis than on an instantaneous one, which is consistent with the above-mentioned observations.

SUMMARY AND RECOMMENDATIONS

This study used a high-resolution, multi-level, primitive equation ocean model to examine the response of an idealized, flat-bottomed oceanic regime along an eastern ocean boundary to an imposed jet. The imposed jet, intended to simulate the mean CCS during the upwelling season, consists of an equatorward CUJ overlying a poleward CUC. Baroclinic instability of the mean flow occurs after ~30 days resulting in the generation of cyclonic and anticyclonic eddies and jets. Examination of the results shows significant evolution of the CUJ and the CUC fields due to the presence of cyclones, anticyclones, meanders, dipole pairs of eddies with strong jets between them, and squirts of cold water with offshore extensions of ~100-200 kilometers. All of these features are consistent with available observational data.

The model results thus support the hypothesis that a possible generation mechanism for eddy and jet fields is baroclinic instability of the coastal, equatorward, near-surface jet (associated with coastal upwelling) and the poleward California Undercurrent. They are also consistent with observations that as these unstable meanders intensify, they carry the cool, upwelled water offshore and are often cutoff, creating vortex pairs of isolated cyclonic and anticyclonic eddies (Bernstein et al., 1977). The model simulations of the CCS demonstrate great variability in time and space and show a classical two-dimensional coastal jet only when the meridional velocity field is time-averaged. This is consistent with recent observations of Kosro (1987) and Huyer and Kosro (1987), who suggest that a classical two-dimensional upwelling-induced coastal jet may not be very representative of instantaneous observations in the CCS.

This study employed twin constraints of a flat-bottom basin and a regular coastline. Future experiments should include bottom topography and/or an irregular coastline which would represent capes and bays.

Future experiments should also include wind forcing instead of initialization of a baroclinic jet representative of the mean CCS during the upwelling season. It is believed that wind forcing of the local ocean domain in the CCS is another important generation process for eddies and jets in the CCS, and may provide important insight into the dynamics, kinematics and energetics of this complex eastern boundary current.

ACKNOWLEDGEMENTS

This work was done in the Department of Oceanography at the Naval Postgraduate School (NPS) while I was holding a National Research Council (NRC) Research Associateship, and continued under the support of the NPS Research Foundation. I wish to thank Dr. C.N.K. Mooers, my NRC sponsor, and Dr. R.L. Haney, who graciously allowed me to use and modify his primitive equation model.

From: R.H. Bourke and A.A. Bird

On: Review and commentary to paper **MODEL SIMULATION OF A COASTAL JET AND UNDERCURRENT IN THE PRESENCE OF EDDIES AND JETS IN THE CCS** by Mary L. Batteen

This is an interesting study which demonstrates, through the use of a high-resolution, multi-level, primitive equation model, the effects on the circulation induced by eddies generated by baroclinic instability of the mean flow. The rotational sense of the eddy determines whether the mean flow is then confined to a narrow inshore region, as in the case of the undercurrent, or forced to meander offshore possibly producing a "squirt." This was also the only model presented during the workshop which showed eddies in the presence of an undercurrent. Since eddies can be observed from remote sensing techniques, knowledge of this interaction of the eddy field and the undercurrent will be useful input for sampling strategies in field programs.

Although the model is formulated for the California Current System, the results have broad implications for other regions as well. In particular, given enough vertical shear to induce baroclinic instability, The same results should be applicable to other eastern boundary current systems.

Of particular importance to observationalists is the finding that the structure of the coastal jet is discernible only after employing a degree of time averaging. Synoptic observations are likely to miss it due to the complex nature of the short-term flow. This points out the need to employ long-term measuring systems if we are to understand the nature of the flow in meandering, boundary-current regimes.

In this study, Batteen used a flat bottom, a regular coastline, and a prescribed initial flow field, including a poleward undercurrent. Future work, as mentioned in the recommendations, should include bottom topography, various coastline geometries, and generation of eddies by wind forcing. Ongoing work by Batteen to incorporate the barotropic component of flow into the model will allow the energetics of the baroclinic and barotropic components of the eddy field to be determined. It is important that each process be included in the model systematically and compared with the results presented here in order to analyze and understand the dynamical processes and their effects. Separation of each of these processes should also help in the interpretation of field observations of eddy and jet fields.

DO NITROGEN TRANSFORMATIONS IN THE POLEWARD UNDERCURRENT OFF PERU AND CHILE HAVE A GLOBALLY SIGNIFICANT INFLUENCE?

L.A. Codispoti, R.T. Barber, and G.E. Friederich

Monterey Bay Aquarium Research Institute

ABSTRACT

Although it represents only about 0.01% of the total oceanic volume, the Poleward Undercurrent off the coasts of Peru and northern Chile is the site of globally significant nitrogen transformations. The co-occurrence of oxygen deficient (<~0.05 ml/l) waters and high fluxes of organic material causes the regional denitrification rate to comprise about 20% of the total marine rate and most of this occurs within the Undercurrent. Theoretical considerations and the existence of extremely low and extremely high nitrous oxide concentrations in and near the Undercurrent suggest that nitrous oxide turnover in this region occurs at much higher rates than in the "average ocean."

The processes of denitrification and nitrous oxide turnover are of more than casual interest. Denitrification converts combined nitrogen into nitrogen gas which is unavailable to most plants. Therefore, this nitrogen transformation can limit the ocean's ability to sequester atmospheric carbon dioxide. Changes in nitrous oxide cycling in regions such as the Poleward Undercurrent could significantly affect the rate at which nitrous oxide is being added to the atmosphere. This gas contributes significantly to the "greenhouse effect" and is the main source of nitrogen oxides in the stratosphere.

The advective field associated with the Poleward Undercurrent exerts a major control on the supply of oxygen to the subsurface waters off the coasts of Peru and Chile, and helps to establish counter-current systems that can create concentrations of organic material. Since the nitrogen transformations described above are sensitive to both oxygen levels and the supply of organic material, understanding these processes requires an understanding of the dynamics of the Poleward Undercurrent.

INTRODUCTION

The Eastern South Pacific Ocean (ESPO) attracts considerable scientific interest because of its high biological productivity and because it displays some of the most intense signals associated with the El Niño-Southern Oscillation phenomenon (e.g., Halpern, 1983). It is generally less well-appreciated that the waters in the upper ~500 m of this region (0-25°S and within 500 km off the coast, Figure 1) have a globally significant impact on nitrogen cycling (e.g., Codispoti and Christensen, 1985) despite the relatively small areas and volumes involved.

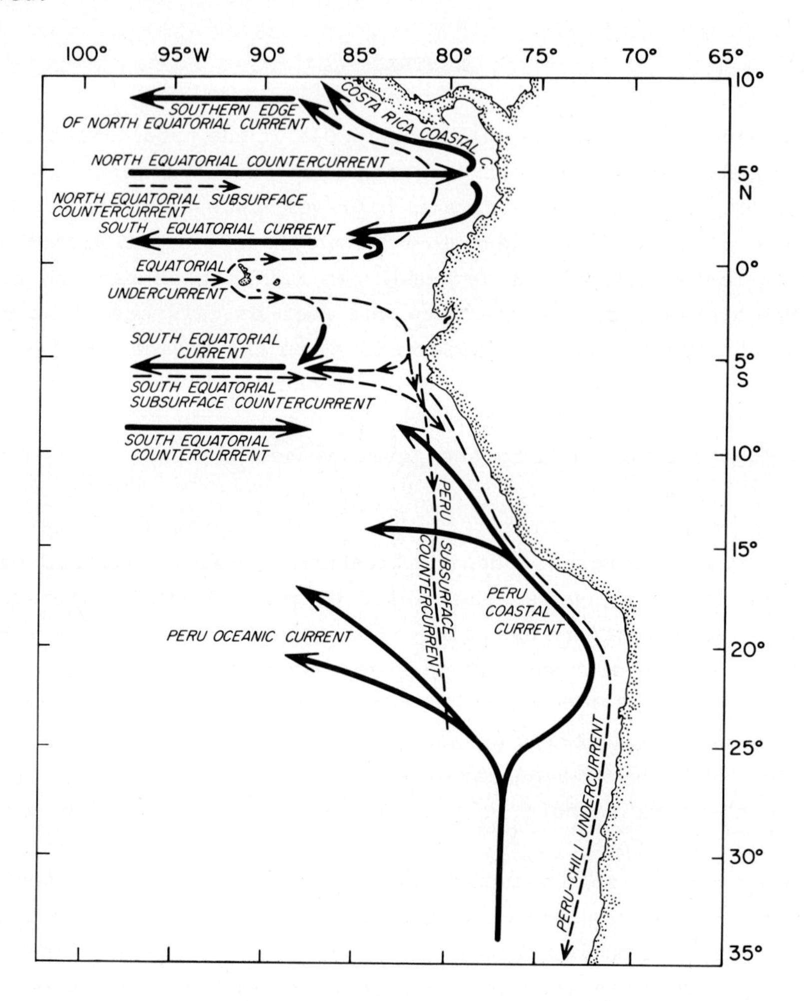

Figure 1: Generalized circulation scheme for the Eastern South Pacific Ocean. (Based on the work of various authors)

In this paper, we will consider the influence of the Poleward Undercurrent in this region on global nitrogen cycling. This current represents less than 0.01% of the total oceanic volume, but because of its position and chemical state, it is a site of particularly intense carbon and nitrogen cycling. We will review data which indicate that the nitrogen transformations that occur within this relatively minor current have globally significant effects that may be important vis-a-vis the issues of climatic change and global habitability.

We think that the most important reasons for studying eastern boundary regions and their associated poleward undercurrents involve providing a better understanding of the factors that govern these extremely interesting systems. Such arguments often fail to convince governments to fund research into these regions, however, and much of the research has been justified because of the relatively rich fishery resources. It should be recognized, however, that the issue of global habitability is becoming increasingly important as mankind's numbers and technological prowess increase. Thus, we believe that the "practical" reasons for studying nitrogen and carbon cycling in the ESPO include the general question of understanding how man is impinging on the environment. This question includes the issues of climatic change and changes in atmospheric chemistry as well as the questions that relate to fishery resources.

METHODS AND DATA

This paper is intended as an overview and review for those who are not intimately familiar with the marine nitrogen cycle. Consequently, some of the data that are presented appear in published papers listed in the references. Other data (including discussions of the methods employed) are presented in the following reports: Friederich et al., 1985; Hafferty et al., 1978; Hafferty et al., 1979.

A parameter called the "nitrate deficit" is used in this paper. This term may be unfamiliar to many readers, so we note here that it estimates the amount of free nitrogen that has been added to a water parcel from the reduction of combined nitrogen during denitrification (see below). Methods for calculating the nitrate deficit are described in Codispoti and Christensen (1985). We used the methods employed by Codispoti and Packard (1980), except for Figure 14 (A) where the values should be multiplied by 1.15 to be comparable.

BACKGROUND

General

Later on, we will show how changes in nitrogen cycling can affect atmospheric carbon dioxide levels, but we note here that both carbon dioxide and nitrous oxide are increasing in the atmosphere, and contribute significantly to the "greenhouse effect" (at present, ~60 and ~10%, of the total, respectively; Lacis et al., 1981; Bolin and Cook, 1983). The present-day rate of increase for atmospheric carbon dioxide is ~3000 Tg C yr^{-1} (Bolin, 1983). For nitrous oxide, the rate of increase is ~2.5 Tg N yr^{-1} (Weiss, 1981; Bolin, 1983; Crutzen, 1983). The consensus view is that increases in these two atmospheric trace gases and in methane and the freons are likely to increase the average temperature of the troposphere by ~5°C within the next ~200 years (Keeling and Bacastow; Lacis et al., 1981). In addition, nitrous oxide plays a significant role in atmospheric chemistry (Crutzen, 1981, 1983) as the main source of NO_x in the stratosphere.

For this discussion, salient features of the ESPO and its Poleward Undercurrent include the following: 1) a high flux of inorganic nutrients into the photic zone (e.g., Codispoti et al., 1982; Codispoti, 1983), 2) high primary production rates that, in turn, support a high downwards flux of organic carbon and nitrogen (e.g, Chavez and Barber, 1987), 3) high subsurface respiration rates arising from "2") (e.g., Packard et al., 1983), and 4) a well-developed oxygen minimum that contains extensive regions between ~75-500 m where oxygen concentrations are almost zero (Codispoti et al., 1986; Wooster et al., 1965; Love, 1970-74; and Zuta and Guillen, 1970).

The well-developed oxygen minimum zone in the eastern South Pacific Ocean is a partial consequence of the ocean-wide fields of advection and biology (e.g., Codispoti et al., 1982), but local processes enhance oxygen-depleted conditions. For example, an upwelling cross-circulation is superimposed on the Poleward Undercurrent. When combined with the high nutrient content of the upwelling waters, this results in high primary production. The resulting flux of organic material means that respiratory rates within the Undercurrent will tend to be high (e.g., Pace et al., 1987; Suess, 1980; Packard et al., 1983). In addition, the counter-current circulations associated with the upwelling circulation and arising in the longshore direction from the presence of the Undercurrent may also lead to the accumulation of nutrients and to enhanced oxygen depletion (Barber and Smith, 1981).

REGIONAL OCEANOGRAPHY

We shall consider the coastal ocean off South America between ~5-25°S (Figure 1). Observations suggest the Poleward Undercurrent in this region exists between depths of ~30 to 200-300 m, and has an average poleward velocity of ~5-10 cm s^{-1} (Brink et al., 1980; Smith, 1981; Fahrbach et al., 1981; Enfield et al., 1978). Its offshore extent is uncertain, but we believe that a choice of 200 km for this dimension is reasonable.

Figures 1 and 2 give highly schematic views of the current system in our study region and show that there are at least two flows that can "feed" the Undercurrent. These are extensions of the Equatorial Undercurrent (EUC) and of the Subsurface South Equatorial Countercurrent (SSECC). Although, oxygen concentrations are relatively low in these two currents (Tsuchiya, 1975 and 1983), they must nevertheless be regarded as sources of oxygen because conditions tend to become even more oxygen depleted within the Undercurrent. This situation arises partially from the high downwards flux of organic material into the Undercurrent. Additional causes are the existence of both cross-shelf and longshore countercurrent systems arising from the upwelling circulation and from the juxtaposition of the shallow equatorward wind drift just above the Poleward Undercurrent (Figures 1 and 2). Such countercurrent circulations form "nutrient traps" that can accumulate more nutrients than originally present in the source waters (Calvert and Price, 1971; Barber and Smith, 1981; Margalef, 1978). Since nutrients accumulate as a result of the breakdown of organic material that is being supplied to the "nutrient trap," these sites are also zones of enhanced respiration and oxygen consumption. Equations (Figure 3), that describe how chemicals can be stripped from an upper layer by the sinking and breakdown of organic matter in either a slower moving lower layer or a lower layer moving in the opposite direction to the surface flow, have been given by Redfield et al. (1963). Accumulation in the lower layer tends to be more extreme when the layers are moving in opposite directions and when the upper layer is highly productive as would be the case for estuarine and upwelling circulations. Such a situation would also apply to the juxtaposition of the Equatorward surface layer and Poleward Undercurrent off Peru and Chile (see Figures 2 and 3).

So far, we have identified two sources of oxygen for the Poleward Undercurrent, namely the waters supplied by the Equatorial Undercurrent between ~3-5°, and the waters suplied by the SSECC at approximately 10°S. A third source of oxygenated and relatively low nutrient water is provided by the Subantarctic waters that move equatorward. The bulk of this water

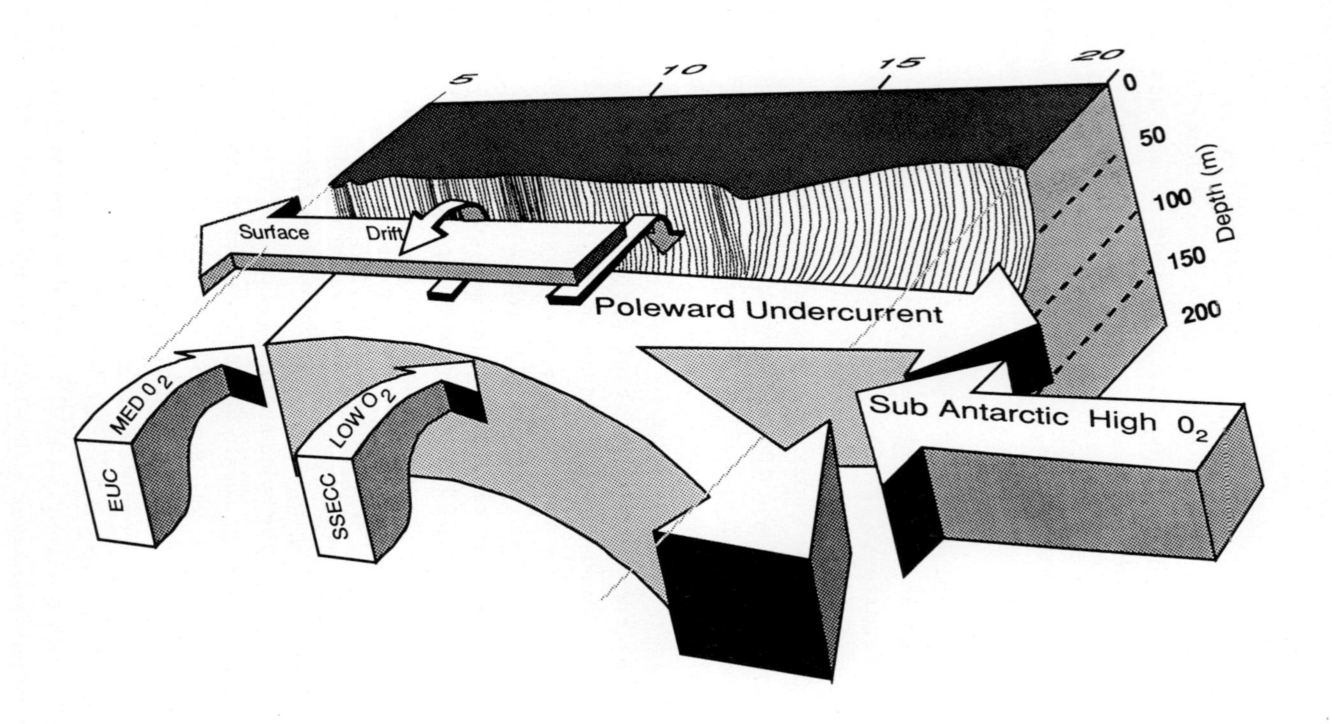

Figure 2: Block diagram showing the Poleward Undercurrent, the surface "wind-drift" and the three advective sources of oxygen for the Undercurrent which are the Equatorial Undercurrent, the Subsurface South Equatorial Countercurrent, and the equatorward flow of Subantarctic waters in the Peru Oceanic Current.

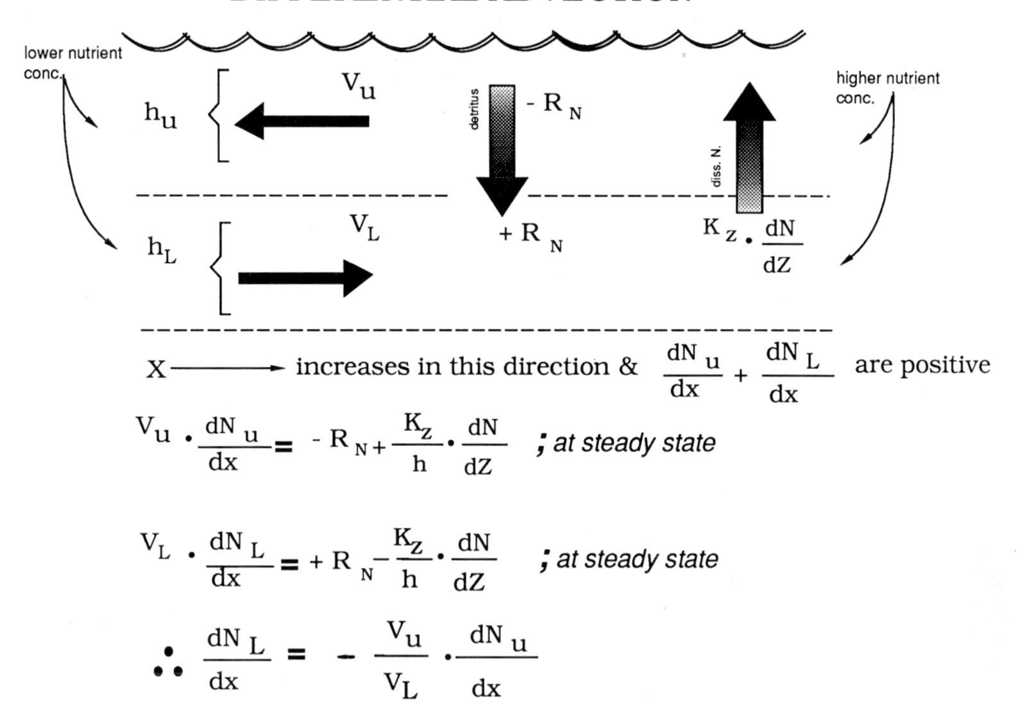

Figure 3: Diagram and equations showing how a countercurrent system can lead to enhanced nutrient levels and oxygen depletion that increase in the downstream direction in the lower current. The equations are taken from Redfield et al. (1963) and assume upper and lower layers of equal depth (h), and a steady-state. It is also assumed that the downward loss of some bioactive particulate chemical (N) is ultimately balanced by upwards diffusion of the N that is released into solution by decomposition and respiration in the lower layer. Nutrients increase towards the right in both layers, but the highest values will be in the lower layer. The upwards transport of N could also be accomplished by upwards vertical advection, so for the upwelling case the re-supply of N to the surface layer (per m^3) would more properly be expressed as the sum of vertical diffusive and advective losses.

tends to turn offshore near 25°S (Wyrtki, 1963, 1977), but the influence of the Subantarctic waters which have higher oxygen concentrations than EUC water within the ESPO (e.g., Friederich and Codispoti, 1981) can be felt within the zone that is dominated by the Undercurrent. This influence extended as far north as 10°S during southern winter 1976, and during March-May 1977, a front between Subantarctic and Equatorial Water occurred near 16°S on the 26.0 sigma-t surface (Figure 4). During February 1985, continuous vertical profiles suggested interleaving of equatorial and subantarctic waters near 15°S (Friederich and Codispoti, 1987).

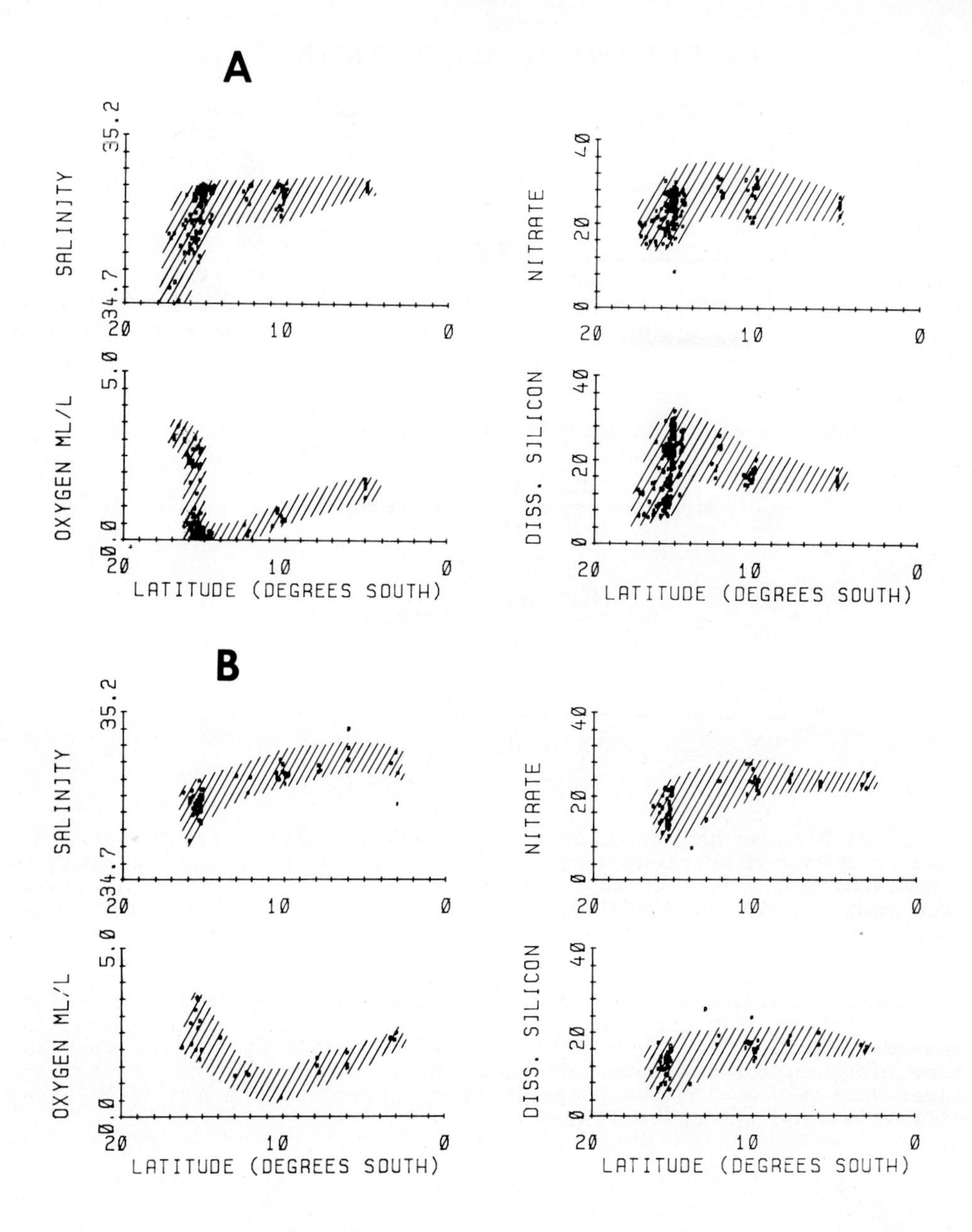

Figure 4: Trends of salinity, dissolved oxygen, nitrate, and dissolved silica on the 26.0 σ_t surface along the coast of Peru during March to May, 1977 (A), and during July to October, 1976 (B). Depth $\geq$ 40m. Nitrate and dissolved silica concentrations are in μM (from Friederich and Codispoti, 1981).

Sections taken along the coasts of Peru and Chile suggest that filaments of low oxygen water penetrate as least as far south as 28°S (Stommel et al., 1973). Some of these low oxygen (<0.5 ml/1) waters are found as far as ~1000 km offshore, but at 28°S the significant nitrate and nitrite consumption appeared to be restricted to the region that one would expect to be influenced by the Poleward Undercurrent (i.e., within 200 km of shore). The offshore zones of low oxygen may arise from the advection of the Peru Countercurrent (Figure 1) which may exist further offshore, but the relationship between the Undercurrent and Countercurrent has not yet been clearly defined. In a similar manner, sections taken near 15°S frequently show horizontal variability in oxygen depletion that could be interpreted as the juxtaposition of equatorward and poleward flows or variations in strength of the poleward flow even though the mean advection may be poleward. In our simplified view (Figure 2) we assume that the Undercurrent and Countercurrent are joined near 5° and "split" near 15°S.

THE NITROGEN CYCLE

Figure 5 outlines the biological nitrogen cycle. Nitrogen is a major constituent of organisms and the supply of available nitrogen is a control on the global carbon cycle. This is because carbon tends to be abundant relative to the needs of organisms, but nitrogen in forms readily available to plants (nitrate, nitrite, ammonium, urea, etc.) is often present in limiting quantities.

Dugdale and Goering (1967) made a distinction between the "new production" supported by the flux of inorganic nitrogen into the photic zone and the total primary production rate which depends on re-cycling of nitrogen within the photic zone. In a steady-state, the export of organic material from the photic zone cannot exceed the rate of "new production." The fundamental distinction is between primary production supported by the transport of available nitrogen into the photic zone vs. the primary production supported by re-cycling of nitrogenous compounds within the photic zone. Although nitrogen fixation (Figure 5) and inputs from the atmosphere and from runoff can support new production, the dominant source of nitrogen arises from the advection and mixing of nutrient-rich subsurface waters into the photic zone. Nitrate is the major form of combined nitrogen supplied by these ascending waters, and Dugdale and Goering (1967) stated that in most cases new production could be estimated by measuring the nitrate uptake rate. They also suggested that the production supported by re-cycling within the photic zone could be estimated by measuring the ammonium uptake rate since ammonium seems to

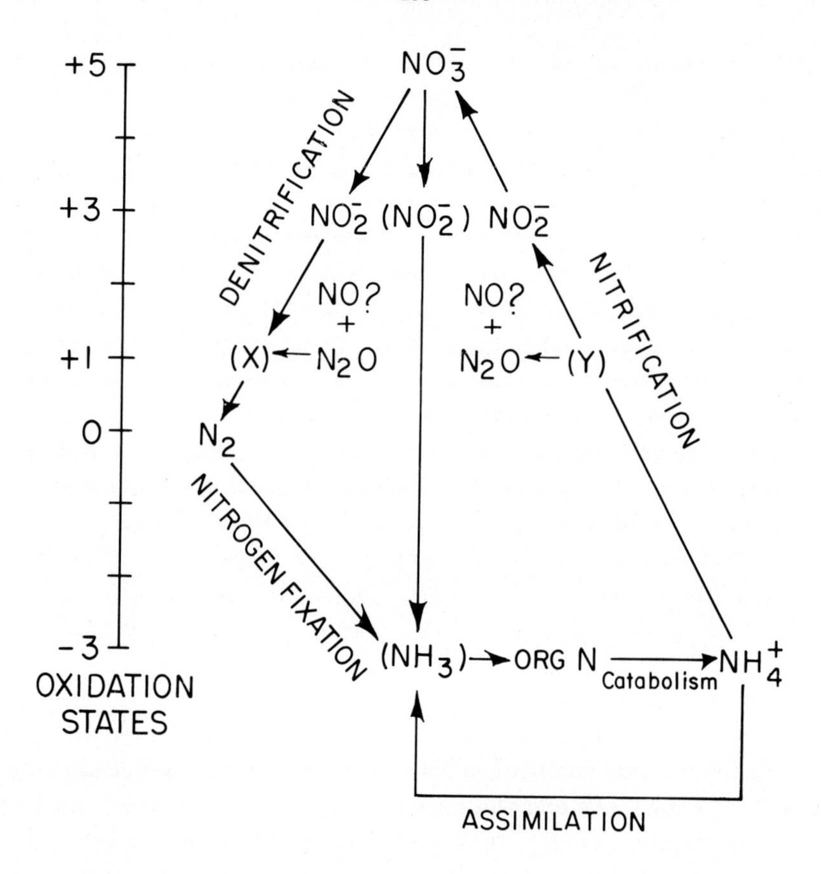

Figure 5: Diagram for the marine nitrogen cycle. This figure is a slightly modified version of a diagram presented by Liu (1979, 1982). (X) and (Y) represent intra-cellular intermediates that do not appear to accumulate in seawater. NO has been added to the original diagram based on recent work (Lipschultz et al., 1981), and the diagram has been modified to suggest that N_2O is produced mainly during nitrification and consumed during denitrification in the sea. Photochemical processes that occur near the sea surface (Zafiriou and True, 1979a,b) are not included in this figure.

be the most abundant form of re-cycled combined nitrogen within the photic zone.

Most new production is exported from the photic zone by a downward flux of particles ($-R_n$ in Figure 3) or dissolved organic matter. Since the supply of nitrate is exhausted by phytoplankton uptake in most of the oceanic photic zone, the downward flux of organic nitrogen and associated biogenic material is essentially equal to the time-averaged upward nitrate transport into the photic zone throughout most of the ocean. The notable exception is at high latitudes where light limitation may be more important than nitrogen limitation. In such cases, waters may leave the surface containing appreciable amounts of unutilized or "preformed nutrients" (Redfield et al., 1963) that behave as conservative tracers. Because of this phenomenon, some models of changing atmospheric carbon dioxide concentration invoke a greater utilization of these preformed nutrients during times when the atmospheric carbon dioxide was lower (e.g., Knox and McElroy, 1984; Sarmiento and Toggweiller, 1984).

The process of nitrification (Figure 5) is of interest to this discussion because it produces nitrate, nitrite and nitrous oxide which are used to oxidize organic matter during denitrification, and because the first nitrification step ($NH_4 \rightarrow NO_2^-$) is thought to be the major in situ source for nitrous oxide (Elkins et al., 1978; Yoshinari, 1976). Although nitrification requires oxygen, nitrifying bacteria can thrive at low oxygen concentrations (Goreau et al., 1980; Carlucci and McNally, 1969). The major end-product of the first stage of nitrification is nitrite, but nitrous oxide is produced as a side-product, and yields of nitrous oxide appear to increase as oxygen concentrations decrease (Goreau et al., 1980). It is possible that nitrous oxide is the major end-product at extremely low (<0.05 ml/1) oxygen concentrations (Codispoti and Christensen, 1985; Delwiche, 1981).

When oxygen concentrations fall below ~0.05 ml/1, nitrate reduction ($NH_4 \rightarrow NO_2^-$) and denitrification become important respiratory modes (Figure 5). For brevity, we will consider nitrate reduction to be an initial step in the denitrification process even though more species of bacteria can perform the former than the latter. Denitrification is classically defined as the bacterially mediated reduction of the ionic nitrogen oxides (nitrate and nitrite) to the gaseous oxides (nitric oxide, and nitrous oxide) or to free nitrogen (N_2) during the breakdown of organic material (Knowles, 1982). Free nitrogen appears to be the major end-product of denitrification in the marine water column, and there appears to be a net consumption of nitrous oxide (e.g., Cohen and Gordon, 1978). The important aspects of oceanic denitrification for the purposes of this paper are as follows:

1) This process converts combined nitrogen in a form readily available to plants (nitrate) to free nitrogen which can only be used by nitrogen fixing plants and bacteria
2) Denitrification appears to be the only in situ sink for nitrous oxide; the rest of the nitrous oxide produced during nitrification probably escapes to the atmosphere
3) Denitrification appears to become a major respiratory process only when oxygen concentrations fall below ~1% of the saturation value (e.g., Devol, 1978).

As a consequence, most marine denitrification may be restricted to oxygen deficient portions of the water column (~50%) which represent only ~0.1% of the total oceanic volume and to shallow and hemipelagic sediments (~50%; Codispoti and Christensen, 1985). Codispoti and Packard (1980) estimated that the denitrification rate in the low oxygen environments of the ESPO was ~25 Tg N/yr $\left(Tg = 10^9 g\right)$ and that most (~80%) of this occurred within 175 km of the coast.

Globally significant variability has been observed in the denitrification rate in the waters off Peru (Codispoti and Packard, 1980; Codispoti et al., 1986). McElroy (1983) has suggested that the variability in marine denitrification could be responsible for some of the variability in atmospheric carbon dioxide concentrations that has been inferred from ice-core data. Periods of increased denitrification could cause the combined nitrogen inventory of the ocean to experience transient decreases, and Codispoti and Christensen (1985) have suggested that the present-day ocean may be in deficit by ~70 Tg N/yr. Such an imbalance is possible because increases in denitrification may not be closely coupled to the source terms for combined nitrogen. Piper and Codispoti (1975) argued, for example, that denitrification and nitrogen fixation may not be closely coupled since the latter process appears to be concentrated in the photic zone and may occur at relatively low rates (Capone and Carpenter, 1982; Carpenter and Capone, 1983).

Figure 6 shows how increases in marine denitrification could lead to increases in atmospheric carbon dioxide as suggested by McElroy. During the transient period when increased denitrification is not balanced by decreases in subsurface respiration or by increased source terms, deeper waters rising to the surface will give off carbon dioxide that was sequestered by biological productivity when the nitrate flux to the photic zone was higher. The photic zone, however, will take up less carbon dioxide from the atmosphere because the nitrate supply is reduced. If nitrogen fixation could increase rapidly, the imbalance would be short-lived, but if we must rely on reduced subsurface respiration arising from reduced new production to gradually decrease the denitrification rate, several ocean mixing times might be required to restore a new steady-state. The equations in Figure 6 are simplified and not stoichiometrically balanced, but they do point out one factor that makes it possible for dentrification rates to vary rapidly; during phytoplankton growth the N/C uptake ratio is ~6 (by atoms) but during denitrification, the $\Delta N/\Delta C$ ratio is ~1. Thus, only 18% of the organic material produced by phytoplankton could remove all of the nitrate associated with this production if respiration proceeded by denitrification. The practical consequence of this is that relatively small changes in the organic matter supply could cause large

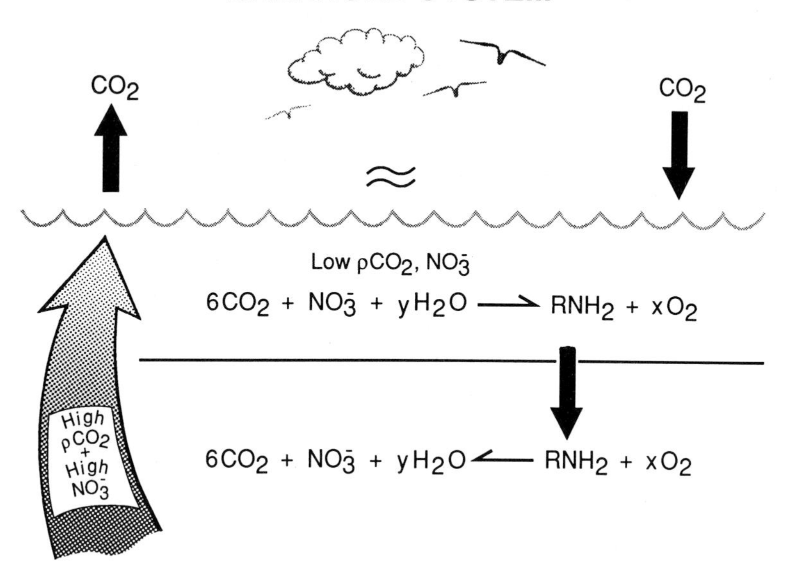

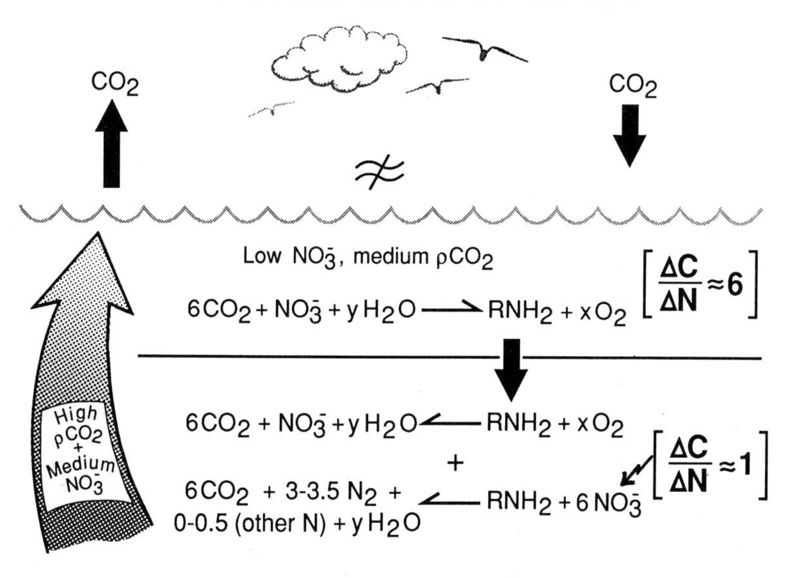

Figure 6: Conceptual models of (upper panel) an ocean with a balanced combined nitrogen budget (combined-N inputs = outputs) and (lower panel) an ocean in which excess denitrification leads to a net of loss of carbon dioxide from the ocean (from Codispoti et al., 1988). The carbon dioxide loss will occur only until a new balance is established for combined nitrogen, and this would eventually have to occur as reduced primary production arising from decreased nitrate levels caused decreases in subsurface respiration (Broecher and Peng, 1982; Piper and Codispoti, 1975). A balance might occur faster if nitrogen fixation rates could increase rapidly in the face of increased denitrification which, at present, is an open question.

Table 1. Mechanisms for Doubling the Denitrification Rate Off Peru and Chile*

A. O_2 Decrease in Source Waters

1. Vol. of MSNM ~1.4×10^{14} m^3

2. Denitrification rate in MSNM ~ 2×10^{13}g N yr^{-1}

3. Nitrate → N_2 Rate ~10 µg-atoms l^{-1} yr^{-1}

4. 10 Nitrate ×5/2 = 25 µg-atoms O_2 l^{-1} yr^{-1}
   [5/2=Nitrate/0 Equivalence Ratio]

5. Therefore 25 µg-atoms O_2 l^{-1} (~0.3 ml/l) decrease in source waters could double the rate.

B. Increase in Carbon Flux

1. Area of MSNM ~1.0×10^{12} m^2 (see Figure 9)

2. "New" productivity of ~400 g C m^{-2} yr^{-1}

3. Dentrification rate of 20 g N m^{-2} yr^{-1}
   (Codispoti and Packard, 1980)

4. This requires ~20 g C m^{-2} yr^{-1} (see Figure 6)

5. Therefore an additional ~5% of "new" primary productivity is all that is required to double the denitrification rate.

C. Decrease in Downward O_2 Flux

1. Range of possible downward O_2 fluxes through the bottom of the photic zone using k_z = 0.3 cm^2 s^{-1} for weak winds and 1.0 cm^2 s^{-1} for strong winds and an average O_2 gradient

   min: 15 g-atoms O_2 m^{-2} yr^{-1}

   max: 50 g-atoms O_2 m^{-2} yr^{-1}

2. Difference max-min = 35 g-atoms $O_2 m^{-2}$ yr^{-1} = 13 g-atoms of C m^{-2} yr^{-1} with a Redfield ratio of O_2 to C of 276:106 by atoms

3. A C:N ratio of 1:1 by atoms during denitrification

4. ∴ 13 g-atoms C m^{-2} yr^{-1} ≈ 13 g-atoms N m^{-2} or 182 g N m^{-2} yr^{-1}

5. Therefore changes in downward flux of O_2 only 10% as large as the one estimated above could double the denitrification rate since the area of the MSNM is ~ 1.0×10^{12} m^2.

*These calculations assume a doubling of the rate that Codispoti and Packard (1980) estimated for the main secondary nitrite maximum (MSNM). This table is from Codispoti et al. (1988).

Table 1:

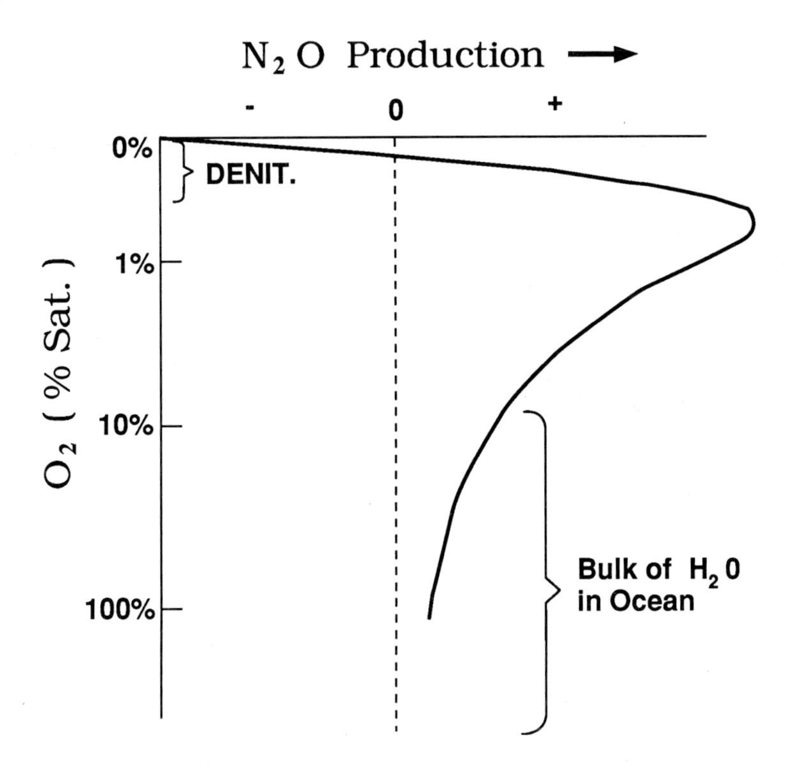

Figure 7: A "cartoon" of nitrous oxide production and consumption rates as a function of dissolved oxygen concentration in the oceanic water column.

changes in denitrification. In addition, since denitrification occurs at low oxygen concentrations and since water column denitrification may be confined largely to 0.1% of the marine volume, very small changes in the global oxygen distribution could also cause large changes in denitrification (see Table 1). Since a large fraction of marine denitrification occurs in eastern boundary regions such as the ESPO, understanding the rates and variability of denitrification in these zones is important.

Because nitrous oxide production is enhanced at low oxygen concentrations and because this gas is consumed by denitrification at even lower concentrations, Codispoti and Christensen (1985) suggested that nitrous oxide turnover in the ocean could be very uneven as shown schematically in Figure 7. Their calculations suggested that nitrous oxide turnover in productive eastern boundary regions and in the Arabian Sea (~1% of the oceanic area) rivaled that in the rest of the ocean and was significant in comparison to the rate at which nitrous oxide is increasing in the atmosphere. Their analysis suggested that small changes in oxygen distribution could have massive changes on nitrous oxide and that a globally significant amount of nitrous oxide was consumed by denitrification in the ESPO. On balance, however, the ESPO may be a net source of N_2O because of the extremely high values occurring at the periphery of the denitrification zone (see below).

DISCUSSION

Advective Controls on Nitrogen Cycling in the Poleward Undercurrent

We have established above that key portions of the nitrogen cycle depend critically on the distribution of dissolved oxygen. In particular, nitrous oxide production is enhanced at lower oxygen concentrations (Goreau et al., 1980 and Figure 7), and both nitrous oxide consumption and denitrification are favored by oxygen concentrations of <~0.05 ml/l (<1% of the saturation value). Scaling calculations (Table 1) suggest, for example, that a 0.3 ml/l decrease in oxygen in the water supplied to the region could cause a doubling of the denitrification rate which in turn would represent a 10-30% increase in the total oceanic rate. The question of the advective supply of oxygen to the Poleward Undercurrent is therefore an important one. Our schematic view of the Undercurrent system suggests two "upstream" sources for the Undercurrent, the EUC and the SSECC (Figure 2). Of these, we believe that the EUC may be a more important source of oxygen. Data from the El Niño Watch experiment (Patzert et al., 1978,), for example suggest that the extension of the SSECC carries waters of lower oxygen content into the region than is carried in by the EUC (Figure 8).

We believe that the relatively greater importance of the EUC is also suggested by the tendency for a monotonic progression of decreasing oxygen followed by increases in nitrite and the nitrate deficit produced by denitrification as the Undercurrent proceeds to the South. In other words, reversals in trend tend to occur only when the Undercurrent travels far enough south to become significantly influenced by Subantarctic waters which can have oxygen concentrations that are higher than those in the EUC within our study area (Figure 4). For example, a high nitrite feature which has been called the Main Secondary Nitrite Maximum (Codispoti and Packard, 1980) is typically found between ~10-25°S, and it has its maximum offshore extent between ~12-15° (Figure 9). This is consistent with the view that total oxygen deficiency (O_2 depletion plus nitrate deficit) in the Undercurrent tends to increase towards the South until influenced by the Subantarctic waters which tend to turn offshore near 25°S.

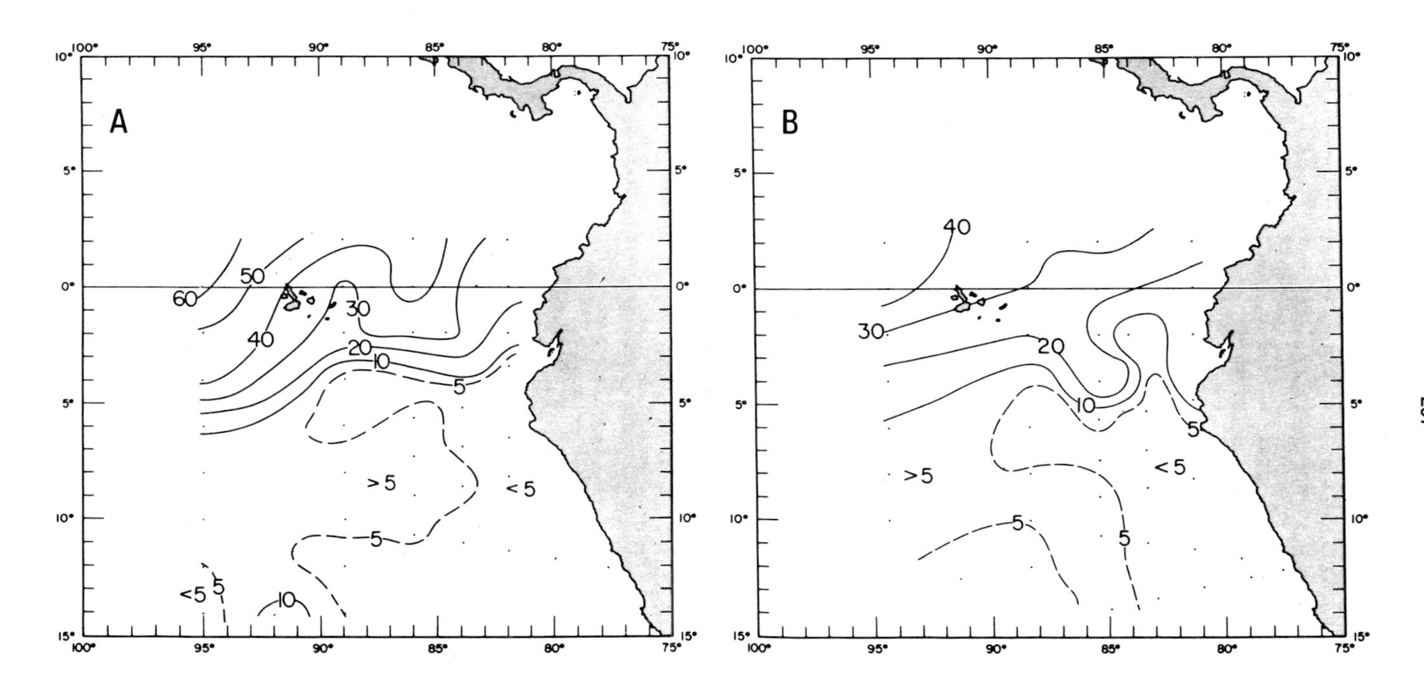

Figure 8: (A) Oxygen (μmoles/kg) on the surface where δ=160 cl/t during 20 February-31 March, 1975, and (B) during 17 April-27 May, 1975 (from Patzert et al., 1978) during the El Niño watch experiment. The 160 cl/t surface is ≈ 26.35 σ_t.

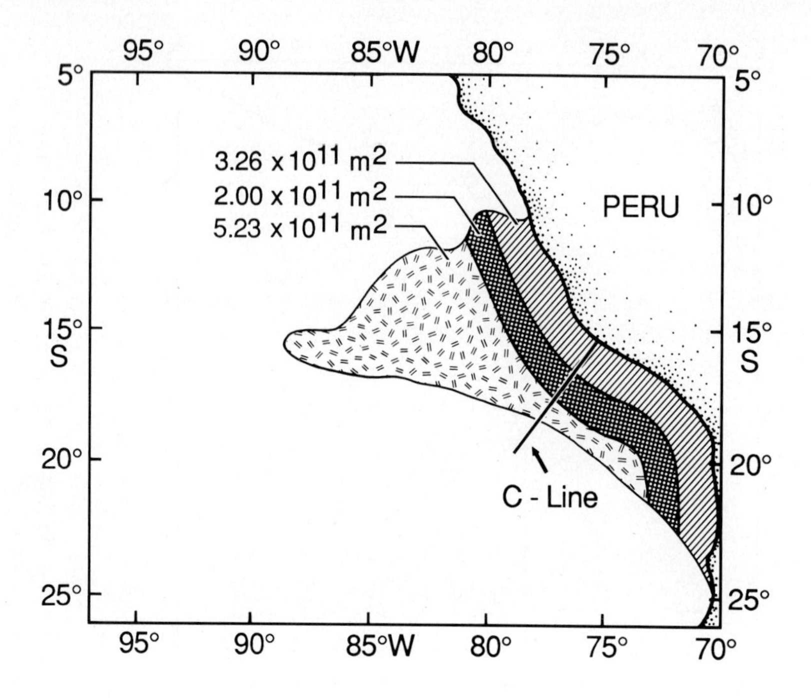

Figure 9: Approximate horizontal extent of the Main Secondary Nitrite Maximum (MSNM) with the position of the "C-line" indicated (From Codispoti and Packard, 1980 as re-drawn from Wooster et al., 1965). The inner band is 175 km wide and Codispoti and Packard suggest that 76% of the denitrification within the MSNM occurred in the inner band. Their calculation also suggested that 9% and 15% of the total occurred in the middle and outer bands, respectively.

A series of sections taken during May-June 1977 (Figures 10-12) also suggests progressively increasing oxygen depletion within the depth range of the Undercurrent (~30 to 250-300 m) as it proceeds from ~5 to 15°S. Similarly, high nitrous oxide concentrations were found within the Undercurrent equatorward of 10° during 1976 (Pierotti and Rasmussen, 1980) while some extremely low values were found poleward of 10° in 1978 (Elkins et al., 1978) suggesting a progression from low oxygen conditions which favor high nitrous oxide production rates to oxygen deficiency which favors net consumption (Figures 7 and 13).

There are times when oxygen deficiency has been observed equatorward of 10° (Sorokin, 1978; Patzert et al., 1978; Zuta and Guillen, 1970), but during such a situation observed in 1985 (Codispoti et al,. 1986) there was still a tendency for increasing oxygen deficiency towards the south at least over the slope and shelf as indicated by nitrate deficit sections taken at 10 and ~15° (Figure 14). Further offshore, a break in high nitrite concentrations near 11° (Codispoti et al., 1986) may have signaled the influence of the SSECC and geostrophic calculations indicated onshore flow in this region. A Coastal Zone Color Scanner Image (supplied by G. Feldman) also indicated an onshore movement in the area of the break in high nitrite concentrations.

The unusual 1985 observations which were made in February and March have been attributed to a region-wide shoaling of the thermocline that may have been an accentuation of a phenomenon that occurs with some frequency during southern summer (Codispoti et al., 1986; Codispoti et al., in press). It would be interesting, however, to know whether episodic denitrification equatorward of 10°S is also influenced by changes in the oxygen supplied by the EUC. An experiment to look at the annual cycle of the EUC extension would be highly instructive since the observations of the oxygen deficiency equatorward of 10°S seem to be concentrated in southern summer.

Minima in Nitrite Concentrations Over the Outer Shelf and Upper Slope

Figures 15 and 16 display minima in nitrite concentrations over the outer shelf and upper slope at 10 and 12°S and a corresponding zone of relatively high oxygen at 12°S. Correlations between light scattering (a measure of the suspended particle load) and poleward velocities using data collected in 1977 near 15°S suggested that the suspended load decreased when poleward velocities increased (G. Kullenberg, personal communication). In addition, Fahrbach et al., (1981) suggest that poleward velocities were strong and reached the surface over the outer

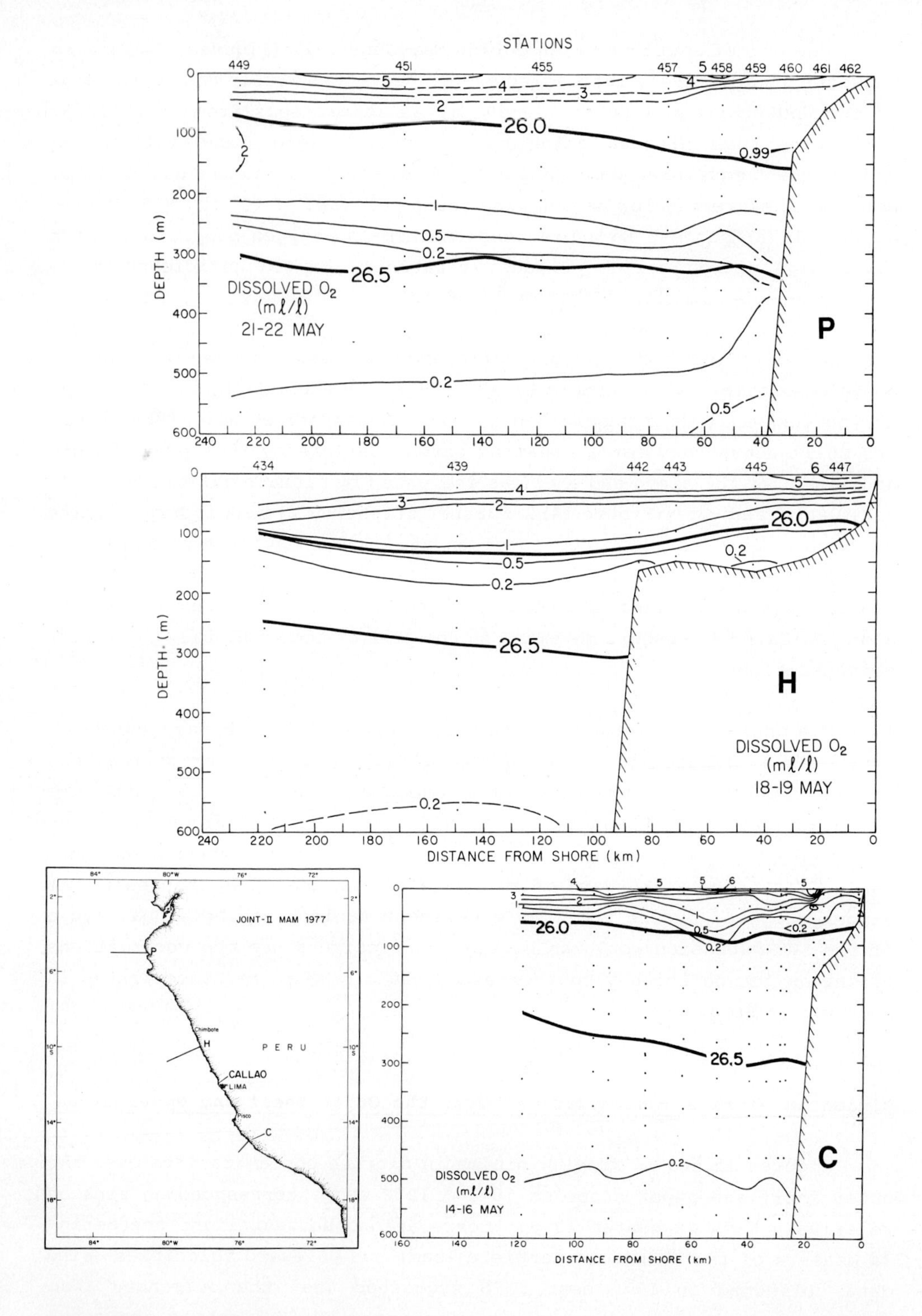

Figure 10: Oxygen distributions at ~5, 10 and 15°S during May 1977 (re-drawn from Hafferty et al., 1979). The 26.0 and 26.5 σ_t surface are indicated by the thick lines.

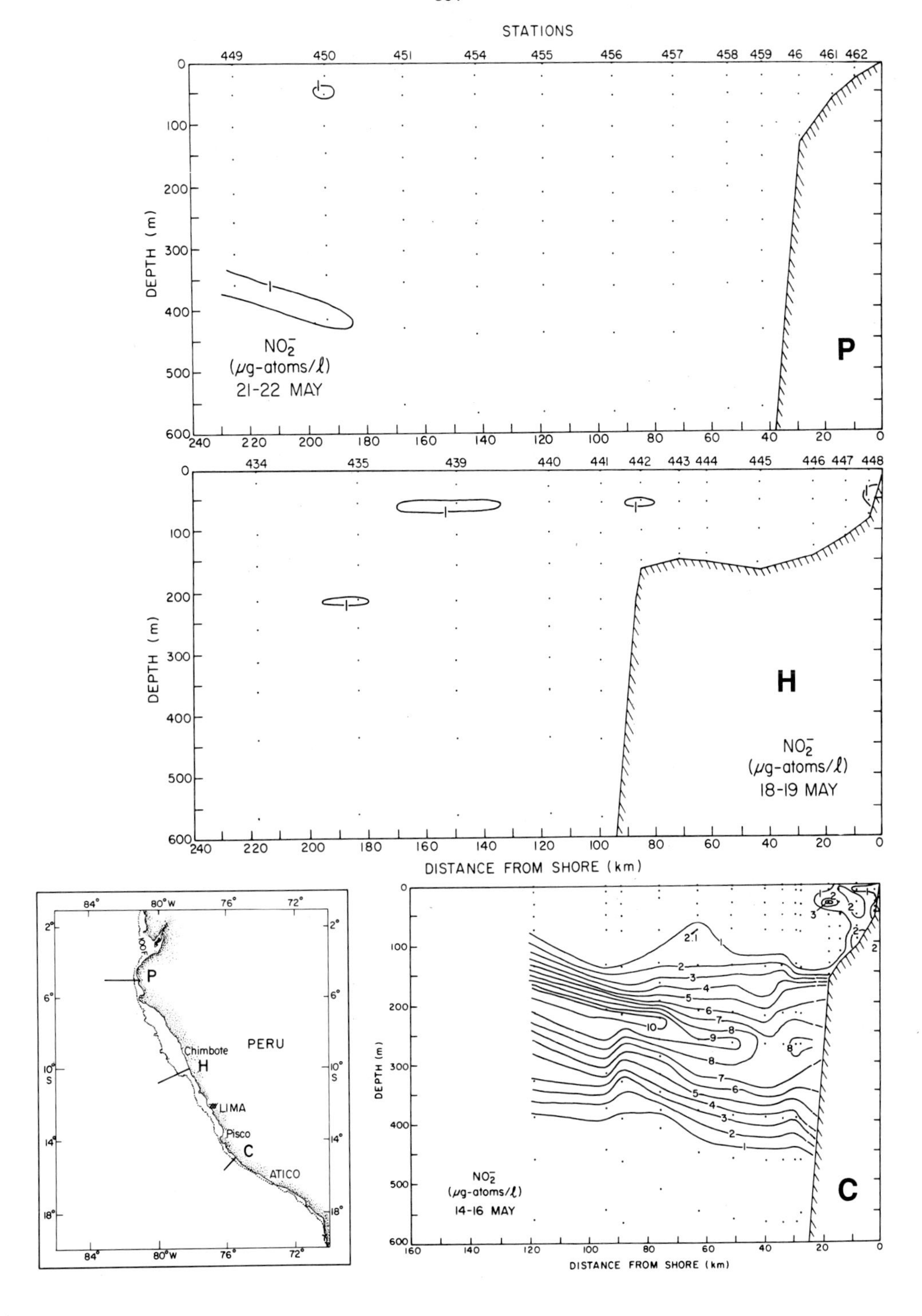

Figure 11: Nitrite distributions at ~5, 10 and 15°S during May 1977 (re-drawn from Hafferty et al., 1979).

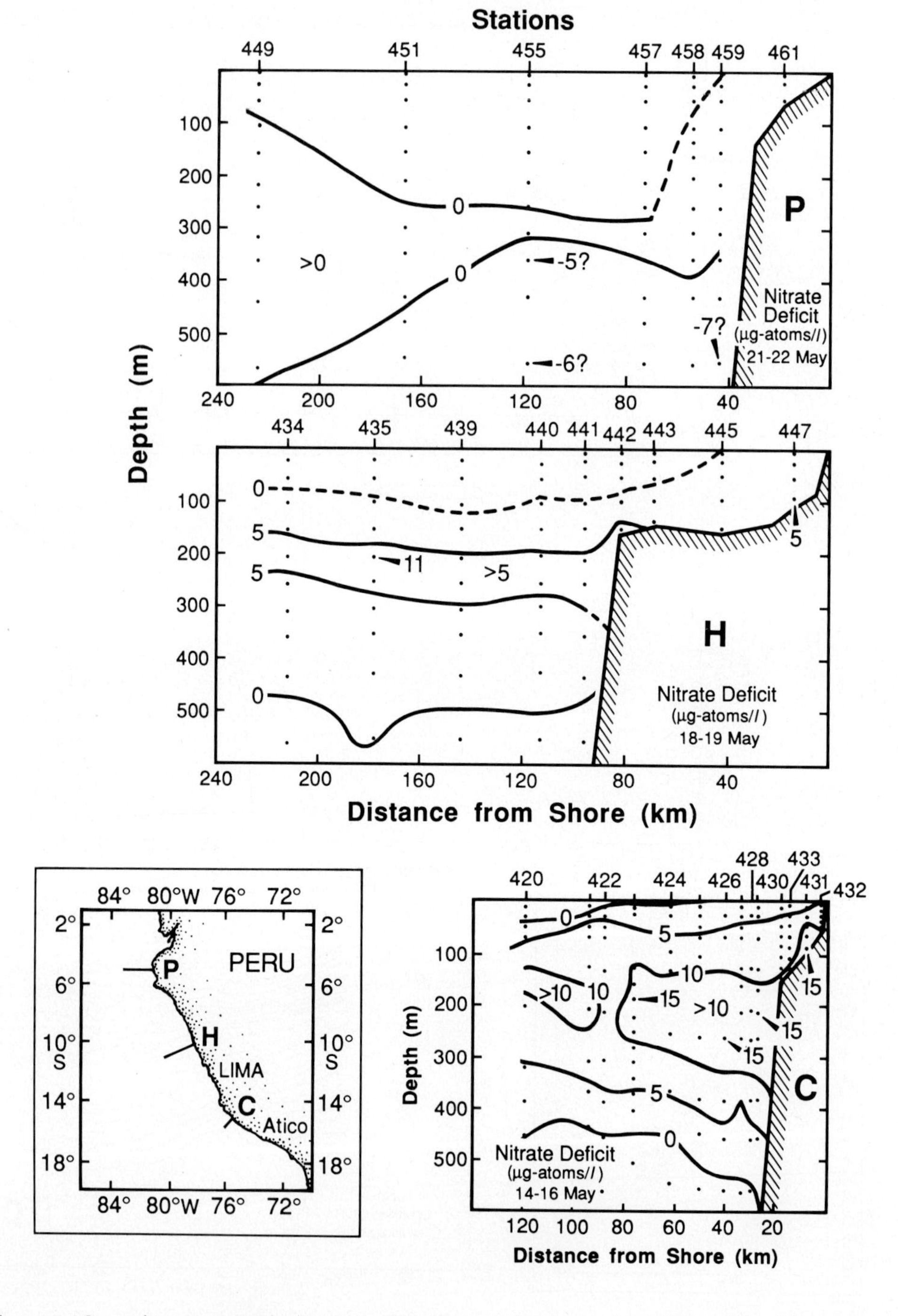

Figure 12: Nitrate deficits at ~5, 10 and 15°S during May 1977 (from data presented in Hafferty et al., 1978).

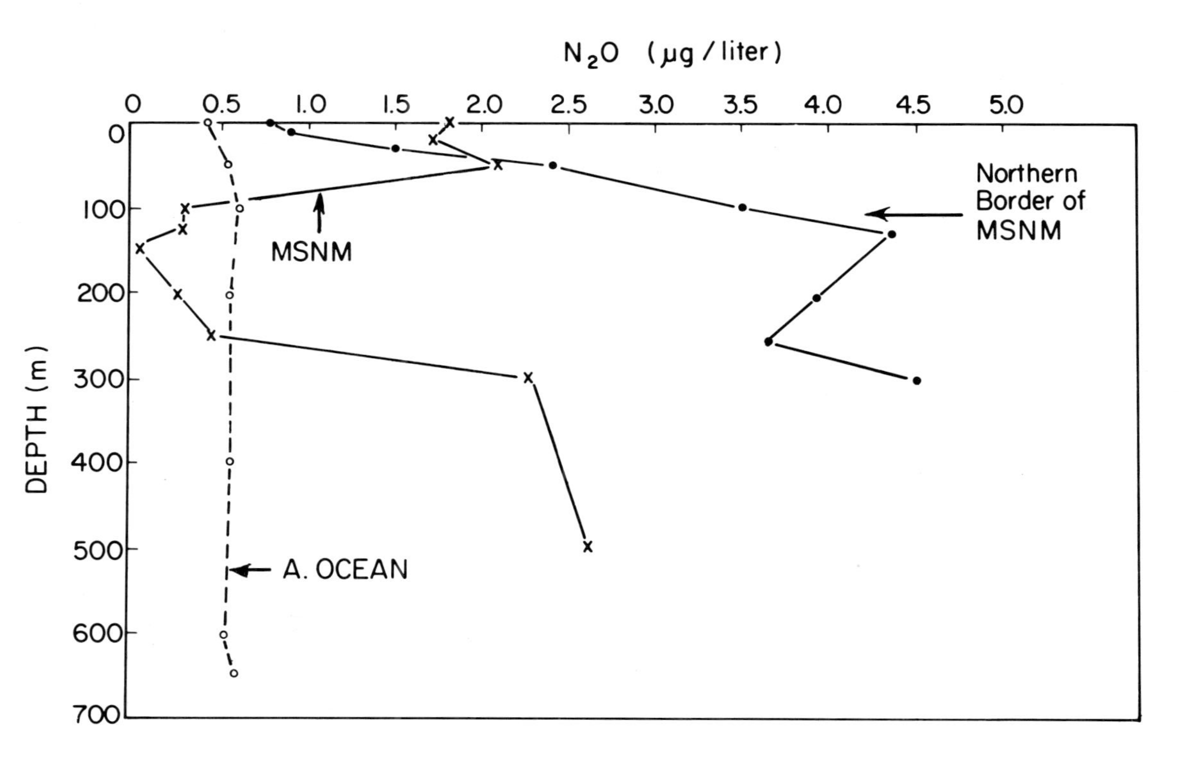

Figure 13: Nitrous oxide distributions in the North Atlantic Ocean (from Hahn, 1981), the Main Secondary Nitrite Maximum that is usually centered at ~200 m (from Elkins, 1978), and just outside of the northern border of the Main Secondary Nitrite Maximum (from Pierotti and Rasmussen, 1980). This figure is taken from Codispoti and Christensen, 1985).

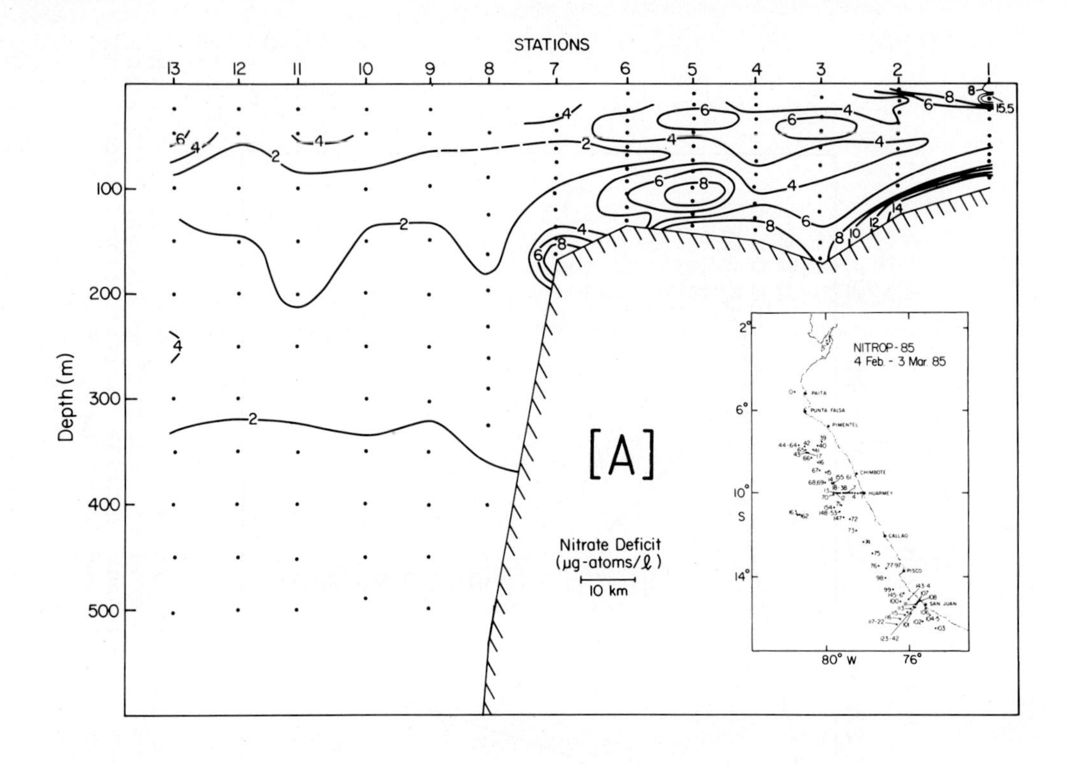

Figure 14a: Nitrate deficit sections at 10° (A) and ~15° (B) during February 1985. The inset shows the station locations. Figure 14B is from Codispoti et al., (1988).

shelf and slope at 5° during observations made in 1977. Consequently, one explanation for these minima is that they signal the "core" of the Undercurrent. Given the three-layered upwelling circulation with onshore flow at mid-depths, nitrite uptake in the photic zone which is about 20m deep, and accumulation of organic material on the bottom, part of this distribution may be attributed to the upwelling circulation. However, one might expect to find higher nitrite concentrations in the bottom layer at the outer shelf if the cross-circulation and local biological processes were the only operative factors. The relationship between poleward velocity and chemical and optical properties is an interesting problem that one can address with existing technology since moored sampling devices exist for acquiring nitrate and nitrite samples (Friederich et al., 1986) as well as current velocity and light transmission data.

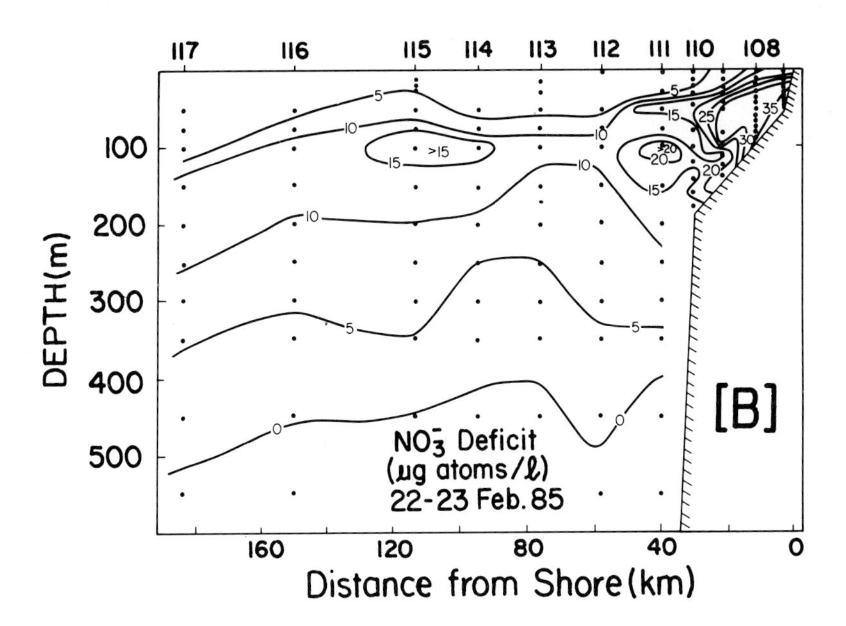

Figure 14b:

Rates of Denitrification and Nitrous Oxide Consumption Within the Undercurrent

If we consider the region between 10 and 25°S, the volume of the Undercurrent, as we have defined it (200 km wide by ~250 m deep), is 8×10^{13} m^3 or ~<0.01% of the total oceanic volume. Codispoti and Packard (1980) have estimated the denitrification within 175 km of shore in this region to be about 20 Tg N yr^{-1} and most of this occurs within the depth range of the Undercurrent since respiration rates tend to decrease exponentially with depth (e.g., Packard et al., 1983). Since estimates for the total marine denitrification rate range from ~50 to ~120 Tg N yr^{-1} (Hattori, 1983; Liu, 1979; Codispoti and Christensen, 1985), it is clear that the Undercurrent has an importance to the oceanic combined nitrogen budget that is immense when compared to the small volume of water that is involved. Similarly, Codispoti and Christensen (1985) have suggested that 0.4 Tg N yr^{-1} of nitrous oxide is consumed by denitrification in the ESPO and, once again, most of this consumption might occur within the Undercurrent as we have defined it. Since the rate of atmospheric increase in nitrous oxide is ~2.5 Tg N yr^{-1}, the Poleward Undercurrent may also be an important factor in the global nitrous oxide distribution.

Existing nitrous oxide data from the Poleward Undercurrent (Figures 13 and 15, Elkins et al., 1978; Friederich et al., 1985) reveal extreme spatial and temporal variability and contain both extremely high and low values when compared with the rest of the ocean. For example, the high near-surface values shown in the inset in Figure 15 are amongst the 10 highest values ever observed in the ocean, and less than 100 m below, extremely low concentrations were observed. Clearly, the relationship between Undercurrent dynamics and nitrous oxide distribution off Peru is a question that may have global significance. Whether or not the Undercurrent is a major source or sink for nitrous oxide or, alternately, be neither because of a balance between enhanced consumption and production is a question that remains to be settled.

Since the Undercurrent per se is defined as a poleward advection, it is interesting to speculate on its importance as an exporter of the products of denitrification and nitrous oxide consumption relative to the total exports which would include diffusion and vertical advection. Taking our simplified definition of the Undercurrent (200 km wide, 250 m deep) and assuming an average net poleward advection of 5-10 cm s^{-1} (Enfield et al., 1978; Fahrbach et al., 1981) gives an average poleward transport of ~2.5-5 Sv. Nitrate deficits within the Undercurrent at the "downstream" end of the MSNM are ~10 M (Figures 12 and 14b) (=140 mg N m^{-3}) and multiplying gives an annual export of nitrate deficit by the Undercurrent of 10-20 Tg N yr^{-1}. Since nitrate deficit estimates the free nitrogen produced by denitrification, this is an approximate estimate of the free nitrogen produced by denitrification that is exported by poleward advection in the Undercurrent. Since the total marine denitrification rate is ~50-150 Tg N yr^{-1} (see above), this calculation suggests that horizontal Undercurrent advection may be a globally important term in calculating denitrification rates.

Importance of the Vertical Advection

Since the waters that upwell within the Rossby radius of deformation in the ESPO come from a depth of ~50-100 m (e.g., Wyrtki, 1963), it is generally agreed that the Undercurrent is the major source of upwelling water. The volume transport of upwelling between 6-24°S has been estimated as ~3.3 Sv (Wyrtki, 1963). If we take 3 Sv to be the upwelling rate between 10-24°S and assume an average nitrate deficit of 5 μM in the ascending waters, an annual nitrate deficit loss due to coastal upwelling would be 7 Tg N yr^{-1} in agreement with a similar calculation by Tsunogai (1971).

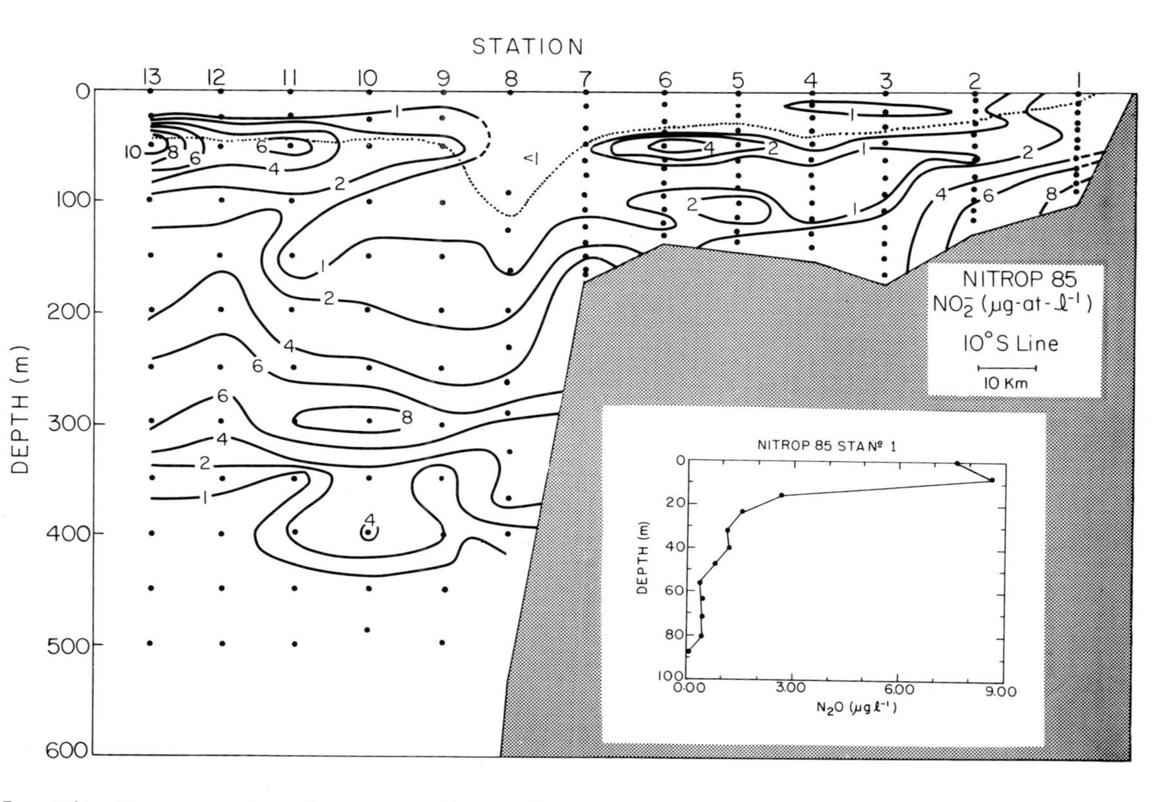

Figure 15: Nitrite concentrations at 10°S during February 1985. The inset shows the nitrous oxide levels distributions at the innermost stations. The high nitrous oxide levels are amoungst the ~10 highest values ever observed in the ocean, but ~50 m below in the oxygen deficient waters the concentrations are extremely low. These data are in line with the concept summarized in Figure 7 (from Codispoti et al., 1986).

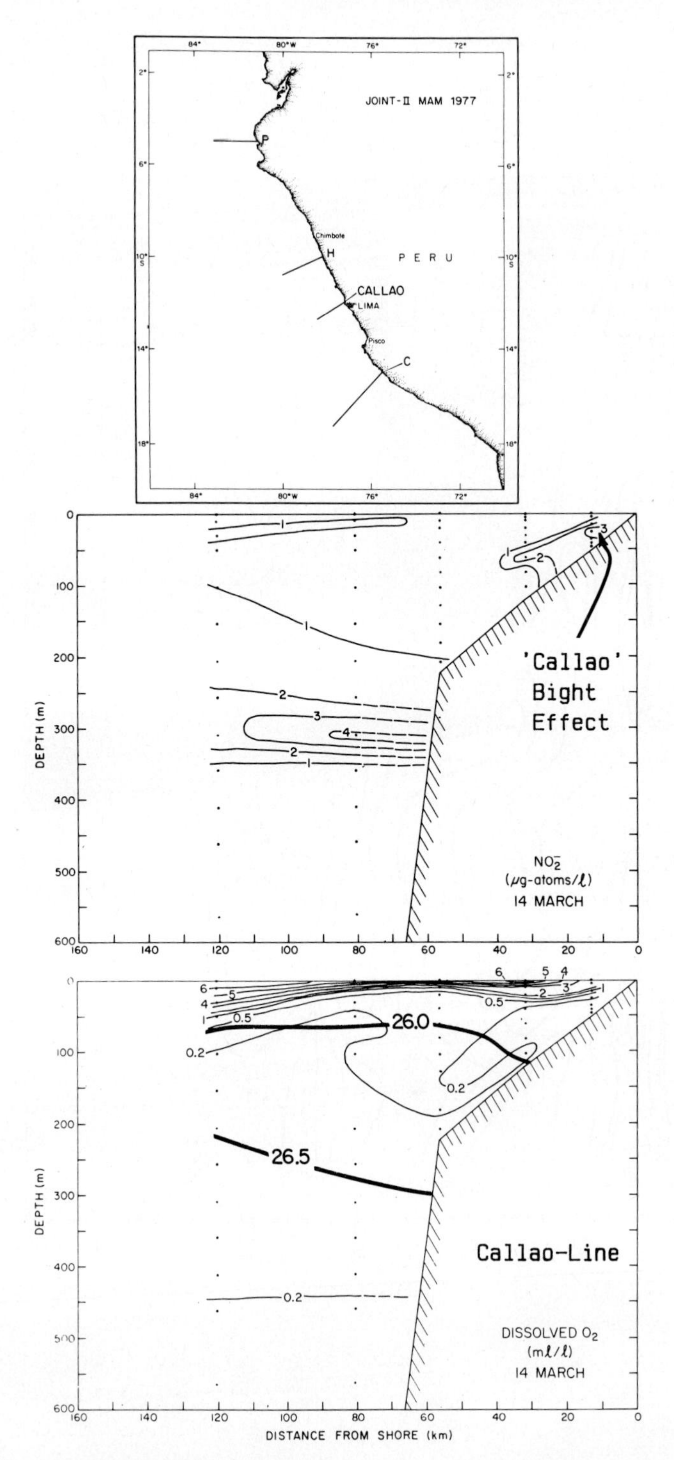

Figure 16: Nitrite and dissolved oxygen concentration on the "Callao Line" during March 1977. The thick lines show the 26.0 and 25.5 σ_t surfaces (re-drawn from Hafferty et al., 1979).

Wyrtki (1963) suggests that the downwelling rate through the 300 m surface is also ~3 Sv, and the nitrate deficits in these waters are also about 5 μM. Thus, the total loss of nitrate-deficit due to vertical advection within the undercurrent regime is ~14 Tg N yr^{-1}. This indicates that vertical advection associated with the Undercurrent waters is another important term for global denitrification.

Carbon Production Supported by Coastal Upwelling Vs. Denitrification

The upwelling waters supplied by the Undercurrent have a potential new production rate of ~200 Tg C yr^{-1} (Chavez and Barber, 1987; Chavez, et al., in press) which is vital to the maintenance of denitrification and nitrous oxide turnover rates close to the coast, but may also be important to the entire regional rate. For example, as indicated by the equations in Figure 6, the ratio of carbon to nitrogen in the denitrification process is approximately 1 by atoms or 0.85 by wt., so the new production supported by coastal upwelling is more than sufficient to support the entire regional denitrification rate. Available data do not permit an evaluation of what percentage of the regional denitrification rate is supported by organic matter production, arising from coastal upwelling (within the Rossby radius) versus production occurring further offshore. It has been suggested, however, that organic material sinking to the bottom over the shelf and outer slope could be transported into the offshore portion of the MSNM because of the net offshore transport in the bottom layer over the shelf and outer slope (Pak et al., 1980; Codispoti and Packard, 1980). This offshore transport should be higher when Undercurrent velocities are higher, and it is interesting to note that data from 1976 during a weak El Niño suggest that Undercurrent velocities were stronger than in 1977 (Codispoti and Packard, 1980; Enfield et al., 1978). It is possible, therefore, that the reduction in the flux of organic material to the Undercurrent due to decreased primary production during an El Niño is compensated to some extent by an increase in the offshore transport of organic matter that may accumulate over the shelf and slope when the Undercurrent is weaker. This is another question that is amenable to investigation with the current generation of moored instruments. Clearly, the organic matter supply and transport processes associated with Undercurrent dynamics is important to the oceanic denitrification rate. This is particularly so since only small changes in the carbon supply can cause large changes in denitrification (Table 1).

CONCLUSION

This paper is speculative and our description of the Undercurrent is certainly over-simplified. Our major intent has been to convince our physical oceanographic colleagues that the Poleward Undercurrent off Peru may be a major "actor" in the global nitrogen cycle, even though it is a dynamic "pipsqueak" in comparison to major currents such as the Gulf Stream. We will be encouraged if this paper engenders a greater appreciation of the relationship between studies of the Poleward Undercurrent in the ESPO and the issue of global habitability.

ACKNOWLEDGEMENTS

Our research in the eastern South Pacific has been supported financially by the National Science Foundation and the Office of Naval Research. We thank Jim Rollins for drafting assistance including the excellent job that he did with the block diagram (Figure 2). We are also indebted to P. Colby and A. Sanico for their skill in turning our rough drafts into readable type.

From: Dr. Theresa Paluszkiewicz, MMS Branch of Environmental Modeling, Reston VA

On: Review and commentary to paper **DO NITROGEN TRANSFORMATIONS IN THE POLEWARD UNDERCURRENT OFF PERU AND CHILE HAVE A GLOBALLY SIGNIFICANT INFLUENCE?"**, by L.A. Codispoti, R.T. Barber, and G.E. Friederich

GENERAL

This paper discusses the nitrogen cycle along the Eastern South Pacific Ocean (ESPO), its possible global significance, and the influence of the circulation characteristic to that region. The results summarized in this paper should be published because they add to the understanding of this complex region and challenge one to consider the importance of considering this region as an ecosystem with globally significant effects. The few comments and suggestions that I offer are meant to clarify the issues for the general reader and strengthen the case for the role of the undercurrent in this ecosystem.

ON THE ISSUE OF THE GLOBAL SIGNIFICANCE OF THE NITROGEN CYCLE IN THIS REGION

As this is a major point in this paper, it probably deserves a little more attention in the introduction. A reader unfamiliar with the nitrogen cycle literature might not be aware of how denitrification can be significant to the carbon dioxide concentrations and just what values are globally significant. The issue of global significance becomes clearer and convincing when it is discussed in more detail [later in the paper].

ON THE POLEWARD UNDERCURRENT AND REGIONAL OCEANOGRAPHY

The description of the undercurrent and the associated dynamics of the region [can be updated]. For example, there are recently published papers which strengthen the case for the influence of the Equatorial Undercurrent on the Peru Undercurrent and are pertinent to the work presented here. These papers are cited below:

Lukas, R. (1986). The Termination of the Equatorial Undercurrent in the Eastern Pacific. Prog. Oceanog., 16, 63-90. See Figure 14.

Rothstein, L.M. (1984). A Model of the Equatorial Sea Surface Temperature Field and Associated Circulation Dynamics. J. Phys. Oceanogr., 14, 1875-1892. This modeling study of the equatorial SST field and associated circulation generates a coastal undercurrent continuous with the EUC.

This section could be improved by a more detailed description of Figure 3 in the context of the ESPO system. For instance, the processes which govern vertical exchange of N are of particular interest to this paper. When is a simple sinking assumed? How are R and Kz parameterized to reflect the various processes which occur in this region? How does a system where there is considerable vertical shear (due to opposing currents in the vertical) effect the magnitude of the vertical exchange? Has this been explored or measured? Can the profiles that exist be used to approximate this term or will new measurements be needed? How will the spatial variability of this term affect the nitrogen cycle in this region?

Consider the difference in the upwelling at 5°S versus that at 10° or 15°S; i.e., regions of less upwelling versus more enhanced upwelling. The paper by Huyer, et al. (JGR, 92, 1987) clearly shows examples of this spatial variability. In fact, Barber and Huyer (Geophysical Research Letters, 6, 1979) indicate that the accumulation of organic material is not a simple passive accumulation of material from sinking from above, but that the variable mixing dynamics of this region are of importance. The interleaving structures found by Codispoti and Friederich may be further indications of the mixing dynamics in the region.

By including issues such as this it may make the "over-simplified" description "less-simplified," but it will point out the dynamics which may be a factor in the uniqueness of regions such as these to the global significance of the nitrogen cycle.

ON THE DISCUSSION

The discussion suggests that there are large temporal variations in the nitrogen cycle and budget that may be related to temporal variations in the strength, persistence or penetration of the EUC and its influence on the ESPO circulation. There is certainly evidence of the waning of the EUC during the 1982-83 El Nino (see Firing, et al., Science, 222, 1983). It would be interesting to see if there were any significant changes in the nitrogen cycle in the ESPO at this time, should you have any coincident data. Dr. Huyer may have hydrographic data during this period. This issue of the temporal and spatial variability of the ESPO circulation is germane also to other points discussed here.

Some useful information may also be gained from the previously cited reference on the variability of the upwelling; namely, that the temporal and spatial variability in the upwelling may be an important factor in the dynamics of this system. This may have added significance when one considers that the most noted case of this variability took place during the 1982-83 El Nino during times of other climatic imbalances.

There are additional references which would aid the discussion of the increased undercurrent velocities. An article by Smith (Science, 221, 1983) describes incidences of higher velocities in the undercurrent associated with the 1982-83 El Nino. In addition, there is a report "Current Measurements Off the West Coast of South America, November 1981-March 1985" (Knoll, et al., 1987, OSU Report 140, 87-31) which shows time series of velocity from current measurements in the undercurrent.

The association of stronger offshore flow associated with stronger alongshore flow may not be entirely correct. Examination of shelf currents at 50 and 100 m from the above mentioned report shows little evidence of the increased offshore flow; this may be due to the insensitivity of the location of the meters and rotation angle, for this particular question. It is difficult to support this statement with data because the location of the measurements may not have been chosen to examine this question. As this is an important question, but one that is not readily answered by available data, it may serve as a worthwhile hypothesis for further study.

AUTHORS' RESPONSE TO THE REVIEWER'S COMMENTS

Because of participation in an extended field program, it was impossible to revise our paper in accordance with the excellent suggestions of the reviewer, and an editorial decision was made to append her remarks to our paper. We agree with her comments, and think that reading the additional references that she cites, and considering some of the complexities that she mentions would be useful for the interested reader. In our defense, we will point out that our main goal was to pique the interest of physical oceanographers in a subject with which most of them are unfamiliar. Consequently, the paper was conceptualized as a lecture to interested non-specialists, and contains a significant amount of review material as well as some previously unpublished material. Glossing over some of the complexities is, of course, a time honored technique when giving introductory lectures, but it can lead to misconceptions which the addendum should help elminate.

TOPOGRAPHIC STRESS IN COASTAL CIRCULATION DYNAMICS

Greg Holloway
Institute of Ocean Sciences
Sidney, B.C., V8L 4B2 Canada

Ken Brink
Woods Hole Oceanographic Institution
Woods Hole, MA 02543

Dale Haidvogel
Chesapeake Bay Institute
Johns Hopkins University
The Rotunda
711 West 40th Street
Baltimore, MD 21211

ABSTRACT

The interaction of nonlinear shelf waves or eddies with longshore variation of topography is addressed. The dynamics are subtle and, perhaps for this reason, have been largely overlooked. Our own results are preliminary; but they point toward an unexpectedly strong influence such that, in many coastal environments, topographic stress may overwhelm other forces. The topographic stress acts to drive undercurrents in the sense of intrinsic shelf wave propagation, hence poleward on eastern boundaries. Potentially, a very important consideration is that a secondary circulation associated with this stress has the sense of upwelling on all boundaries in either hemisphere.

THE PROBLEM

A host of forces act on the coastal ocean, viz. steady and fluctuating wind, windstress curl, buoyancy sources/sinks, longshore pressure gradients, and encounters with open ocean eddies. No doubt this list could be extended. Many of the forces cannot be measured accurately and/or with adequate space-time resolution to obtain confident statistics. Response of the coastal ocean is correspondingly complicated and difficult to observe with sufficient resolution, whatever "sufficient" means! It is an understatement to acknowledge the difficulty to verify by observation a causal relation between forcing and response. Under these circumstances, one may not wish to consider yet another force acting on the coastal ocean. Nonetheless, that is our purpose.

We are concerned with the role of topographic form drag, i.e., the force arising from correlations of pressure fluctuations with bottom topographic slopes. In studies of global atmospheric angular momentum, these are the correlations that support mountain torque (Newton, 1971). A new result is to find that this "drag" really is not a drag. Rather it is a systematic stress with a definite tendency which, among other effects, drives poleward undercurrents on eastern boundaries. More generally the topographic form stress tends to drive currents in the sense of intrinsic shelf wave propagation, i.e., cyclonically with respect to ocean basins.

Not only does topographic stress appear to drive longshore flow, but a secondary circulation has the sense of upwelling on all boundaries in either hemisphere. With significant consequences in terms of biological productivity, regional climate, waste disposal, etc., as well as concern for large scale, longshore circulation, it would seem surprising that an effect such as this has gone unnoticed. The key question is the quantitative one: Is the process so weak as to be ineffective? It is probably too early to answer this question definitely, but our preliminary results are exciting or alarming, depending upon one's predispositions.

In the following sections we trace certain precedents provide a synopsis of a statistical, theoretical approach, then provide results from numerical experimentation.

PREMONITIONS AND EXTREMAL CONJECTURES

We will try to show that there are three necessary ingredients:

a. A large scale gradient of potential vorticity.
b. Topographic variation along the large scale contours of potential vorticity.
c. Nonlinearity.

These ingredients were present in early ocean circulation models under simplifying assumptions of quasigeostrophy and β-plane with small amplitude bottom roughness. Among the results of such modeling was the observation by Bretherton and Karweit (1975) that topographic interaction tended to propel westward flow. Flow around individual topographic features tended also to mean circulation of anticyclonic sense about elevations, and cyclonic about depressions. Thus, the flow everywhere tended "westward" in the deformed f/H contours, as discussed also by Holloway and Hendershott (1974).

Interpreting these numerical experimental results, two quite different extremal conjectures were advanced. Bretherton and Haidvogel (1976, hereafter BH) proposed that such flows approach a condition of minimum total enstrophy at given total energy. Let the overall depth be written $H = H_o\left(1-h\left(\underline{x}\right)/f\right)$ where H_o is mean depth, h/f is fractional elevation of the bottom above the mean depth, $\underline{x} = \left(x,y\right)$ is spatial coordinate with x east and y north, and f is Coriolis parameter. In a barotropic, quasigeostrophic flow with rigid upper lid, motion is defined by a velocity streamfunction $\underline{u} = z \times \underline{\nabla}\psi$ where $\underline{u} = \left(u,v\right)$, z is unit vertical and $\underline{\nabla} = (\partial_x, \partial_y)$. Omitting forcing and dissipation, and taking f constant the equation of motion is

$$\frac{D}{Dt}\left[\nabla^2 \psi + h\right] = 0 \tag{1}$$

with

$$\frac{D}{Dt} = \partial_t + J\left(\psi, \cdot\right)$$

and

$$J\left(A,B\right) = \partial_x A \partial_y B - \partial_y A \partial_x B.$$

In a closed domain or in a domain satisfying periodicity conditions on $\nabla^2\psi$ and h, integral invariants of (1) are energy E and total enstrophy Q, where

$$E = \frac{1}{2}\int dxdy\ |\underline{\nabla}\psi|^2 \tag{2}$$

and

$$Q = \int dxdy\left(\nabla^2\psi + h\right)^2 \tag{3}$$

In brief, the argument of BH is that turbulent shearing will draw out contours of potential vorticity $q = \nabla^2\psi + h$, producing steep gradients of q. If one hypothesizes potential vorticity mixing at short length scales, then q-variance (i.e., Q) will be dissipated. E is at larger scales and, moreover, will be retained at large scales by any reverse cascading tendency in 2D turbulence. The result is to suggest that the flow approach a state of miminum Q for given E. Maps of ψ calculated from this extremal method were found to approximate numerical experimental results also obtained by BH and to agree in sense with Bretherton and Karweit (1975) and Holloway and Hendershott (1974).

Extended to the ß-plane for which $q = \nabla^2\chi + h + \beta y$, BH obtained

$$\psi\left(\underline{x}\right) = \Sigma_{\underline{k}}\ \frac{\hat{h}_k e^{i\underline{k}\cdot\underline{x}}}{k^2 + \mu} + \frac{\beta y}{\mu} \tag{4}$$

where $h(\underline{x}) = \sum \hat{h}_{\underline{k}} e^{i\underline{k}\cdot\underline{x}}$ and μ is a positive constant which depends upon E. Interest in (4) is directed to the term $\beta y/\mu$ which provides a westward flow of magnitude $\beta y/\mu$. If f/H contours are dominated by the topography of a continental margin, the "westward" flow of BH is the poleward flow on an eastern boundary.

While the extremal principle of BH was advanced to explain these results, a very different extremal approach was given by Salmon et al. (1976, hereafter SHH). The problem considered was again restricted to quasigeostrophy, with one or two layers in the vertical. For the case of one layer on a ß-plane with bottom topography, $q = \nabla^2\psi + h + \beta y$ as previously. Without forcing or dissipation, integral invariants are E and Q, including contributions from βy in the latter. In a closed basin, boundary conditions may be taken as $\Psi = 0$ along the boundary. SHH supposed that $\Psi(\underline{x},t)$ is expanded upon a complete, orthogonal set of eigenfunctions

$$\psi = \sum_i \psi_i(t)\, \phi_i(\underline{x}) \tag{5}$$

where

$$\nabla^2\phi_i + k_i^2\, \phi_i = 0$$

with $\phi_i = 0$ on the boundary. Truncating the sum in (5) at some large N, the vector $\{\psi_i\}$ describes the motion of a point in N-dimensional phase space. SHH then inquired what is the probability distribution of $\{\psi_i\}$ for fixed expectations of E and Q. The extremal solution is that which maximizes entropy subject to E and Q. For the ith expansion coefficient, SHH (eqn. 4.3) obtains

$$P_i\left(\psi_i\right) = \left(\frac{a_i + a_2 k_i^2}{\pi}\right)^{1/2} \exp\left\{-\left(a_1 + a_2 k_i^2\right) k_i \left(\psi_i - \langle\psi_i\rangle\right)^2\right\} \tag{6}$$

with

$$\langle\psi_i\rangle = \frac{a_2 k_i^2}{a_1 + a_2 k_i^2}\, h_i \tag{7}$$

where h_i is the expansion coefficient of $h(\underline{x}) + \beta y$ and a_1 and a_2 are determined by E and Q.

The solution (7) in fact contains (4) in the limit that one choose Q at its minimum for given E, in which case a_1 and a_2 collapse to the single parameter μ. In common with (4), (7) expresses a westward flow $\langle\psi\rangle \cong \frac{a_2}{a_1}\beta y$ which is valid for all Q including the case where Q is at its minimum. (At unrealistically large Q for given E, the sign of a_2/a_1 reverses to give eastward flow). The closed basin solutions include eastward return flows in narrow boundary currents. More recent investigation by Carnevale and Frederiksen (1987) confirms the maximum entropy result that mean flow is westward, relating this also to nonlinear stability theory. In re-entrant ß-channel geometry, the mean is westward at all latitudes without the eastward return flows required by closed basin geometry.

One is left in a quandary that is not resolved to date. Two quite different extremal conjectures have been advanced: that flows tend either to minimize enstrophy or to maximize entropy. A question remains as to which is "right," with some uncertainty what constitutes "right." For present purposes we may leave this question open, noting only that each conjecture leads to a tendency for westward flow on a ß-plane corresponding to poleward flow on an eastern boundary where topography controls the gradient of f/h.

MECHANISM: ASYMMETRICAL FORM DRAG

In their interpretation of model results, BK recognized the important role of asymmetrical topographic form drag. When a flow was forced toward the east on the ß-plane, large topographic form drag

$$D_x = -\frac{H_o}{f}\int_A dA\, p\partial_x h \qquad (8)$$

arose, where p is pressure at the bottom and $\int_A dA$ is an integral over a fluid region of interest. When a flow was toward the west, BK remarked that the form drag vanished. This asymmetry is due to the possibility of stationary Rossby in case of eastward flow, whereas only freely propagating Rossby waves were possible in the westward flow. Qualitatively, these considerations appeared to support the numerical experimental result that westward flow was "favored."

Analytical results for the form drag due to an isolated topographic bump on a ß-plane were obtained by Johnson (1977) for cases of uniform upstream flow from the west or from the east. In cases of flow from the west, Rossby wave radiation of energy resulted in drag which could be expressed in terms of length and height of the bump and amplitude of the upstream flow. No drag resulted in cases of flow from the east. Johnson's results thus supported the account given by BK. Although the analytical results were available only for cases of steady flow with uniform upstream conditions and for bump heights of such small amplitude that closed recirculation above the bump would not occur, Clarke (1982) has argued that Johnson's drag is broadly consistent with inferred topographic resistance to the eastward flowing Antarctic Circumpolar Current.

Brink (1986) considers the theoretical analysis of topographic drag on a continental shelf and slope, extending previous analyses which were limited to quasi-geostrophic dynamics on the ß-plane. Steady, inviscid, barotropic motion is assumed while the topography is permitted to have small amplitude departure from a mean continental margin which is uniform in the longshore coordinate. Except for a small amplitude restriction, the topography is permitted to have extensive spatial structure with the provision (for analytical convenience) that topographic variations vanish beyond some large distance from the region of interest.

Brink's procedure is to expand perturbatively in the amplitude of the topographic variation, scaled by a small quantity ε. At zeroth order the steady flow is directed alongshelf with arbitrary shear in the cross-shelf direction. To first order in ε, the transport streamfunction perturbation is expanded on eigenfunction bases which depend upon the zeroth order flow and mean continental margin profile. The topographic drag (8) can then be evaluated at order ε^2, resulting in Brink's eqn. 2.26. For details of this calculation, one should refer to Brink's article. Importantly, the drag is expressed in terms of integral statistics of the topographic variations.

Brink identifies three possibilities:

a. Zeroth order flow is in the direction of free shelf wave propagation. Drag vanishes.

b. Zeroth order flow opposes free shelf wave propagation and is slower than the first mode intrinsic phase speed. This is the situation which supports non-zero drag due to standing shelf lee waves.

c. Zeroth order flow opposes free shelf wave propagation with amplitude greater than the maximum intrinsic phase speed. Termed "supersonic" by Brink, drag vanishes in this case since the mean flow is so swift that all perturbations are swept downstream.

Using a gridded topographic data set off Oregon (Peffley and O'Brien, 1976) and assuming a southward zeroth order flow with values from 6.3 cm s^{-1} inshore to 2.2 cm s^{-1} offshore, Brink (1986, Figure 6) obtains the topographic drag shown here as Figure 1. Averaged across the shelf, this stress is 3.36 dyne cm^{-2}. Thus Brink remarks that if one were to ignore questions of validity of the perturbation expansion and use actual, unscaled topography ($\varepsilon = 1$), the apparent northward stress of 3.36 dyne cm^{-2} would overwhelm the coastal momentum balance. However, validity of the perturbation expansion is in doubt and the assumption of steady flow is not satisfied. As well, Brink remarks that the actual near bottom flow is persistently northward (Kundu and Allen, 1976) which, if assumed for the sense of the zeroth order flow, would result in no topographic drag at all, corresponding to case a. The important point here is to realize that a topographic drag can arise with, possibly, quite large values. To go further than Brink's analysis requires numerical simulation such as reported by Haidvogel and Brink (1986, hereafter HB).

HB treats the fully nonlinear, time-dependent problem of barotropic flow above a continental margin with finite amplitude topographic irregularities. A flat rigid upper lid is assumed. In coordinates such that x is alongshelf and y is cross-shelf, HB solves numerically for evolution of transport streamfunction

$$\left(-\partial_y \phi,\ \partial_x \phi\right) = \left(Hu,\ Hv\right) \tag{9}$$

with H given by

$$H = H_o \exp\left\{\alpha y + \sin\frac{\pi y}{W} \cdot \Sigma d_m \sin\left(k_m x + \theta_m\right)\right\} \tag{10}$$

The flow is in a channel of width W bounded by straight free-slip sidewalls. H and ϕ are assumed to be periodic over some x-interval L.

The study by HB is motivated by a thought experiment based upon the topographic drag analysis of Brink. It was seen that mean flow opposed to the sense of shelf wave propagation generates large drag due to lee wave formation, whereas mean flow in the sense of shelf wave propagation encounters no topographic drag. Suppose then that an x-directed wind stress is applied which is periodic in time about zero mean stress. Although the alongshelf wind stress is symmetric with respect to periods of positive and negative x-driving, the resulting flow will encounter strongly asymmetrical topographic drag and result in a time averaged flow

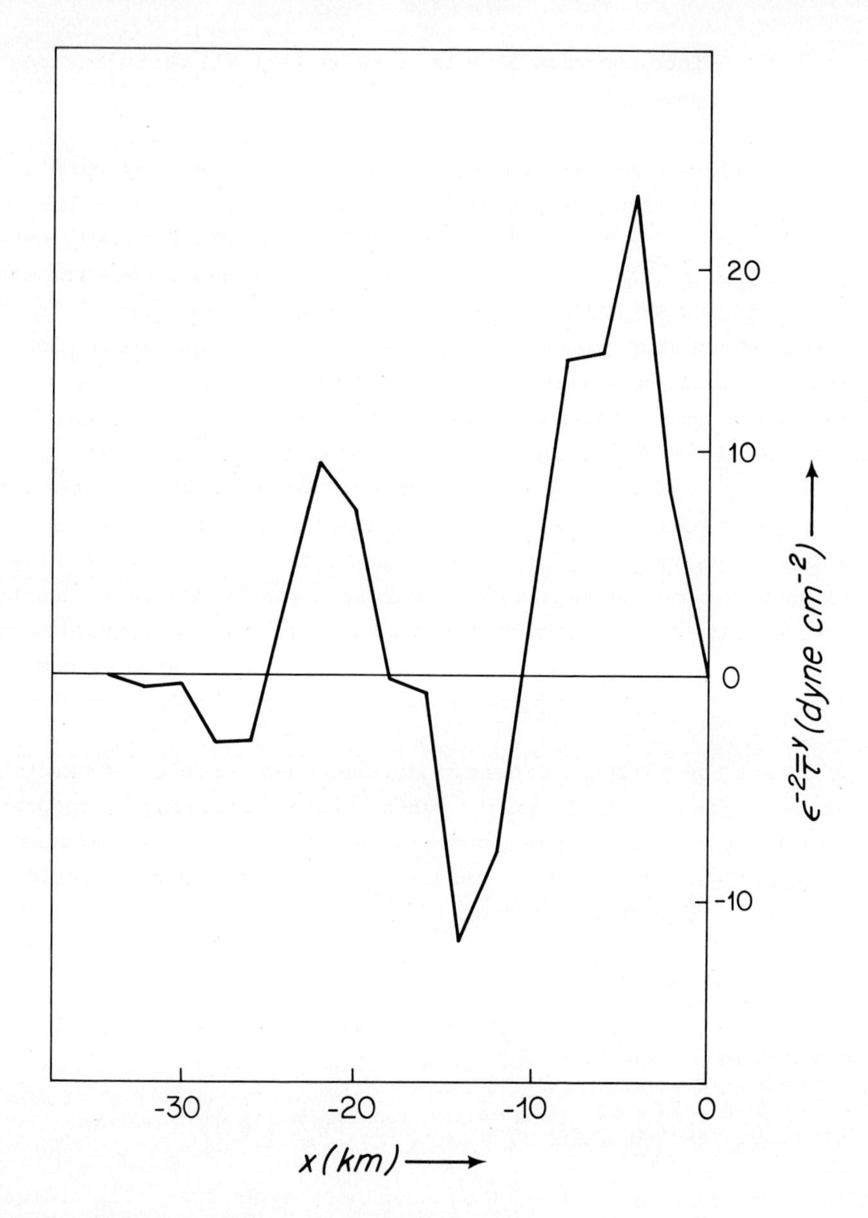

Figure 1: Bottom stress as a function of cross-shelf distance for a case of topography resembling the Oregon shelf.

directed toward positive x (the direction of intrinsic wave propagation in the coordinate scheme of HB).

Numerical simulation results of HB tend to bear out the thought experiment over a range of parameters including wavelength and amplitude

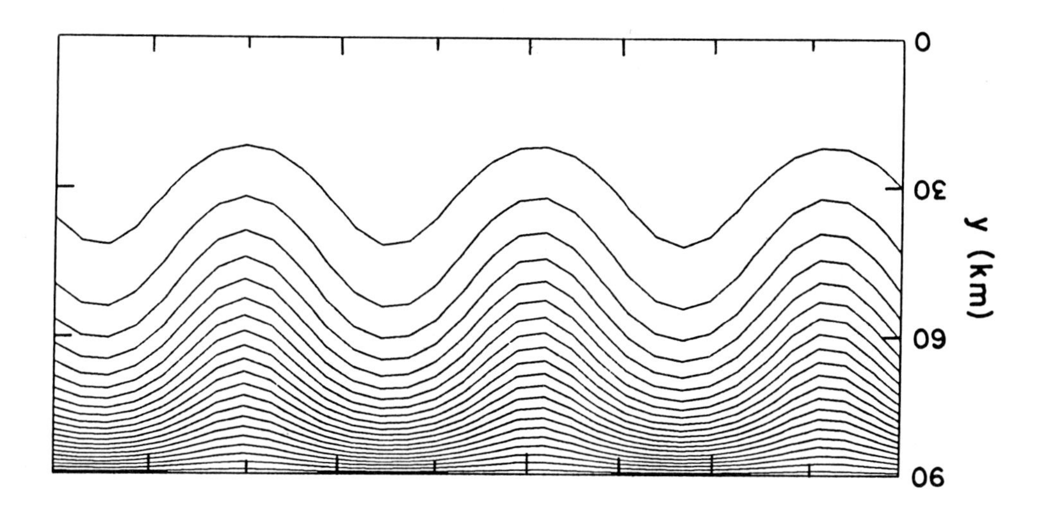

Figure 2a: Bottom topographic contours for the central experiment studied by HB. The contour interval is 100 meters.

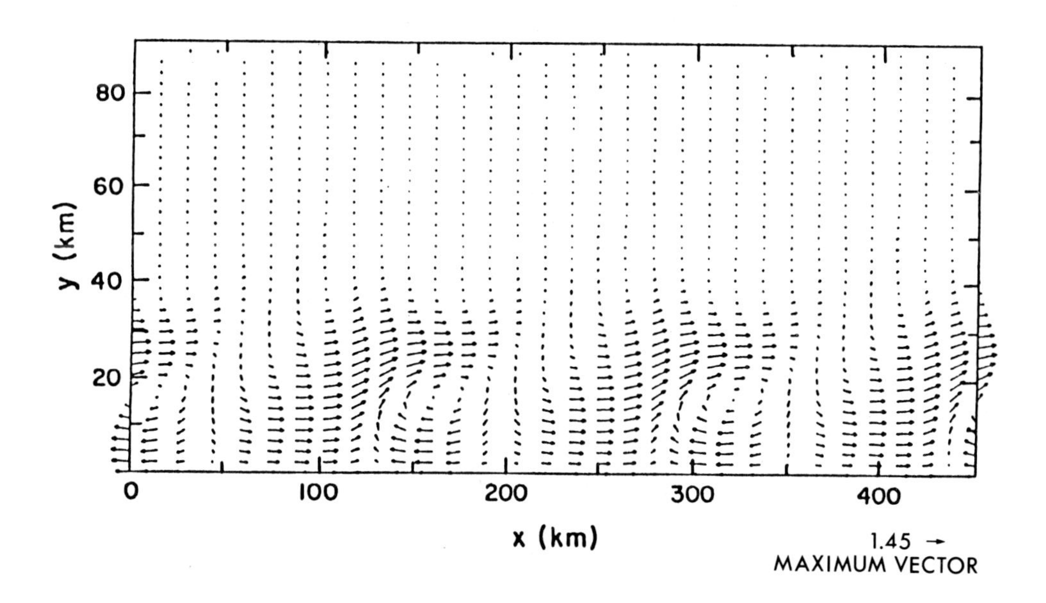

Figure 2b: Time-mean Eulerian Velocity vectors for the central experiment of HB. The maximum vector length is 1.45 cm/sec. The circulation is forced by a spatially broad, along-channel wind stress which oscillates in time with a ten day period and an amplitude of 1 dyne/cm**2.

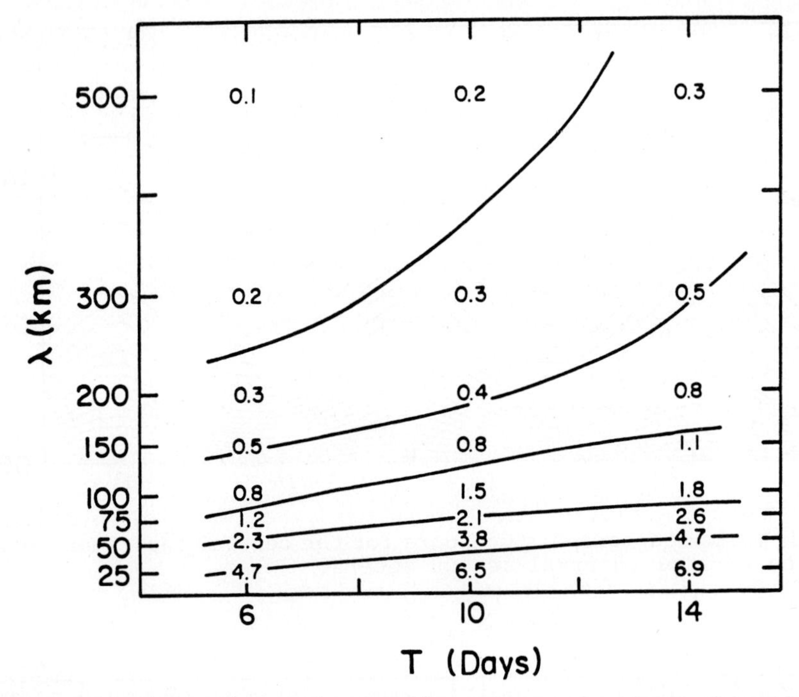

Figure 3: Maximum time and zonally averaged along-channel currents (9 cm/sec) observed by HB, as a function of bottom bump wavelength (km) and wind stress periods (days).

of topographic variation, strength and period of wind stress forcing, and coefficient of linear bottom friction. Isobaths of topography and a time averaged flow field are shown in Figure 2 (from HB Figure 2 and 5). In the illustration shown, the maximum mean flow is 1.45 cm s^{-1}. For other experiments summarized in Figure 3 (from HB Figure 10), mean flows as large as 6.9 cm s^{-1} were obtained.

WHEN A DRAG'S NOT A DRAG: NEPTUNE EFFECT

It is clear from the numerical experiments just described that substantial mean flows can result from zero mean but time-periodic applied stress. What is not so clear is that this is due to asymmetrical topographic drag. In particular, cases where the mean flow is persistently in the sense of shelf wave propagation should be characterized by absence of topographic drag. The question is re-opened: What drives such flows?

A different theoretical approach was developed by Holloway (1986, 1987, hereafter H86 H87) following analytical methods from turbulence theory. The cumbersome calculation is made tractable by returning to barotropic quasigeostrophic flow on a ß-plane. Nonlinear time-dependent motion is considered in the presence of a spectrum of statistically homogeneous topographic variation $H_{\underline{k}} = \left|h_{\underline{k}}\right|^2$. Relative vorticity $\zeta(\underline{x},t) = \hat{z} \cdot \underline{\nabla} \times \underline{u}$ is likewise expanded in a wavenumber spectrum $Z_{\underline{k}} = \left|\zeta_{\underline{k}}\right|^2$. The problem is posed as follows:

Given a spectrum H_k, given a statistical distribution of wind stress torques as well as any mean zonal (along-shelf) wind stress, and given values for ß and for frictional parameters, one seeks the statistical expectation, for vorticity-topography cross-spectra

$$C_{\underline{k}} = R_{\underline{k}} + iI_{\underline{k}} = \xi_{\underline{k}}^{*} h_{\underline{k}}$$

where asterisk denotes complex conjugation, and for a component of spatially uniform zonal (along-shelf) flow.

Due to nonlinearity of the equation of motion, equations for the evolution of second moment quantities Z_k and C_k involve unknown third order correlations $\zeta\zeta\zeta$, $\zeta\zeta h$, ζhh, while evolution of third and higher order correlations involve successively higher order correlations. This unclosed hierarchy is then "closed" by reference to an entropy argument related to eqn. (6): it is required that in the absence of any external forcing or dissipation, flow statistics tend toward maximum entropy. This is accomplished at the level of fourth cumulant statistics, i.e., the portion of fourth order correlations which represent departures from random phase configurations. It is assumed that fourth cumulants have the effect of providing a linear relaxation of third order correlations. The result (H87, eqn. 16) is

$$C_{\underline{k}} = \frac{ik_x UH_{\underline{k}} - \gamma_{\underline{k}} H_{\underline{k}}}{-i\omega_{\underline{k}} + \eta_{\underline{k}} + \nu_k} \tag{11}$$

where k_x is the x-component wavenumber, $\gamma_{\underline{k}}$ and $\eta_{\underline{k}}$ are expressed (H87, eqn. 12 and 13) in terms of weighted sums over all wavevectors of the vorticity and topography variance of the vorticity and topography variance spectra, ν_k is the wavevector representaton of any explicit dissipation law, and $\omega_{\underline{k}} = k_x\left(U - ß/k^2\right)$ is the Doppler shifted Rossby wave frequency.

The importance of C_k is that the total x-directed force F which the topography exerts upon the flow can be expressed in terms of the imaginary part of C_k as $F = \Sigma_{\underline{k}} F_{\underline{k}} = -\Sigma_{\underline{k}} k_x k^{-2} I_{\underline{k}}$. From (11)

$$F_k = -\frac{\eta_k + \nu_k}{\omega_k^2 + \left(\eta_k + \nu_k\right)^2} \frac{k_x^2}{k^2} UH_{\underline{k}} + \frac{\omega_k k_x}{\omega_k^2 + \left(\eta_k + \nu_k\right)^2} \frac{\gamma_k}{k^2} H_{\underline{k}} \tag{12}$$

The first term on the right side of (12) is an asymmetrical drag, i.e., opposed in sense to U. (Both $\eta_{\underline{k}}$ and $\gamma_{\underline{k}}$ are positive.) However, the derivation is not restricted to steady ideal flow nor to lowest order in an amplitude expansion nor to cases where recirculation above bumps does not occur. More significantly, the result differs from previous analyses insofar as drag arises on either sign of U and has no large $|U|$ cutoff. The drag is asymmetrical in that flow in the sense of intrinsic wave propagation, i.e., $U < 0$ in ß-plane geometry, results in large $\omega_{\underline{k}}^2 = 0$ thereby suppressing contribution to $F_{\underline{k}}$. Flow opposed to intrinsic propagation, i.e., $U > 0$, admits large contributions near $k^2 = ß/U$ for which $\omega_{\underline{k}}^2 = 0$. It is noteworthy that the drag contributions exist for all values of U even if one assumes inviscid flow for which $\nu_k = 0$.

Although topographic drag according to the first term on the right side of (12) differs from the results of Johnson or of Brink, a qualitative thought experiment such as suggested by HB is not sensitive to these differences. This leaves open the question whether asymmetry of drag is sufficient to account for rectified flow. Especially one should be puzzled if flow is persistently in the sense of intrinsic wave propagation, for which it was previously argued that drag vanishes and it is now argued that the drag does not vanish but in fact propels the flow when eddy activity is present.

An answer, perhaps, and a novel effect is given by the second term on the right side of (12). This term takes the sign of $\omega_{\underline{k}} k_x$ and so, in particular, for $U < 0$ this term tends to further accelerate U negatively. Apparently the two terms in (12) may oppose each other. It is suggested in H87 that the net force can be estimated, albeit roughly, if $\eta_{\underline{k}} \approx \gamma_{\underline{k}}$ and $\nu_{\underline{k}}$ plays a minor role. Then the first term acts to eliminate the Doppler shift from $\omega_{\underline{k}} k_x$ in the second term so that the net force takes the sense of the intrinsic phase propagation. A more quantitative evaluation of $\eta_{\underline{k}}$ and of $\gamma_{\underline{k}}$ is required to determine the net force quantitatively. However, H87 provides estimates that overall F could correspond to stress of order 1 dyne cm^{-2} in many "typical" coastal environments.

The surprise in this section is to see that topographic stress may not be a drag at all but rather tends to be a force of definite sign, namely that which acts in the sense of intrinsic shelf wave propagation. There is no obvious name for the phenomenon leading to the second term in (12). Following a cartoon in H86, we call it the 'Neptune effect.'

A mathematical result (12) which runs against intuition begs for some physical interpretation. H87 offers the following account:

a. There is a tendency for relative vorticity to correlate with topographic variation in the sense (N. hemisphere) for negative vorticity above topographic elevation. This tendency was seen in numerical experiments by BK and by HH and has been discussed theoretically by BH and by SHH. One can also say in a loose way that if fluid columns were to relocate "randomly" then a column which finds itself over a local elevation will be foreshortened and hence, on average, exhibit negative vorticity.

b. A geostrophic tendency, which need not be complete, will associate positive pressure anomaly with negative vorticity (N. hemisphere) and therefore, with a., tend to associate positive pressure anomaly on geopotential surfaces with positive elevation of topographic variation (in either hemisphere).

c. Larger scale gradient of potential vorticity, whether by β or by gradient of f/h on continental margin, will induce a tendency for vorticity and pressure anomalies to propagate in the sense of Rossby or shelf wave propagation. Nonlinearity and topographic variations tend to scatter the propagating waves.

d. Competition between tendency b. to establish pressure anomaly above elevation and c. to propagate as waves until scattered leads to an equilibrium tendency for positive pressure to lie on the side of wave propagation relative to a topographic bump. Negative pressure anomaly tends similarly to lie on the side of wave propagation relative to topographic depression. The resulting pressure-slope correlation (in either hemisphere) supports a transfer of momentum from solid earth to fluid which has the same sense as Rossby or shelf wave propagation.

Finally, in case the physical account is no more compelling than the mathematical one, numerical experiments are also reported by H87 as shown here in Figure 4. A steady mean stress $\hat{U}$ is applied in a number of cases ranging from poleward to equatorward stress (in eastern boundary geometry). As well, random wind stress torques (with zero mean momentum) are applied in two of three cases for each choice of $\hat{U}$. The fluid is of nominal depth 300 m and the bottom is characterized by random topographic variations of rms height 30 m. Dashed and dash-dot curves show the response of mean U to increasing amplitudes of random wind stress

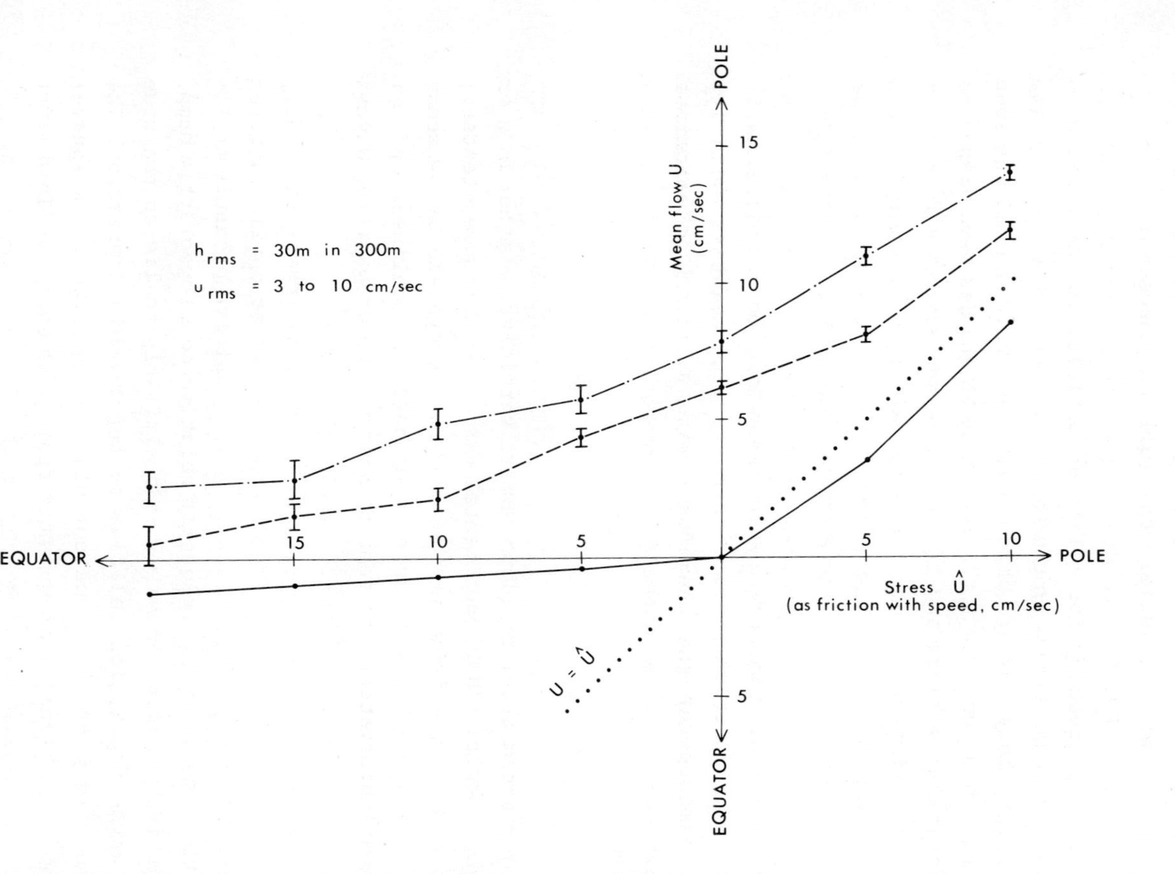

Figure 4: Mean (area averaged) flow U is plotted against mean applied stress $\hat{U}$. A dotted curve shows the purely frictional solution $U = \hat{U}$. When topographic roughness of r.m.s. amplitude 30 m with a reference depth of 300 m is included, U departs from $\hat{U}$. A solid curve shows the response when only the mean stress $\hat{U}$ is applied. Dashed and dash-dot curves show the response when random wind stress torques of different r.m.s. amplitudes are also present. Under the influence of random torques, U is unsteady. Vertical bars show one standard deviation of fluctuating U about its time averaged value.

torques. Of particular interest are experiments in which $\hat{U}$ is equatorward with a strength that would produce equatorward U of 5 or 10 cm s^{-1} in absence of topography. When wind stress torques are sufficient to maintain eddy energy levels corresponding to several cm s^{-1} rms speeds, mean U is poleward at some few to several cm s^{-1}. Values of U fluctuate in time and vertical bars shows one standard deviation from time averaged U. Although these mean currents are persistently poleward, the net topographic stress is also poleward resisting both bottom friction and applied $\hat{U}$. Indeed, eddy energy is being withdrawn by working against the mean applied stress $\hat{U}$.

Results presented in this section and in the preceding section are so recent that they should be regarded tentatively. Indications are that flow interaction with longshore variation of topography may indeed have very substantial consequences. Much closer investigation is warranted.

SECONDARY CIRCULATIONS AND ROLE OF BAROCLINICITY

To this point both theories and numerical experiments have been based upon barotropic models. (Not quite. Johnson (1977) included effects of stratification). In this final section we address two further aspects.

First, even within the context of barotropic theory, we may observe that there is an important secondary recirculation within the vertical planes normal to the coastline. At (11) we calculated a longshore momentum source $F = -\Sigma k_x k^{-2} I_{\underline{k}}$ due to pressure-bottom slope correlations. Let us now consider the correlation of onshore velocity $v = \partial_x \psi$ (in the notation of H86 and H87) with topographic elevation h. That is

$$\overline{vh} = \mathrm{Re}\Sigma\left(ik_x \Psi_{\underline{k}}\right)^* h_{\underline{k}} = -\Sigma k_x k^{-2} I_{\underline{k}} \tag{13}$$

Contributions to the sum in (13) are just those already evaluated at (12). Arguments following (12) indicated a strong tendency for negative sign, hence for $\overline{vh} < 0$, i.e., onshore motion tends to occur above negative topography such as canyons or other depressions. If, for example, we suppose a situation with no longshore gradient of mean pressure, then we expect no average off/onshore motion of water in a nearly geostrophic portion of water column. However, within a depth range spanned by the range of topographic variation about the mean depth, longshore pressure differences are supported by the topography. It is within this depth range that negative $\overline{vh}$ supports net onshore (upslope) transport due to Neptune effect.

A return transport is required to avoid net onshore mass transfer—else Coriolis reaction to such transfer would need to be included in the longshore momentum budget. Three return paths are available. There may be an adverse mean pressure gradient $\overline{\partial_x p} \cong F < 0$, supporting net offshore geostrophic transport in the water column above depths spanned by topographic variation. A mean adverse wind stress may drive an offshore transport in a surface Ekman layer, thereby closing an upwelling cell. Thirdly, the mean flow may be driven so strongly in the direction of shelf wave propagation that a benthic Ekman layer supports the offshore transport. The proportions in which these paths are used will depend upon external forcing of wind stress and longshore mean pressure gradients. The key remark here is that the transport associated with the topographic interaction has a definite preferred sense, favoring onshore motion or upwelling of the deeper water.

Discussion thus far has not addressed truly baroclinic effects due to stable stratification. In a qualitative way, one anticipates stratification shielding much of the water column from the influence of topography. Thus equatorward wind stress at an eastern boundary might drive equatorward surface circulation while topographic stress drives a poleward undercurrent. Exploration of these circumstances would appear to be the natural province of numerical modeling. Very preliminary stratified simulations, similar to those described by HB but utilizing a primitive equation coastal model, appear to support these conjectures. We hope to report these calculations in the near future.

ACKNOWLEDGEMENTS

We are grateful to Billie Mathias for typing and to Patricia Kimber for assistance with the figures. We appreciate support received from the Office of Naval Research (contracts N00014-C-85-0440, -87-G-0262, and -87-K-0092.)

PART IV:

EPILOGUE

EPILOGUE

We sent to each workshop participant a set of questions about Eastern Boundary Poleward Undercurrents (EBPUC). From their answers we assembled a consensus of scientific opinion on the nature of flows along eastern ocean boundaries.

1. WHAT DOES YOUR OWN STUDY SUGGEST AS THE APPROPRIATE SPATIAL-TEMPORAL SCALES OF THE EBPUC?

Rationale: Smith posed the thesis that the EBPUC must have some kind of persistence - spatial, temporal or both; Mittelstaedt suggests that the EBPUC of the North Atlantic is "connected" from Africa to Norway, say 15 to 65°N; Neshyba pointed out that the salinity-oxygen characteristic of the Peru-Chile EBPUC is traced through more than a dozen cross-shelf hydrographic sections from 10 to 48°S, from a data set spanning many years; Yamazaki and Lueck state that the time scale of dissipation of total mechanical energy in the California Undercurrent compares to its transit time from 33 to 50°N; Freeland finds the North Pacific EBPUC is persistent at Vancouver Island, and shows a similar generic feature in the Canadian Arctic Basin.

Opinion: **HUTHNANCE.** There are several aspects to the notion of spatial persistence or continuity: (i) transport volume, (ii) advection of identifiable water, (iii) propagation of slope current fluctuations, (iv) cross-slope structure.

For quasi-geostrophic flow along depth contours, (iv) is determined by (i) in detail between individual depth contours so that the persistence is similar. The persistence of (ii) may be much less, however, if there is strong lateral exchange but the volume is maintained by large scale driving. Persistence of (iii) may be different again, being the frictional decay distance for coastal trapped waves (Huthnance 1987) which also corresponds to the along-shelf evolution/persistence scale for establishing (i) or (iv) after a barrier is encountered. From 48 to 60°N in the Atlantic, (i) is quite well conserved along the slope (with values) between 1.2 and 2.2 Sv (Huthnance 1986), but rapidly increases NE of the Wyville-Thomson Ridge as upper layer flows from across the Ridge and Faeroe-Shetland Channel recirculation are added. Thus the implied greater than 1000 km scale south of 60°N is upset by the Ridge. Statistics for (ii) are poor but water characteristics and a few drogue tracks suggest 500 km or so from Rockall Trough northwards. Evidence regarding (iii) comes from our CONtinental SLope EXperiment (CONSLEX) in winter 1982-3; along-shore coherences are significant, but not good in general, up to 500 km at periods of two days, but are upset by the Wyville-Thomson Ridge intersecting the slope; coherences are poorer at shorter periods except actually on the shelf where the scale exceeds 500 km. I hesitate to comment on (iv) with the poor cross-slope resolution hitherto.

McCLIMANS. I can say that the Norwegian Coastal Current (NCC) and the Norwegian Current (NC) have seasonal and mesoscale variability. The NCC exhibits a more direct response to variable winds, primarily in the Skagerrak near its source, but also along the coast, especially with gale or stronger winds and transport less than 1/4 Sv. Pressure fluctuations over Scandinavia make the Baltic Sea behave like a bellows, pressing out large amounts of buoyant water during the fall and winter months.

MITTELSTAEDT. The climatological mean boundary flow along the Eastern Atlantic Margin is permanent and exhibits a pronounced seasonal signal.

HICKEY. Data from both Washington and the Southern California Bight almost always clearly indicate a poleward undercurrent (as defined by a subsurface maximum in poleward flow) over the slope during summer and early fall. I feel confident that the Undercurrent is a very large scale feature of about 4000 km length whose principal time scale is seasonal. If we could remove smaller spatial scales and shorter time scales, the Undercurrent would emerge as a continuous longshore feature.

MOOERS. The seasonal aspects of the EBPUCs need to be determined and related to the seasonal aspects of coastal upwelling; eastern boundary currents *per se*, including planetary Rossby waves; and related mesoscale variablility.

2. IS THERE A "NORMAL" PATTERN TO THE EBPUC?

Rationale: Can you associate numbers with descriptive characteristics: width? thickness? relation to topography? steady-state? seasonal/ transient aspects?

Opinion: **HUTHNANCE.** The slope current west of Britain, 48 to 60°N, appears to be confined to the steep slope bordering the continental shelf (Booth and Ellett 1983; Huthnance 1986) and, therefore, has a width of 20 to 50 km. It appears to decrease downward but is essentially barotropic in the depths where it occurs, to 2000 m or more at 48°N decreasing to 1000 m or less at 60°N and beyond. It closely follows the depth contours along the slope. Although varying, it is steady in comparison with most obvious local atmospheric forcing, i.e., rarely reversing. There is some evidence for larger transports in winter (Gould, et al. 1985; Huthnance 1986).

BATTEEN. My model results suggest that the Undercurrent can change significantly if mesoscale eddies are present. My paper in this volume shows typical seasonal vertical (Figure 2a) and horizontal (Figure 2b) profiles and vertical cross-sections (Figure 8) of an equatorward coastal jet overlying a poleward undercurrent during the upwelling season and before the onset of eddies. These are used as initial conditions. After about 30 days, eddies and jets are generated, due to the baroclinic instability associated with the vertical shear between the coastal upwelling jet (CUJ) and the California Undercurrent (CU). Results show significant changes from the imposed initial conditions in the CUJ and CU fields due to the presence of eddies and jets. For example, the CU, instead of being a strictly northward flow, as prescribed in the initial conditions, now meanders around nearshore anticyclone eddies, resulting in offshore extensions of about 100 km. In the presence of cyclonic eddies, the CU flows poleward in a very narrowly confined region of 8 to 20 km inshore of the eddies.

McCLIMANS. In our work we rely heavily on models, from very simple dimensional arguments to 1, 2, and 3 dimensional numerical models and laboratory models. Laboratory models are quite valuable for studies of instabilities due to their high spatial resolution, but are limited to regional dynamics over which $\underline{f}$ is reasonably constant, about 1000 km. Remote driving mechanisms must be simulated as boundary conditions. This is satisfactory for the Norwegian Coastal Current and the Norwegian Current, but may prove to be a problem for the undercurrents if it turns out that they are part of the basin-wide dynamics.

HICKEY. Temporal scale: Off Washington and Southern California, the vertical shear associated with the Undercurrent has a seasonal cycle, such that the maximum poleward flow occurs in summer and early fall. The poleward flow itself at a depth of 100 to 200 m appears to have a semi-annual signal with maxima in summer and in winter. I believe that the second maximum, which is generally not associated with a strong subsurface maximum, may have a different driving mechanism than the first. Width: The off Washington appears to be about 20 km or about the width of the continental slope. In the California Bight, an undercurrent occurs across the entire Santa Monica basin, about 40 km, i.e., it does not appear to be trapped to one side or the other as might be expected. Thickness: This is about 200 m to the 5 cm s^{-1} contour. The depth of the subsurface maximum is shallower in the Southern California Bight, about 100 m, than off

Washington, about 200 m. Transients: There is clearly much high frequency variability. The one Washington data set which had four moorings across the undercurrent did not indicate that meandering was significant, in spite of suggestions by Mysak and others.

MITTELSTAEDT. The EBPUC's observed off Northwest Africa and off the Iberian Peninsula represent the "boundary signal" of a much broader and thicker subsurface circulation than say 100 km in the horizontal and a few hundred meters in the vertical. The domain of the poleward subsurface flow in the eastern basin seems to increase with latitude.

When/where the regional/seasonal external driving forces like the winds dominate the large scale pressure gradient, the surface flow is different from the flow at depths. This is the case in winter and spring at latitudes south of about 23°N and throughout the year north thereof up to about 50°N in the Eastern Atlantic. This is about the latitudinal zone where I expect an EBPUC to become disconnected from the surface layer. There are, of course, local and transient perturbations due to boundary eddies and waves.

Topography is important. It locally determines the flow direction and may generate eddies/waves that in turn interfere with the boundary flow.

MOOERS. The instabilities of EBPUCs seem to be linked to the mesoscale variability of eastern boundary current regimes. The interactions between the EBPUCs and the mesoscale eddies and meandering jets observed in those regimes may play an important role in cross-shore exchange processes and, thus, ecosystem dynamics. Similarly, the interactions of the EBPUCs with the coastal upwelling centers found equatorward of coastal capes may also play an important role in cross-shore exchange processes and, thus, ecosystem dynamics.

3. ARE THERE ESSENTIALLY TWO EBPUCs IN EACH BASIN?

Rationale: One EBPUC may be coupled to the eastern boundary of the sub-tropical gyre, while another is coupled to the eastern boundary current of the sub-polar gyre, the two being discontinuous?

Opinion: **HICKEY.** The seasonal variation of the NE Pacific poleward undercurrent of about 50°N suggests that this EBPUC is tied to the sub-tropical gyre. I was not aware that the feature has been observed north of Vancouver Island. By the "feature", I define a flow pattern with a subsurface maximum. Thus, I continue to differentiate in my mind between flows that may be directly driven by local poleward wind stress and flows that generally oppose the local wind stress and hence usually have a subsurface maximum of some sort, even though the depth of the maximum may be a function of wind stress or other factors.

MITTELSTAEDT. I believe there is a connection of the EBPUC off (North) Africa and Europe. The dynamics of the various branches of the boundary flow along the eastern margin of the Atlantic are different. But still, they all might be a common manifestation of the boundary dynamics to the large scale processes in the interior of the ocean.

HUTHNANCE. Confinement over the slope distinguishes the EBPUC west of Britain from a broader drift (hundreds of kilometers) northwards in the upper layers of the Rockall Trough and Plateau, and not tied to topography. To my mind, the most pertinent respective theories are poleward density gradients with the shelf/slope/ocean topography (Huthnance 1984) and a thermal-diffusion eastern boundary layer (McCreary 1981). These account for the two widths. However, the poleward density gradient applies from the tropics to high latitudes and is not directly coupled to the two gyres separately; essentially one (not two) slope currents would be expected in each basin. More observations in Biscay and between Portugal and Africa are required to establish this in view of the aforementioned persistence scale of 1000 km or less. On the other hand, the eastern boundary layer would adjoin two gyres separately. This boundary layer form of the EBPUC will be quasi-steady, but will be transient with season owing to its association with slow baroclinic Rossby waves decaying westwards. In fact, Rockall Trough is so occupied by eddies that consistent estimates of any northward drift are hard to determine.

NESHYBA. Having traced the continuity of the Peru-Chile Undercurrent from 10 to 48°S using its salinity/oxygen characteristic, we noted that this PUC transcends the zone centered on 43°S where the West Wind Drift of the Southeast Pacific impinges the continent. It appears that the PUC proceeds through this the zone of stagnation. To the south of the impingement zone, along the Chilean coast in what constitutes the eastern boundary of a subpolar flow system (not really a gyre because the Drake Passage removes the eastern boundary south of about 53°S), it gradually loses its subsurface geostrophic signature by blending into the surface Cape Horn Current that gains volume via cumulative runoff from the Cordillera. These data suggest that there is but one EBPUC in the Southeast Pacific.

4. WHAT IDEAS DO YOU HAVE ON HOW TO PARTITION EBPUC FORCING INTO LOCAL AND DISTANT EFFECTS?

Rationale: Some boundary phenomena are remotely driven, e.g., coastal trapped waves; are EBPUCs also?

Opinion: **HUTHNANCE.** My thoughts on this are expounded in Huthnance (1987) for simple barotropic contexts. Any eastern boundary has a set of coastal trapped waves which allow a more or less complete description of the motion on the slope from top to bottom. A forced problem has the solution

Particular Integral
\+ Complementary Function (viz a combination of coastal trapped waves to fit initial conditions in the along-shore coordinate and decaying as they propagate along the shelf/slope.)

The slope current is the Particular Integral (corresponding to density gradient forcing, say) but evolves through the decay of coastal trapped waves (e.g., northwards from the equator or a cross-slope barrier). These are initially present in the solution to enable the initial conditions to be satisfied in the presence of the Particular Integral. The evolution/decay distance is the persistence scale (iii) discussed in the first question above.

HICKEY. I believe that the EBPUC is driven by the seasonal along-shore pressure gradient; some of the pressure gradient signal is remotely driven and some locally driven by the coastal wind field. South of Point Conception, other processes must drive the along-shore pressure gradient. Other driving mechanisms for the poleward flow have been suggested and these are decoupled from the pressure gradient mechanism. I suspect these mechanisms will not be as significant as those related to the pressure gradient field.

MITTELSTAEDT. I believe that the EBPUC we observed off Northwest Africa, as well as the one off Europe, is essentially regionally driven.

BATTEEN. One can gain ideas on how to partition local and distant forcing of EBPUCs by running numerical experiments which use local but no remote forcing, then remote but no local forcing, and comparing the results with observations.

5. HOW CAN YOU BETTER USE THEORETICAL STUDIES?

Rationale: Could you use model diagnostics to interpret results? How are model results used for experimental design? —for setting priorities in field work proposed for large scale science like WOCE? What can concretely be done to bring theoretical and observed mechanisms together?

Opinion: **HICKEY.** Topography and wind stress climates vary enough along the West Coast that I believe the models could be used to partition the variance due to specific forcing mechanisms. The various models should be coupled more tightly with the observations. Perhaps this workshop is a step in the right direction.

BATTEEN. Model diagnostics can be used to interpret results, particularly if you use a model to run a number of controlled experiments, such as varying the type of forcing or topography, and then obtain model diagnostics for each case. Close comparison of the results should contribute to a deeper understanding of physical mechanisms responsible for observed oceanic behavior.

A model's structural detail could be used to help plan field investigations. For example, a model's horizontal and vertical cross-sections of horizontal velocity, temperature, and pressure can provide information on the surface and subsurface structure of jets and undercurrents which would influence the design of observational grids and arrays.

HUTHNANCE. Theory is used, as implicit in the above discussions, in a distinction between wind forcing (fluctuating currents on the shelf), density gradient forcing (steady currents along the slope), and the broader diffusive boundary layer. Because coastal trapped waves underlie the first two, their study in context can confirm their form and elucidate shelf and slope flows generally (Huthnance 1986; Freeland et al. 1986). A knowledge of coastal trapped wave dispersion (for wavelength-frequency relations) and decay (for evolution-persistence scales) can assist experiment design, specifically along-slope spacing. Care is required, however, for the degrading effect of higher modes with shorter scales; they may be more prevalent in chaotic reality than in idealized models.

MOOERS. The capabilities and limitations of oceanic observations and models suggest using them jointly to make optimal estimates of ocean fields in general. Considering the intense spatial and temporal variability of EBPUCs relative to our ability to observe or model them, the methods of four-dimensional data assimilation may be essential for dynamical interpolation of limited observations on one hand, and for observational constraints to be imposed on numerical models on the other hand. Time sequences of optimal field estimates are necessary for diagnostic studies of the EBPUC regime.

5. WHICH TECHNOLOGIES WOULD BE USEFUL IN MODERN EBPUC STUDIES?

Rationale: See Jane Huyer's paper What's Next?

Opinion: **MITTELSTAEDT.** Among others, we require multi-ship programs, satellite data products, extensive ADCP application, and both surface and subsurface Lagrangian devices in depths ranging from 100 to 1500 m for periods of 1 to 2 years.

BATTEEN. All the above plus long-term current meter deployment, and high-resolution multi-level primitive equation models such as developed by Haidvogel, Haney, and Batteen.

HICKEY. Pop-up drifters may be important for investigating the continuity of the EBPUC. Shipboard and in-situ acoustic Doppler current meters would provide additional detail on spatial scales needed to properly examine questions of instability, etc.

HUTHNANCE. We used current meters extensively to good effect for variability in CONSLEX, but they are vulnerable to fishing activity in many productive shelf and slope areas. The proposed five years or more of current meter measurements should be within WOCE Core Project 3.

For many years SMBA have worked a CTD section across the Hebrides Shelf and Rockall Trough to Rockall, at two-monthly intervals. For a narrow slope current (not the previous purpose of this CTD work) the spacing would need to be closer, with several casts within the steep slope area. We have relied on remote wind and atmospheric pressure observations from the many land and island recording stations, together with some information about their relation to more open offshore records.

Bottom pressure and coastal sea level records in CONSLEX gave some information on temporal variations, relating best to the currents in 'events' extending 500 km or more along-shore. Pressure gradients were difficult to estimate, however; motion took place on scales of 100 km or less over which pressure differences approach instrumental noise level, especially along-shore.

The role of EBPUCs in transporting (say) heat can be estimated properly only by budgeting for a closed 'box'. Even a section across a channel between two closed boundaries gives only a difference between transports in opposite directions, with no test of their reliability. Estimates from CONSLEX current meters with thermistors were very scattered. Long records are needed but also some preliminary work on how long and on the spatial representativeness of estimates from one mooring. In CONSLEX, cross-slope current variations were a minimum over the upper slope, which makes this location strategic for cross-slope fluxes.

NESHYBA. New Technology: There is no doubt in my mind that a pop-up Lagrangian drifter, preferably air-launched, is the technological key to reveal along-shore continuity within subsurface EBPUCs. This way we can obtain more answers more quickly and cost-effectively than with other technology now possible.

Old Plus New Technology: As all the above commentators point out, long-term records are needed, particularly to evaluate transports and the coherency links between EBPUC transports and other major gyre features. We might set up long-term sections in certain places where each terminus has a sea-level station, and the section is as short as need be to capture eddies, squirts and jets as well as steady flows. The section between Costa Rica and Cocos Island, about 500 km, could monitor the North Equatorial Countercurrent transport. Cocos Island has already been positioned by the Global Positioning System (GPS). When repeated enough times, this device also yields vertical leveling between section end stations. A similar setup is available between Valparaiso and J. Fernandez Island to monitor differential transport in the Humboldt Current system and the strength of the South Pacific subtropical gyre. We can now see far enough ahead technologically to plan for definitive long-term data sets needed to resolve our basic questions about EBPUCs. All we require is resolve, a green light, and about 1/1000th part of NASA's funding for outer space!

MOOERS. The Lagrangian methods, together with long-term current meter moorings on the slope, seem essential. There may be tracer methods which could be used to study the water mass continuity issues. Adroit use of satellite IR and ocean color imagery could prove useful, though indirectly, in monitoring the state of the EBPUCs. Satellite radar altimetry should help determine the offshore boundary conditions and the spatial structure of the pressure gradient field, especially on the mesoscale.

7. SHOULD WE RECOMMEND THAT AN EBPUC PANEL BE CHARTERED WITHIN THE WOCE, FOR THE PURPOSE OF DEFINING THE PRIORITY OF EBPUCs IN OCEAN SCIENCE FOR THE COMING DECADE?

Rationale: The continuity of EBPUCs can be examined as part of the division of West Wind Drifts as they impinge on eastern boundaries, at which they become an integral part of gyre behavior. Included also are the coherency analyses between the EBPUC and basin-wide phenomena, as well as between EBPUC and EBC.

Opinion: **HUTHNANCE**. EBPUCs are relevant to WOCE specifically. The WOCE Science Plan envisages that sea-level recording will carry on, after the five-year special observational period, as a continuing monitor of ocean circulation. It is therefore important that sea-level circulation relationships be understood, implicating all boundary currents across which sea-level changes occur.

Core Project 3 appears to be the relevant part of WOCE. I support the formation of an EBPUC panel or, even wider, a Lateral Boundary Conditions panel, within Core Project 3, by whatever means are appropriate, presumably through the WOCE Science Steering Committee.

HICKEY. I think WOCE may be the one chance to examine the forcing of the EBPUC relative to the Pacific basin forcing, as well as to resolve questions of continuity within these features of eastern boundary flows. The formation of an EBPUC panel in WOCE should be a high priority.

MITTELSTAEDT. Yes.

NESHYBA. EBPUCs ought to be a definitive target in WOCE gyre experiments because of their role in poleward transports of heat and water. For example, it is difficult to foresee success in budget studies of the South Pacific that do not explicitly account for EBPUC transports.

Beyond the contribution that EBPUC research holds for physical science, it is critical to success in fisheries. Eastern boundaries are sites for four out of five major coastal upwelling-fishery zones of the world ocean: (1) Oregon and vicinal states, (2) Peru-Chile, (3) Namibia-Angola, and (4) North Africa. The fifth, Somalia-Arabia, differs in that it is monsoon driven and not an eastern ocean boundary. We now know empirically that ecosystems evolve with integral dependency on a coherency with these physical systems often in the form of migration patterns of nekton. Physical scientists gain new insight when patterns of biological behavior can be blended into physics, and our results are essential to describing life cycles within eastern boundary zones. Although WOCE itself excludes biology explicitly, physical ocean. Science in general would be remiss if its results are not blended into related coastal research.

PART V

REFERENCES

Aagaard, K. and L.K. Coachman (1968). The East Greenland Current North of the Denmark Strait, II, Arctic, 21, 267-290.

Aagaard, K. and P. Greisman (1975). Toward New Mass and Heat Budgets for the Arctic Ocean, J. Geophys. Res., 80, 3821-3827.

Aagaard, K., C. Darnall, A. Foldvik, and T. Torresen (1985a). Fram Strait Current Measurements 1984-1985, Report No. 63, Department of Oceanography, Geophysical Institute, University of Bergen, Bergen, Norway.

Aagaard, K., J.H. Swift, and E. Carmack (1985b). Thermohaline Circulation in the Arctic Mediterranean Seas, J. Geophys. Res., 90 (C3), 4833-4846.

Aagaard, K., A. Foldvik, and S.R. Hillman (1987). The West Spitsbergen Current: Disposition and Water Mass Transformation. J. Geophys. Res., 92 (C4), 3778-3784.

Aas, E. (1977). The Mean Seasonal Variation in the Transport of Atlantic Water Through the Faeroe-Shetland Channel. Institute of Geophysics Report Series, No. 18, 20 pp. University of Oslo.

Ahumada, R., A. Rudolph, and V. Martinez (1983). Circulation and Fertility of Waters in Concepcion Bay. Coastal and Shelf Science, 16, 95-105.

Allain, Ch. (1970). Observations hydrologiques sur le talus du Banc d'Arguin en Decembre 1962, Rapp. Process-Verbaux Reunion ICES, 159, 86-89.

Allan, R.J. and J.I. Pariwono (1987). Aspects of Large-Scale Ocean-Atmosphere Interactions in Low-Latitude Australia. Tropical Ocean-Atmosphere Newsletter, 38, 6-10.

Allen, J.S. and R.L. Smith (1981). On the Dynamics of Wind-Driven Shelf Currents. Philosophical Transactions, the Royal Society of London, A302, 617-634.

Alvarez, L.G., R. Durazo, J. Perez, L.F. Navarro, and R. Hernandez (1984). Observaciones de corrientes costeras superficiales mediante trazadores Lagrangeanos. II. Erendira, Baja California, 1979-1981. CICESE, Dept. of Oceanography, (Unpublished Report).

Alvarez, L.G. (1986). Estudio de corriente litoral y difusion frente a la costa Noroeste de Baja California, con relacion al transporte de desechos vertidos al mar. Departamento de Oceanografia, CICESE. 96 pp. (Unpublished Report).

Alvarez, L.G., A. Badan-Dangon, and A. Valle (1988). A Note on Coastal Currents off Tehuantepec. Estuarine, Coastal, and Shelf Science in press.

Ambar, I. and M.R. Howe (1979). Observations of the Mediterranean Out-flow 1. Mixing in the Mediterranean Outflow, Deep-Sea Res., 26, 535-554.

Ambar, I. (1983). A Shallow Core of Mediterranean Water Off Western Portugal, Deep-Sea Res., 30, 677-680.

Ambar, I. (1984). Seis meses de medicoes de correntes, temperaturas e salinidades na vertente continental ao largo da costa alentejana, Relatorio tecnico, 1/84, Grupo de Oceanografia, Universidade de Lisboa, 47 pp. (Unpublished Manuscript).

Ambar, I. (1985). Seis meses de medicoes de correntes, temperaturas e salinidades na vertente continental Portuguesa a 40°N, Relatorio tecnico, 1/85, Grupo de Oceanografia, Universidade de Lisboa, 40 pp. (Unpublished Manuscript).

Ambar, I. and A. Fiuza (1987). Observations of a Poleward Current Off Portugal.

Anderson, D.L.T. and A.E. Gill (1975). Spin-up of a Stratified Ocean, with Application to Upwelling. Deep-Sea Res., 22, 583-596.

Andrews, J.C. (1977). Eddy Structure and the West Australian Current. Deep-Sea Res., 24, 1133-1148.

Andrews, J.C. (1983) Ring Structure in the Poleward Boundary Current Off Western Australia in Summer. Aust. J. Mar. Freshwat. Res., 34, 547-561.

Arakawa, A. and V.R. Lamb (1977). Computational design of the basic dynamical processes of the UCLA general circulation model. Methods in Computational Physics. (J. Chang, Ed.), Academic Press, 17, 173-265.

Arakawa, A. and M.J. Suarez (1983). Vertical differencing of the primitive equations in sigma coordinates. J. Atmos. Sci., 40, 34-45.

Arana, P. and A. Nakanishi (1971). La pesqueria del camaron nylon (Heterocarpus reedi), frente a la costa de Valparaiso. Inv. Mar. 2(4), 78-88.

Arfi, R. (1987). Variabilite interannuelle de l'hydrologie d'une region d'upwelling (Cap Blanc, Mauritanie), Oceanologica Acta, 10 (2), 151-159.

Austin, T.S. (1960). Oceanography of the East Central Equatorial Pacific as Observed During Expedition EASTROPIC. Fisheries Bulletin, 60, 257-282.

Badan-Dangon, A. (1972). Interpretacion de los datos obtenidos por medio de botellas de deriva en el sistema de la Corriente de California. Tesis de Licenciatura. UABC.Ensenada, 30 pp.

Badan-Dangon, A., K.H. Brink, and R.L. Smith (1986). On the Dynamical Structure of the Midshelf Water Column Off Northwest Africa. Continental Shelf Research, 5, 629-644.

Bainbridge, A.E. (1981). GEOSECS-Atlantic Expedition. Sections and Profiles. Vol. 2, Natl. Sci. Foundation, Washington, D.C., 198 pp..

Bakun, A. and R.H. Parrish (1980). Environmental inputs to fishery population models for eastern boundary current regions, in Intergov. Oceanogr. Comm., UNESCO, Workshop Rep. 28, G.D. Sharp, ed., 67-104.

Bakun, A. (1988). Applications of Maritime Data to the Study of Surface Forcing of Seasonal and Interannual Variability in Eastern Boundary Regions. PhD. Dissertation, Oregon State University, College of Oceanography.

Bang, N.D. (1973). Characteristics of an Intense Ocean Frontal System in the Upwell Regime West of Cape Town. Tellus, 25(3), 256-265.

Barber, R.T. and R.L. Smith (1981). Coastal Upwelling Ecosystems In: Analysis of Marine Ecosystems, A.R. Longhurst (Ed.), Academic Press, 31-68.

Barton, E.D., R.D. Pillsbury, and R.L. Smith (1975). A Compendium of Physical Observations From JOINT-I, Reference 75-17, School of Oceanography, Oregon State University, 60 pp.

Barton, E.D. and M.L. Argote (1980). Hydrographic Variability in an Upwelling Area Off Northern Baja California in June, 1976. J. Mar. Res., 38, 631-649.

Barton, E.D., J.M. Robles, A. Amador, and C. Morales (1980). A Year of Current and Temperature Observations Off Baja California North. Data Report. CICESE, Ensenada, 162 pp.

Barton, E.D. and P. Hughes (1982). Variability of Water Mass Interleaving Off NW Africa, J. Mar. Res., 40 (4), 963-984.

Barton, E.D. (1985). Low-Frequency Variability of Currents and Temperatures on the Pacific Continental Shelf Off Northern Baja, California, 1978 to 1979. Continental Shelf Research, 4, 425-443.

Barton, E.D. (1987). Meanders, Eddies, and Intrusions in the Thermohaline Front Off Northwest Africa, Oceanologica Acta, 10(3), 267-283.

Barton, E.D., This Volume. The Poleward Undercurrent on the Eastern Boundary of the Subtropical North Atlantic.

Batteen, M.L. (1988). On the use of sigma coordinates in large-scale ocean circulation models. Ocean Modelling, 77, 3-5.

Baumgartner, T.R. and N. Christensen, Jr. (1985). Coupling of the Gulf of California to Large-Scale Interannual Climatic Variability. J. Mar. Res., 43, 825-848.

Beardsley, R.C., C.E. Dorman, C.A. Friehe, L.K. Rosenfeld, and D.D. Winant (1987). Local Atmospheric Forcing During the Coastal Ocean Dynamics Experiment 1, A Description of the Marine Boundary Layer and Atmospheric Conditions Over Northern California.

Bennett, E.B. (1963). An Oceanographic Atlas of the Eastern Tropical Pacific Ocean, Based on Data from the ESTROPIC Expedition, October - December 1955. Inter-American Tropical Tuna Commission, Bulletin 8, 33-165.

Bernal, P.A. (1981). A Review of the Low-Frequency Response of the Pelagic Ecosystem in the California Current. CalCOFI Report XXII, 49-62.

Bernal, P.A. and J.A. McGowan (1981). Advection and Upwelling in the California Current. In: Coastal Upwelling, F.A. Richards (Ed.), American Geophysical Union, Washington, 381-399.

Bernstein, R.L., L.C. Breaker and R. Whritner (1977). California current eddy formation: Ship, air, and satellite results. Science, 195, 353-359.

Blackburn, M. (1962). An Oceanographic Study of the Gulf of Tehuantepec. U.S. Fish and Wildlife Service, Special Scientific Reports, Fisheries, No. 404, 28 pp.

Blanco, J. (1984). Caracteristicas de la Circulacion sobre la plataforma conteinental de Talcahuano. Thesis. Esc de Ciencias del Mar U.C.V., 41 pp.

Blumberg, A.F. and G.L. Mellor (1987). A description of a three-dimensional coastal ocean circulation model. Three-dimensional Coastal Ocean Models, (N. Heaps, Ed.), American Geophysical Union, 4, 1-16.

Bolin, B. (1983). C, N, P, and S Cycles: Major Reservoirs and Fluxes. In: The Major Biogeochemical Cycles and Their Interactions, Bolin, B. and R.B. Cook Eds., Wiley, N.Y., 41-45.

Bolin, B. and R.B. Cook (1983). The Major Biogeochemical Cycles and Their Interactions, Wiley, New York, 532 pp.

Booth, D.A. and D.J. Ellett (1983). The Scottish Continental Slope Current. Continental Shelf Research, 2, 127-146.

Booth, D.A. and D.T. Meldrum (1987a). Drifting Buoys in the Northeast Atlantic. Journal due Conseil International pour l'Exploration de la Mer, 43, 261-267.

Booth, D.A. and D.T. Meldrum (1987b). Northeast Atlantic Satellite-Tracked Buoy Drifts. Offshore Technology Report OTH 87 270, HMSO London, 95 pp.

Bourke, R.H., J.L. Newton, R.G. Paquette, and M.D. Tunnicliffe (1987). Circulation and Water Masses of the East Greenland Shelf. J. Geophys. Res., 92, (C7), 6729-6740.

Braathen, B.R. and R. Saetre (1973). Farming Salmon in Norwegian Coastal Waters. Environment and Equipment. Fisken og Havet. Ser. B. (9) (in Norwegian).

Brandhorst, W. (1963). Descripcion de las condiciones oceanograficas de las aguas costeras entre Valparaiso y el golfo de Arauco, con especial referencia al contenido de oxigeno y su relacion con la pesca. (Resultados de la expedicion AGRIMAR, 1959). Min. Agricultura Dir. Agr. y Pesca, pp. 3-55, Dantiago, Chile.

Brandhorst, W. (1971). Condiciones oceanograficas estivales frente a la costa de Chile, Rev. Bio. Mar., Valparaiso, 14(3), 45-84.

Bretherton, F.P. and M.J. Karweit (1975). Mid-Ocean Mesoscale Modeling. In: Numerical Models of the Ocean Circulation, Ocean Affairs Board, National Research Council, National Academy of Sciences, Washington, DC, 237-249.

Bretherton, F.P. and D.B. Haidvogel (1976). Two-dimensional Turbulence above Topography. J. Fluid Mech., 78, 129-154.

Brink, K.H. and J.S. Allen (1978). On the Effect of Bottom Friction on Barotropic Motion Over the Continental Shelf. J. Phys. Oceanogr., 8, 919-922.

Brink, K.H., D. Halpern, and R.L. Smith (1980). Circulation in the Peruvian Upwelling System Near 15°S. J. Geophys. Res., 85, 4036-4048.

Brink, K.H., D. Halpern, A. Huyer, and R.L. Smith (1983). The Physical Environment of the Peruvian Upwelling System. Progress in Oceanography, 12, 285-305.

Brink, K.H. and D.C. Chapman (1985). Programs for Computing Properties of Coastal-Trapped Waves and Wind-Driven Motions Over the Continental Shelf and Slope. WHOI Tech. Rep., 85-17.

Brink, K.H. (1986). Topographic Drag Due to Barotropic Flow Over the Continental Shelf and Slope. J. Phys. Oceanogr., 16, 2150-2158.

Brink, K.H. and R.D. Muench (1986). Circulation in the Point Conception-Santa Barbara Channel Region. J. Geophys. Res., 91 (C1), 877-895.

Brockmann, C., E. Fahrgach, A. Huyer and R.L. Smith (1980). The Poleward Undercurrent Along the Peru Coast: 5 to 15°S. Deep-Sea Res., 27, 847-856.

Broecker, W.S. and T.H. Peng (1982). Tracers in the Sea. Eldigio Press, New York, 690 pp.

Brown, P.C. and L. Hutchings (1985) - Phytoplankton Distribution and Dynamics in the Southern Benguela Current. In: International Symposium on the Most Important Upwelling Areas Off Western Africa (Cape Blanco and Benguela){Barcelona, 1983.} Bas, C., Margelef, R. and P. Rubias (Eds.) Barcelona, Institute de Investigaciones Pesqueras, 319-344.

Brown, W.S., J.D. Irish, and M.R. Erdman (1983). CODE-1: Bottom Pressure Observations. Chapter V, CODE Technical Report No. 21, WHOI Tech. Report, 83-23, 117-138.

Brundrit, G.B. (1984). Monthly Mean Sea Level Variability Along the West Coast of Southern Africa. S. Afr. J. Mar. Sci., 2, 195-203.

Brundrit G.B., B.A. De Cuevas, and A.M. Shipley (1987). Long-Term Sea-Level Variability in the Eastern South Atlantic and a Comparison With That in the Eastern Pacific. In: The Benguela and Comparable Ecosystems. Payne, A.I.L., J.A. Gulland, and K.H. Brink (Eds.). S. Afr. J. Mar. Sci., 5, 73-78.

Calvert, S.E. and N.B. Price (1971). Upwelling and Nutrient Regeneration in the Benguela Current, October 1968. Deep-Sea Res., 18, 505-523.

Camerlengo, A.L. and J.J. O'Brien (1980). Open boundary conditions in rotating fluids. J. Comput. Physics, 35, 12-35.

Cannon, G.A., N.P. Laird, and T.V. Ryan (1975). Flow Along the Continental Slope Off Washington, Autumn 1971. J. Mar. Res., 33, Suppl., 97-107.

Capone, D.G. and E.J. Carpenter (1982). Nitrogen Fixation in the Marine Environment. Science, 217, 1140-1142.

Carlucci, A.F. and P.M. McNally (1969). Nitrification by Marine Bacteria in Low Concentrations of Substrate and Oxygen. Limnol. Oceanogr., 14, 736-739.

Carmack, E. (1972). On the Hydrography of the Greenland Sea, PhD Thesis, University of Washington, Seattle, WA.

Carmack, E. and K. Aagaard (1973). On the Deep Water of the Greenland Sea, Deep-Sea Res., 20, 687-715.

Carnevale, G.F. and J.S. Frederiksen (1987). Nonlinear Stability and Statistical Mechanics of Flow Over Topography. J. Fluid Mech., 175, 157-181.

Carpenter, E.J. and D.G. Capone (1983). Nitrogen in the Marine Environment. Academic Press, N.Y., 900 pp.

Chapman D.C. and K.H. Brink (1987). Shelf and Slope Circulation Induced by Fluctuating Offshore Forcing. J. Geophys. Res., 92, 741-759.

Chapman, P. and L.V. Shannon (1985). The Benguela Ecosystem. II Chemistry and Related Processes. In: Oceanography and Marine Biology. An Annual Review. 23, Barnes, M. (Ed)., Aberdeen; University Press, 183-251.

Chavez, F.P. and R.T. Barber (1987). An Estimate of New Production in the Equatorial Pacific. Deep-Sea Res., 34(7), 1229-1243.

Chavez, F.P., R.T. Barber and M.P. Sanderson. The Potential Primary Production of the Peruvian Upwelling Ecosystem: 1953-1984. In: ICLARM Conference Proceedings Series. Pauly, D., H. Salzwedel, P. Muck, and J. Mendo, (Eds.), ICLARM/PROCOPA/IMARPE. In Press.

Chelton, D.B. (1981). Interannual Variability of the California Current - Physical Factors. CalCOFI Reports, XXII, 34-48.

Chelton, D.B. (1982). Large-Scale Response of the California Current. CalCOFI Report XXIII, 130-148.

Chelton, D.B., P.A. Bernal, and J.A. McGowan (1982). Large-Scale Interannual Physical and Biological Interaction in the California Current. J. Mar. Res., 40, 1095-1125.

Chelton, D.B. and R.E. Davis (1982). Monthly Mean Sea Level Variability Along the West Coast of North America. J. Phys. Oceanogr., 12, 757-784.

Chelton, D.B. (1984). Seasonal Variability of Alongshore Geostrophic Velocity Off Central California. J. Geophys. Res., 89, 3473-3486.

Chelton, D.B., R.L. Bernstein, A. Bratkovitch, and P.M. Kosro (1987). The Central California Coastal Circulation Study. EOS, 68, 1-13.

Chelton, D.B., A.W. Bratkovitch, R.L. Bernstein, and P.M. Kosro (1988). Poleward Flow off Central California During the Spring and Summer of 1981 and 1984. J. Geophys. Res., 93(10), 604-10,620.

Christensen, N., Jr. and N. Rodriguez (1979). A Study of the Sea Level Variations and Currents Off Baja California. J. Phys. Oceanogr., 9, 631-638.

Christensen, N., Jr., R. de la Paz, and G. Gutierrez (1983). A Study of Sub-Inertial Waves Off the West Coast of Mexico. Deep-Sea Res., 30, 835-850.

Clarke, A.J. (1977). Observational and Numerical Evidence for Wind-Forced Coastal-Trapped Long Waves. J. Phys. Oceanogr., 7, 231-247.

Clarke, A.J. (1982). The Dynamics of Large-Scale, Wind-Driven Variations in the Antarctic Circumpolar Current. J. Phys. Oceanogr., 12, 1092-1105.

Clarke, A.J. and S. Van Gorder (1986). A Method for Estimating Wind-Driven Frictional, Time-Dependent, Stratified Shelf and Slope Water Flow. J. Phys. Oceanogr., 16, 1013-1028.

Coachman, L.K. and K. Aagaard (1974). Physical Oceanography of Arctic and Subarctic Seas, In: Marine Geology and Oceanography of the Arctic Seas, Y. Herman (Ed.), Springer-Verlag, New York, 1-72.

Codispoti, L.A. and F.A. Richards (1976). An Analysis of the Horizontal Regime of Denitrification in the Eastern Tropical North Pacific. Limnol. Oceanogr., 21, 379-383.

Codispoti, L.A. and T.T. Packard (1980). On the Denitrification Rate in the Eastern Tropical South Pacific. J. Mar. Res., 38, 453-477.

Codispoti, L.A. (1981). Temporal Nutrient Variability in Three Different Upwelling Regions. In: Coastal Upwelling, F.A. Richards (Ed.), American Geophysical Union, Washington, 209-220.

Codispoti, L.A., R.C. Dugdale, and H.J. Minas (1982). A Comparison of the Nutrient Regimes Off Northwest Africa, Peru, and Baja, California. Rapports et Proces-Verbaux Reunions, International Council for the Exploration of the Sea, 180, 184-201.

Codispoti, L.A. (1983). Nitrogen in Upwelling System. In: Nitrogen in the Marine Environment, E. Carpenter and D. Capone (Eds.), Academic Press, N.Y.

Codispoti, L.A. and J.C. Christensen (1985). Nitrification, Denitrification and Nitrous Oxide Cycling in the Eastern Tropical South Pacific Ocean. Mar. Chem., 16, 277-300.

Codispoti, L.A., G.E. Friederich, T.T. Packard, H.E. Glover, P.J. Kelley, R.W. Spinrad, R.T. Barbver, J.W. Elkins, B.B. Ward, F. Lipschultz, and N. Lostaunau (1986). High Nitrite Levels Off Northern Peru: A Signal of Instability in the Marine Denitrification Rate. Science, 233, 1200-1202.

Codispoti, L.A., G.E. Friederich, T.T. Packard, and R.T. Barber (1988). Remotely Driven Thermocline Oscillations and Denitrification in the Eastern South Pacific: The Potential for High Denitrification Rates During Weak Coastal Upwelling. Science of the Total Environment, 75, 301-318.

Cohen, Y. and L.I. Gordon (1978). Nitrous Oxide in the Oxygen Minimum of the Eastern Tropical North Pacific. Deep-Sea Res., 25, 509-524.

Colborn, J.G. (1975). The Thermal Structure of the Indian Ocean. International Indian Ocean Expedition Oceanographic Monograph No. 2.

Connary, S.D. (1972). Investigations of the Walvis Ridge and Environs. Ph.D. Thesis, Columbia University, 228 pp.

Crawford, W.R. and T.R. Osborn (1980). Energetics of the Atlantic Equatorial Undercurrent. Deep-Sea Res., 26, Suppl. II, 309-323.

Cresswell, G.R. (1977). The Trapping of Two Drifting Buoys by an Ocean Eddy. Deep-Sea Res., 24, 1203-1209.

Cresswell, G.R. and T.J. Golding (1980). Observations of a South Flowing Current in the Southeastern Indian Ocean. Deep-Sea Res., 27A, 449-466.

Cresswell, G.R. (1986). The Role of the Leeuwin Current in the Life Cycles of Several Marine Creatures. Proceedings International Conference on Pelagic Biogeography, Amsterdam, 1985.

Cromwell, T. and E.B. Bennett (1959). Surface drift Charts for the Eastern Tropical Pacific. Inter-American Tropical Tuna Commission Bulletin, 3, 217-237.

Crowe, F.J. and R.S. Schwartzlose (1972). Release and Recovery Records of Drift Bottles in the California Current Region 1955 through 1971. CalCOFI Atlas No. 16, 140 pp.

Crutzen, P.J. (1981). Atmospheric Chemical Processes of the Oxides of Nitrogen, Including Nitrous Oxide. In: Denitrification Nitrification and Atmospheric Nitrous Oxide, C.C. Delwiche (Ed.), Wiley, N.Y., 17-44.

Crutzen, P.J. (1983). Atmospheric Interactions - Homogeneous Gas Reactions of C, N, and S Containing Compounds. In: The Major Biogeochemical Cycles and Their Interactions, B. Bolin and R.B. Cook (Eds.), 67-112.

Csanady, G.T. (1978). The Arrested Topographic Wave. J. Phys. Oceanogr., 8, 47-62.

Csanady, G.T. (1980). Longshore Pressure Gradients Caused by Offshore Winds. J. Geophys. Res., 85, 1076-1084.

Dakin, W.J. (1919). The Percy Sladen Trust Expeditions to the Abrolhos Islands (Indian Ocean). Report I. Journal of the Linnean Society, 34, 127-180.

Davis, R.E. (1985). Drifter Observations of Coastal Surface Currents During CODE: The Method and Descriptive View. J. Geophys. Res., 90, 4741-4755.

De Decker, A.H.B. (1970). Notes on an Oxygen-depleted Subsurface Current Off the West Coast of South Africa. Investl. Rep. Div. Sea Fish. S. Afr., 84, 24 pp.

Defant, A. (1941). Die Relative Topographie Einzelner Druckflachen im Atlantischen Ozean. "Meteor" Rep., 6(2:4), Teil., X-XVIII, 183-190 +.

Defant, A. (1941). Die absolute Topographie des physikalischen Meeresniveaus und der Druckflachen im Atlantischen Ozean, Wiss. Ergebn. Dt. Atlant. Exped. "Meteor" 1925-7, VI, 2, 191-260.

Delwiche, C.C. (1981). The Nitrogen Cycle and Nitrous Oxide. In: Denitrification, Nitrification and Nitrous Oxide, C.C. Delwiche (Ed.), Wiley, N.Y., 1-16.

Denbo, D.W. and J.S. Allen (1983). Mean Flow Generation on a Continental Margin by a Periodic Wind Forcing. J. Phys. Oceanogr., 13, 78-92.

Denbo, D.W., K. Polzin, J.S. Allen, A. Huyer, and R.L. Smith (1984). Current Meter Observations Over the Continental Shelf Off Oregon and California, February 1981-January 1984. College of Oceanography, Oregon State University, Data Rep. 112, Ref. 84-12.

Devol. A.H. (1978). Bacterial Oxygen Uptake Kinetics as Related to Biological Processes in Oxygen Deficient Zones of the Oceans. Deep-Sea Res., 25, 137-146.

Dickson, R.R. (1972). Variability and Continuity within the Atlantic Current of the Norwegian Sea, Rapp. Cons. Explor. Mer., 162, 167-183.

Dietrich, G., K. Kalle, W. Drauss, and G. Siedler (1980). General Oceanography: An Introduction. InterScience.

Dooley, H.D. and J. Meincke (1981). Circulation and Water Masses in the Faroese Channels During Overflow '73. Deutsche Hydrographische Zeitschrift, 34, 41-54.

Dugdale, R.C. and J.J. Goering (1967). Uptake of New and Regenerated Forms of Nitrogen in Primary Productivity. Limnol. Oceanogr., 12, 196-206.

Dunlop, J.N. and R.N. Wooller (1986). Range Extensions and the Breeding Seasons of Seabirds in Southwestern Australia. Rec. West. Aust. Mus., 12, 389-394.

Elkins, J.W., S.C. Wofsy, M.B. McElroy, C.E. Kolb and W.A. Kaplan (1978). Aquatic Sources and Sinks for Nitrous Oxide. Nature, 275, 602-606.

Ellett, D.J. and D.A. Booth (1983). Some Oceanographic Applications of ODAS Data. COST-43 Technical Document 100, 83-97.

Ellett, D.J., A. Edwards and R. Bowers (1986). The Hydrography of the Rockall Channel--An Overview. Proceedings of the Royal Society of Edinburgh, 88B, 61-81.

Embley, R.W. and J.J. Morley (1980). Quaternary Sedimentation and Paleoenvironmental Studies Off Namibia (South-West Africa). Mar. Geol., 36, 183-204.

Emery, W.J. and L.A. Mysak (1980). Dynamical Interpretations of Satellite-sensed Thermal Features Off Vancouver Island. J. Phys. Oceanogra., 10, 961-970.

Emery, W.J., W.G. Lee and L. Magaard (1984). Geographic and seasonal distributions of Brunt-Vaisala frequency and Rossby radii in the North Pacific and North Atlantic. J. Phys. Oceanogr., 14, 294-317.

Enfield, D.B., R.L. Smith and A. Huyer (1978). A Compilation of Observations from Moored Current Meters. Data Report 70, Oregon State University.

Enfield, D.B. and J.S. Allen (1980). On the Structure and Dynamics of Monthly Mean Sea Level Anomalies Along the Pacific Coast of North and South America. J. Phys. Oceanogr., 10, 557-578.

Enfield, D.B. and J.S. Allen (1983). The Generation and Propagation of Sea Level Variability Along the Pacific Coast of Mexico. J. Phys. Oceanogr., 13, 1012-1033.

Espinoza, R., S. Neshyba and Z. Maoxiang (1983). Surface Water Motion off Chile with Satellite Images of Surface Chlorophyll and Temperature. In: Recursos Marinos del Pacifico. Arana (Ed): 41-57.

Fahrbach, E. (1976). Einige Beobachtungen zur Erzeugung und Ausbreitung interner Gezeitenwellen am Kontinentalabhang vor Sierra Leone, "Meteor" Forsch.-Ergebn., A, 18, 64-77.

Fahrbach, E., C. Brockmann, N. Lostaunau and W. Urquizo (1981). The Northern Peruvian Upwelling System During the ESACAN Experiment. In: Coastal Upwelling, F.A. Richards (Ed.), Am. Geophys. Union, Washington, D.C., 134-145.

Feliks, Y. (1985). Notes and Correspondence on the Rossby Radius of Deformation in the Ocean. J. Phys. Oceanogr., 15, 1605-1607.

Fiuza, A. (1982). The Portuguese Coastal Upwelling System, In: Present Problems of Oceanography in Portugal, Junta Nacional de Investigacao Cientifica e Tecnologica, Lisbon, 45-71 (Unpublished Manuscript).

Fiuza, A. and D. Halpern (1982). Hydrographic Observations of the Canary Current between 21°N and 25.5°N During March-April 1974, Rapp. Proces-Verbaux Reunion ICES, 180.

Flament, P., L. Armi and L. Washburn (1985). The Evolving Structure of an Upwelling Filament. J. Geophys. Res., 90, 11765-11778.

Fonseca, T.R. (1977). Proceso de surgencia en Punta Curaumilla con especial referencia a la Circulacion-Thesis. Esc. de Pesquerias y Alimentos U.C.V., 90 pp.

Fonseca, T. (1985). Fisica de las aguas costeras de la zona Central de Chile. Tralka Vol. (4), 337-354.

Fraga, F. (1974). Distribution des masses d'eau dans l'upwelling de Mauretanie, Tethys, 6, (2), 5-10.

Freeland, H.J. and K.L. Denman (1982). A Topographically Controlled Upwelling Center of Southern Vancouver Island. J. Mar. Res., 40(4), 1069-1093.

Freeland, H.J. (1983). Low Frequency Currents Observed Off Southern Vancouver Island, 1979-1981. Can. Data Rep. Hydrogr. Ocean Sci., 7, 80 pp.

Freeland, H.J., W.R. Crawford, and R.E. Thomson (1984). Currents along the Pacific Coast of Canada. Atmos.-Oc., 22(2), 151-172.

Freitag H.P. and D. Halpern (1981). Hydrographic Observations Off Northern California During May 1977. J. Geophys. Res., 86, 4248-4252.

Friederich, G.E. and L.A. Codispoti (1981). The Effects of Mixing and Regeneration on the Nutrient Content of Upwelling Waters Off Peru. In: Coastal Upwelling, F.A. Richards (Ed.), Am. Geophys. Union, Washington, D.C. 221-227.

Friederich, G.E., P.M. Kelley, L.A. Codispoti, R.W. Spinrad, G. Kullenberg, J.W. Elkins, J. Kogelschatz, T.T. Packard, F. Lipschultz, H.E. Glover, A.E. Smith, and B.B. Ward (1985). Microbial Nitrogen Transformations in the Oxygen Minimum Zone Off Peru. NITROP-85 Data Report, R/V Wecoma Cruise, Feb 1-Mar 5, 1985, Bigelow Lab Tech. Rep. No. 59.

Friederich, G.E., P.J. Kelly, and L.A. Codispoti (1986). An Inexpensive Moored Water Sampler for Investigating Chemical Variability. In: Tidal Mixing and Phytoplankton Dynamics, M. Bowman, C.M. Yentsch, and W.T. Peterson, (Eds.), Springer-Verlag, N.Y.

Friederich, G.E. and L.A. Codispoti (1987). An Analysis of Continuous Vertical Nutrient Profiles Taken During a Cold-Anomaly Off Peru. Deep-Sea Res., 34(5/6), 1049-1065.

Garcia, J. (1983). Variaciones Hidrograficas y eventos de surgencia frente a Punta Colonet, B.C., en junio de 1980. Tesis de licenciatura. UABC. 116 pp.

Gardner, D. (1977). Nutrients as Tracers of Water Mass Structure in the Coastal Upwelling of Northwest Africa, In: A Voyage of Discovery, M. Angel (Ed.), Supplement to Deep-Sea Res., 24, 305-325.

Gentilli, J. (1972). Thermal Anomalies in the Eastern Indian Ocean. Nature, 238, Physical Sciences, 93-95.

Gill A.E. and E.H. Schumann (1974). The Generation of Long Shelf Waves by the Wind. J. Phys. Oceanogr., 4, (1), 83-90.

Gill, A.E. (1982). Atmosphere-Ocean Dynamics, International Geophysics Series, 30, Academic Press, 662 pp.

Gladfelter, W.H. (1964). Oceanography of the Greenland Sea, USS ATKA (AGB-3) Survey Summer 1962, Informal Report 0-64-63, U.S. Naval Oceanographic Office, Washington, D.C., 154 pp.

Godfrey, J.S. and K.R. Ridgway (1985). The Large-Scale Environment of the Poleward-Flowing Leeuwin Current, Western Australia: Longshore Steric Height Gradients, Wind Stresses, and Geostrophic Flow. J. Phys. Oceanogr., 15, 481-495.

Godfrey, J.S., D.J. Vaudrey, and S.D. Hahn (1986). Observations of the Shelf-Edge Current South of Australia, Winter 1982. J. Phys. Oceanogr., 16, 668-679.

Gomez Valdez, J. (1984). Estructura hidrografica promedio frente a Baja California. Ciencias Marinas, 9, 75-85.

Gordon, R.L. (1978). Internal Wave Climate Near the Coast of Northwest Africa During JOINT-1. Deep-Sea Res., 25, 625-643.

Goreau, T.J., W.A. Kaplan, J.C. Wofsy, M.B. McElroy, F.W. Valois, and S.W. Watson (1980). Production of NO2 and N2O by Nitrifying Bacteria at Reduced Concentrations of Oxygen. Appl. Environ. Microbiol., 40, 526-532.

Gould, W.J., J. Loynes, and J. Backhaus (1985). Seasonality in Slope Current Transports NW of Shetland. ICES Hydrography Committee, C.M. 1985/C:7.

Gregg, M.C. and C.S. Cox (1972). The Vertical Microstructure of Temperature and Salinity. Deep-Sea Res., 19, 355-376.

Griesman, P. (1976). Current Measurements in the Eastern Greenland Sea, PhD Thesis, University of Washington, Seattle, WA.

Greisman, P. and K. Aagaard (1979). Seasonal Variability of the West Spitsbergen Current, Ocean Modelling, 19, 3-5.

Griffiths, R.C. (1965). A Study of Ocean Fronts Off Cape San Lucas, Lower, California. U.S. Dept. of the Interior. Special Scientific Report. Fisheries. No. 499.

Griffiths, R.C. (1968). Physical, Chemical, and Biological Oceanography of the Entrance to the Gulf of California, Spring 1960. Special Scientific Report. U.S. Fish and Wildlife Service. Fisheries, 573, 47 pp.

Griffiths, R.W. and A.F. Pearce (1985). Instability and Eddy Pairs on the Leeuwin Current South of Australia. Deep-Sea Res., 32, 1511-1534.

Gunn, J.T., P. Hamilton, H.J. Herring, L.H. Kantha, G.S.E. Lagerloef, G.L. Mellor, R.D. Muench, and G.R. Stegan (1986). Santa Barbara Channel Circulation Model and Field Study, Appendix C: Observational Data. Report Number 92 (C2), Dynalysis of Princeton, 221 pp.

Gunther, E.R. (1936). A Report on Oceanographical Investigations in the Peru Coastal Current. Discovery Rep., 13, 107-276.

Hafferty, A.J., L.A. Codispoti, and A. Huyer (1978). JOINT-II R/V Melville Legs I, II, and IV, R/V Iselin Leg II Bottle Data, March 1977-May 1977. Coastal Upwelling Ecosystem Analysis Data Rep. 45, 779 pp.

Hafferty, A.J., D. Lowman, and L.A. Codispoti (1979). JOINT-II Melville and Iselin Bottle Data Sections, March-May 1977. Coastal Upwelling Ecosystems Analysis Data Rep. 38, 130 pp.

Hahn, J. (1981). Nitrous Oxide in the Oceans. In: Denitrification, Nitrification, and Nitrous Oxide, C.C. Delwiche (Ed.), Wiley, N.Y., 191-240.

Haidvogel, D.B. and K.H. Brink (1986). Mean Currents Driven by Topographic Drag over the Continental Shelf and Slope. J. Phys. Oceanogr., 16, 2159-2172.

Halpern, D. (1983). Special Issue III; Update 1982/83 Equatorial Pacific Warm Event. Tropical Ocean-Atmosphere Newsletter, 21, 1-34.

Halpern, D., R.L. Smith, and R.K. Reed (1978). On the California Undercurrent Over the Continental Slope Off Oregon. J. Geophys. Res., 83, 1366-1372.

Hamann, I., H.-Ch. John, and E. Mittelstaedt (1981). Hydrography and its Effect on Fish Larvae in the Mauretanian Upwelling Area, Deep-Sea Res., 28A, 561-575.

Hamilton, P. and M. Rattray, Jr. (1978). A Numerical Model of the Depth-Dependent, Wind-Driven Upwelling Circulation on a Continental Shelf. J. Phys. Oceanogr., 8, 437-457.

Hamon, B.V. (1965). Geostrophic Currents in the South-Eastern Indian Ocean. Aust. J. Mar. Freshwat. Res., 16, 255-271.

Haney, R.L. (1985). Midlatitude Sea Surface Temperature Anomalies: A Numerical Hindcast. J. Phys. Oceanogr., 15, 787-799.

Hanzlick, D.J. (1983). The West Spitsbergen Current: Transport, Forcing, and Variability, PhD Thesis, University of Washington, Seattle, WA.

Harris, T.F.W. and L.V. Shannon (1979). Satellite-Tracked Drifter in the Benguela Current System. S. Afr. J. Sci., 75(7), 316-317.

Hart, T.J. and R.I. Currie (1960). The Benguela Current. Discovery Reports, 31, 123-182.

Hastenrath, S. and P.J. Lamb (1977). Climatic Atlas of the Tropical Atlantic and Eastern Pacific Oceans. University of Wisconsin Press, Madison.

Hattori, A. (1983). Denitrification and Dissimilatory Nitrate Reduction. In: Nitrogen in the Marine Environment, E. Carpenter and D. Capone (Eds.), Academic Press, N.Y., 191-232.

Helland-Hansen, B. and F. Nansen (1909). The Norwegian Sea, Its Physical Oceanography. Rep. Norweg. Fish. Invest., 2.

Hickey, B.M. (1979). The California Current System-Hypotheses and Facts. Prog. Oceanogr., 8, 191-279.

Hickey, B.M. and Hamilton (1980). A Spin-Up Model as a Diagnostic Tool for Interpretation of Current and Density Measurements on the Continental Shelf of the Pacific Northwest. J. Phys. Oceanogr., 10, 12-24.

Hickey, B.M. (1981). Along-Shore Coherence on the Pacific Northwest Continental Shelf (January-April 1975). J. Phys. Oceanogr., 11, 822-835.

Hickey, B.M. and N. Pola (1983). The Seasonal Along-Shore Pressure Gradient on the West Coast of the United States. J. Geophys. Res., 88, 7823-7633.

Hickey, B.M. (1984). The Fluctuating Longshore Pressure Gradient on the Pacific Northwest Shelf: A Dynamical Analysis. J. Phys. Oceanogr., 14, 276-293.

Hickey, B.M. (1988). Poleward Flow Near the Northern and Southern Boundaries of the U.S. West Coast. This Volume.

Hickey, B.M. (1989a). Patterns and Processes of Circulation In and Over the Continental Shelf Off Washington. In: Coastal Dynamics of the Pacific Northwest. M. Landry and B. Hickey, Wiley.

Hickey, B.M. (1989b). Seasonal Variability Over the Santa Monica/San Pedro Basin and Shelf. J. Geophys. Res. (to be submitted).

Hill, J.W. and A.J. Lee (1957). The Effect of Wind on Water Transport in the Region of the Bear Island Fishery, Proc. Roy. Soc., B, 148, 104-116.

Holden, C.J. (1987). Observations of Low-frequency Currents and Continental Shelf Waves Along the West Coast of South Africa. In: The Benguela and Comparable Ecosystems. Payne, A.I.L., J.A. Gulland and K.H. Brink (Eds). S. Afr. J. Mar. Sci., 5, 197-208.

Holloway, G. and M.C. Hendershott (1974). The Effects of Bottom Relief on Barotropic Eddy Fields. MODE Hotline News, No. 65 (Unpublished Manuscript).

Holloway, G. (1986). A Shelf Wave/Topographic Pump Drives Mean Coastal Circulation. Parts I and II. Ocean Modeling, No. 68 and 69 (Unpublished Manuscripts).

Holloway, G. (1987). Systematic Forcing of Large Scale Geophysical Flows by Eddy-Topographic Interaction. J. Fluid Mech. (To Appear).

Holloway, G., K. Brink and D. Haidvogel (1988). Topographic Stress in Coastal Circulation Dynamics. (This volume.)

Holloway, P.E. and H.C. Nye (1985). Leeuwin Current and Wind Distributions on the Southern Part of the Australian North West Shelf between January 1982 and July 1983. Aust. J. Mar. Freshwat. Res., 36, 123-137.

Horn, W. and J. Meincke (1976). Note on the Tidal Current Field in the Continental Slope Area Off Northwest Africa. Memoires de la Societe' Royale des Sciences Liege, 6(10), 31-42.

Houghton, R. (1976). Circulation and Hydrographic Structure Over the Ghana Continental Shelf During the 1974 Upwelling, J. Phys. Oceanogr., 6, 909-924.

Hsiung, J. (1985). Estimates of the Global Oceanic Meridional Heat Transport. J. Phys. Oceanogr., 15, 1405-1413.

Huber, K., E. Mittelstaedt and G. Weichart (1977). Zur Hydrographie der Gewasser vor Marokko. Physikalische und chemische Daten. Meereskundliche Beobachtungen und Ergebnisse, Nr. 46, Deutches Hydrographisches Institut, Hamburg, 131 pp.

Hughes, P. and E.D. Barton (1974). Stratification and Water Mass Structure in the Upwelling Area Off NW Africa in April/May 1969, Deep-Sea Res., 21, 611-628.

Hurlbcurt, H.E. and J.D. Thompson (1973). Coastal Upwelling on a B-Plane. J. Phys. Oceanogr., 3, 16-32.

Huthnance, J.M. (1984). Slope Currents and "JEBAR." J. Phys. Oceanogr., 14, 795-816.

Huthnance, J.M. (1986). The Rockall Slope Current and Shelf-Edge Processes. Proceedings of the Royal Society of Edinburgh, 88B, 83-101.

Huthnance, J.M. (1987). Along-Shelf Evolution and Sea Levels Across the Continental Slope. Continental Shelf Research, (To Appear).

Huyer, A., B.M. Hickey, J.D. Smith, R.L. Smith and R.D. Pillsbury (1975a). Along-Shore Coherence at Low Frequencies in Currents Observed Over the Continental Shelf Off Oregon and Washington. J. Geophys. Res., 80 (24), 3495-3505.

Huyer, A., R.D. Pillsbury and R.L. Smith (1975b). Seasonal Variation of the Along-Shore Velocity Field Over the Continental Shelf Off Oregon. Limnol. Oceanogr., 20(1), 90-95.

Huyer, A. (1976). A Comparison of Upwelling Events in Two Locations: Oregon and Northwest Africa, J. Mar. Res., 34, 531-546.

Huyer, A. and R.L. Smith (1976). Observations of a Poleward Undercurrent Over the Continental Slope Off Oregon May-June 1975. Transactions of the Amer. Geophys. Union, 57, 263 pp.

Huyer, A. (1977). Seasonal Variation in Temperature, Salinity, and Density Continental Shelf Off Oregon. Limnol. Oceanogr., 22, 442-453.

Huyer, A., R.L. Smith and E.J.C. Sobey (1978). Seasonal Differences in Low-Frequency Current Fluctuations Over the Oregon Continental Shelf. J. Geophys. Res., 83 (C10), 5077-5089.

Huyer, A., E.J.C. Sobey and R.L. Smith (1979). The Spring Transition in Currents Over the Oregon Continental Shelf. J. Geophys. Res., 84, 6995.

Huyer, A. (1983). Upwelling in the California Current System. Prog. in Oceanog., 12, 259-284.

Huyer, A., R.L. Smith and B.M. Hickey (1984). Observations of a Warm-Core Eddy Off Oregon, January to March 1978. Deep Sea Res., 31 (2), 97-117.

Huyer, A. and R.L. Smith (1985). The Signature of El Nino Off Oregon, 1982-83. J. Geophys. Res., 90, 7133-7142.

Huyer, A. and P.M. Kosro (1987). Mesoscale Surveys Over the Shelf and Slope in the Upwelling Region Near Point Arena, California. J. Geophys. Res., 92, 1655-1681.

Huyer, A., P.M. Kosro, S.J. Lentz and R.C. Beardsley, This Volume. Poleward Flow in the California Current System.

Ikeda, M. and W.J. Emery (1984). Satellite Observation and Modeling of Meanders in the California Current System Off Oregon and Northern California. J. Phys. Oceanogr., 7, 1434-1450.

Inostroza, H. (1972). Some Oceanographic Features of Northern Chilean Waters in July, 1962. In: Oceanography of the South Pacific, New Zealand, 37-46.

Johannessen, O.M., J.A. Johannessen, J. Morison, B.A. Farrelly and E.A.S. Svendsen (1983). Oceanographic Conditions in the Marginal Ice Zone North of Svalbard in Early Fall 1979 with an Emphasis on Mesoscale Processes, J. Geophys. Res., 88 (C5), 2755-276

Johnson, D. (1980). The Gunther Undercurrent and Upwelling Along the North and Central Coast of Chile (Unpublished Manuscript).

Johnson, D., T. Fonseca and H. Sievers (1980). Upwelling in the Humboldt Coastal Current Near Valparaiso, Chile. J. of Marine Res., 38(1), 1-16.

Johnson, D.R., E.D. Barton, P. Hughes and C.N.K. Mooers (1975). Circulation in the Canary Current Region Off Cabo Bojador in August, 1972, Deep-Sea Res., 22, 547-557.

Johnson, D.R. and C.N.K. Mooers (1981). Internal Cross-Shelf Flow Reversals During Coastal Upwelling. In: Coastal Upwelling, F.A. Richards, (Ed.), American Geophysical Union, Washington, D.C.

Johnson, E.R. (1977). Stratified Taylor Columns on a Beta-plane. Geophys. Astrophys. Fluid Dyn., 9, 159-177.

Jones, E.L. (1918). The Neglected Waters of the Pacific Coast. Department of Commerce, U.S. Coast and Geodetic Survey Special Publication No. 48, Washington, D.C.

Jury, M. (1985). Case Studies of Along-Shore Variations in Wind-Driven Upwelling in the Southern Region. In: South Africa Ocean Colour and Upwelling Experiment. L. Shannon (Ed): 29-47.

Kamstra F. (1985). Environmental Features of the Southern Benguela With Special Reference to the Wind Stress. In: South African Ocean Colour and Upwelling Experiment. L.V. Shannon (Ed). Sea Fisheries Research Institute, Cape Town, p. 13-27.

Kamstra F. (1987). Interannual Variability in the Spectra of the Daily Surface Pressure at Four Stations in the Southern Hemisphere. Tellus 39A, 509-514.

Keeling, C.D. and R.B. Bacastow (1977). Impact of Industrial Gases on Climate, Energy and Climate. Geophysics Research Board, National Academy of Science National Research Council, Washington, D.C. 110-160.

Kislyakov, A.G. (1960). Fluctuations in the Regime of the Spitsbergen Current, Translation: Soviet Fisheries Investigations in North European Seas, 39-49.

Kitani, K. (1977). The Movements and Physical Characteristics of the Water Off Western Australia in November 1975. Bulletin of Far Seas Fisheries Research Laboratory, 15, 13-19.

Knowles, R. (1982). Denitrification. Microbiol. Rev., 46, 43-70.

Knox, F. and M.B. McElroy (1984). Changes in Atmospheric CO2: Influence of the Marine Biota at High Latitude. J. Geophys. Res., 89, 4629-4637.

Kosro, P.M. (1987). Structure of the Coastal Current Field off Northern California During the Coastal Ocean Dynamics Experiment. J. Geophys. Res., 92, 1637-1654.

Kremling, K. and A. Wenck (1984). Chemical Data From the NW African Upwelling Region. Berichte aus dem Institut fur Meereskunde, Kiel, Nr. 124, 130 pp.

Krey, J. and B. Babenerd (1976). Phytoplankton Production: Atlas of the International Indian Ocean Expedition. (Kiel: Institut fur Meereskunde).

Kundu, P.J. and J.P. McCreary (1987). A Model of the Throughflow from the Pacific to the Indian Ocean. J. Mar. Res., 45, In press.

Kundu, P.K. and J.S. Allen (1976). Some Three-Dimensional Characteristics of Low-Frequency Current Fluctuations Near the Oregon Coast. J. Phys. Oceanogr., 181-199.

Lacis, A., J. Hansen, P. Lee, T. Mitchell and S. Lebedeff (1981). Greenhouse Effect of Trace Gases, 1970-1980. Geophys. Res. Lett., 8, 1035-1038.

Lamberth, R. and G. Nelson (1987). Field and Analytical Drogue Studies Applicable to the St. Helena Bay Area Off South Africa's West Coast. In: The Benguela and Comparable Ecosystems. Payne, A.I.L., Gulland, J.A. and K.H. Brink (Eds). S. Afr. J. Mar. Sci.

LeFloch, J. (1973). Hydrologie et dynamique de Secteur ouest Marocain. Resultat de la campagne CINECA-Charcot I, Publications du CNEXO, Nr. 06, 27-63.

LeFloch, J. (1974). Quelques aspects de la dynamique et de l'hydrologie des couches superficielles dans l'ouest Marocain Campagnes CINECA-I et III. Tethys. 6, 53-68.

Legeckis, R. and G.R. Cresswell (1981). Satellite Observations of Sea Surface Temperature Fronts Off the Coast of Western and Southern Australia. Deep-Sea Res., 28A, 297-306.

Lenz, J., G. Schneider, M. Elbrachter, P. Fritsche, H. Johannsen and T. Weibe (1985). Hydrographic, chemical and planktological data from the Northwest African Upwelling area obtained from February to April 1983. Berichte aus dem Institut fur Meereskunde, Kiel, Nr. 140, 105 pp.

Lentz, S.J. (1987). A Description of the 1981 and 1982 Spring Transitions Over the Northern California Shelf. J. Geophys. Res., 92, 1545-1567.

Lipschultz, F., O.C. Zafiriou, S.C. Wofsy, M.B. McElroy, F.W. Valois and S.W. Watson (1981). Production of NO and N2O by Soil Nitrifying Bacteria. Nature (London), 294, 642-643.

Liu, K.-K. (1979) Geochemistry of Inorganic Nitrogen Compounds in Two Marine Environments: The Santa Barbara Basin and the Ocean Off Peru. Ph.D. Thesis, University of California, Los Angeles, 354 pp.

Liu, K.-K and I.R. Kaplan (1982). Nitrous Oxide in the Sea Off Southern California. In: The Environment of the Deep Sea, W.G. Ernst and J.G. Marin (Eds.), Prentice Hall, N.Y., 74-92.

Loder, J. (1980). Topographic Rectification of Tidal Currents on the Sides of Georges Bank. J. Phys. Oceanogr., 10, 1399-1416.

Love, C.M. (1970-1974). EASTROPAC Atlas, Cruise Data February, 1967 to March 1968. Circular 330, Volumes 1-11, U.S. Dept. Commerce, Washington, D.C.

Lueck, R.G., W.R. Crawford and T.R. Osborn, (1983). Turbulent Dissipation Over the Continental Slope Off Vancouver Island. J. Phys. Oceanogr., 13, 1809-1818.

Lueck, R.G. and T.R. Osborn, (1986). The Dissipation of Kinetic Energy in a Warm-Core Ring. J. Geophys. Res., 91, 803-818.

Lukas, R. (1986). The Termination of the Equatorial Undercurrent in the Eastern Pacific. Prog. Oceanogr., 16, 63-90.

Lynn, R.J. (1983). The 1982-83 Warm Episode in the California Current. Geophys. Res. Lett., 10, 1093-1095.

Lynn, R.J. and J.J. Simpson (1987). The California Current System: The Seasonal Variability of its Physical Characteristics. J. Geophys. Res., 92, 12,947-12,966.

Madelain, F. (1970). Influence de la topographie du fond sur l'ecoulement Mediterraneen entre le detroit de Gibraltar et le Cap Saint-Vincent, Cahiers Oceanographiques, 22, 43-61.

Margalef, R. (1978). What is an Upwelling Ecosystem? In: Upwelling Ecosystems, R.Boje and M. Tomczak (Eds.), Springer-Verlag, N.Y., 12-14.

Markina, N.P. (1976). Biogeographic Regionalization of Australian Waters of the Indian Ocean. Oceanology, 15, 602-604.

Marmen, H.A. (1926). Coastal Currents Along the Pacific Coast of the United States. Spec. Publ. U.S. Cst. Geodetic. Surv., 121, 91 pp.

Martell, C.M. and J.S. Allen (1979). The Generation of Continental Shelf Waves by Along-Shore Variations in Bottom Topography. J. Phys. Oceanogr., 9, 696-711.

Maxwell, J.G.H. and G.R. Cresswell (1981). Dispersal of Tropical Marine Fauna to the Great Australian Bight by the Leeuwin Current. Aust. J. Mar. Freshwat. Res., 32, 493-500.

McCartney, M.S. and L.D. Talley (1984). Warm-to-Cold Water Conversion in the Northern North Atlantic Ocean. J. Phys. Oceanogr., 14, 922-935.

McCreary, J.P. (1976). Eastern Tropical Ocean Response to Changing Wind Systems, with Application to El Nino. J. Phys. Oceanogr., 6, 632-645.

McCreary, J.P. (1981). A Linear Stratified Ocean Model of the Coastal Undercurrent. Philosophical Transactions, the Royal Society of London, A302: 385-413.

McCreary, J.P., Jr. and S.F. Chao (1985). Three-Dimensional Shelf Circulation Along an Eastern Ocean Boundary. J. Mar. Res., 43, 13-36.

McCreary, J.P., S.R. Shetye and P.J. Kundu (1986). Thermohaline Forcing of Eastern Boundary Currents: With Application to the Circulation Off the West Coast of Australia. J. Mar. Res., 44, 71-92.

McCreary, J.P., Jr., P.K. Kundu and S-Y. Chao (1987). On the Dynamics of the California Current System. J. Mar. Res., 45, 1-32.

McElroy, M.B. (1983). Marine Biological Controls on Atmospheric CO2 Climate. Nature, 302, 328-329.

McLain, D. and D. Thomas (1983). Year-to-Year Fluctuations of the California Counter Current and Effects on Marine Organisms. CalCOFI Rep., Vol. XXIV.

McLain, D.R. and D.H. Thomas (1983). Year-to-Year Fluctuations of the California Countercurrent and Effects on Marine Organisms. CalCOFI Reports, XXIV, 165-181.

McNider, R.T. and J.J. O'Brien (1973). A Multi-Layer Transient Model of Coastal Upwelling. J. Phys. Oceanogr., 3, 258-273.

Meincke, J., G. Siedler and W. Zenk (1975). Some Current Observations Near the Continental Slope Off Portugal, "Meteor" Forsch.-Ergebn., A, 16, 15-22.

Meincke, J. (1978). On the Distribution of Low Salinity Intermediate Waters Around the Faeroes. Dt. Hydrogr. Z., 31, 50-64.

Mejia, A. (1985). Efecto de una cuenca de forma arbitraria en la propagacion de una onda de Kelvin costera. MS Thesis, CICESE, Ensenada, 82 pp.

Midtun, L. (1985). Formation of Dense Bottom Water in the Barents Sea. Deep Sea Res., 32, 1233-1241.

Mittelstaedt, E. (1972). Der hydrographische Aufbau und die zeitliche variabilitat der Schichtung und Stromung im nordwest Afrikanischen Auftriebsgebeit im Frujahr 1968, "Meteor" Forsch.-Ergebn., A, 11, 1-57.

Mittelstaedt, E., R.D. Pillsbury and R.L. Smith (1975). Flow Patterns in the Northwest African Upwelling Area, Dt. Hydrogr. Z., 28, 145-167.

Mittelstaedt, E. (1976). On the Currents Along the Northwest African Coast South of 22°N, Dt. Hydrog. Z., 29, 97-117.

Mittelstaedt, E., D. Halpern, R.L. Smith and J.S. Bottero (1980). Tidal Currents Off Manritania at 21°40'N. Dt. Hydrogr. Z., 33(6), 223-235.

Mittelstaedt, E. and I. Hamann (1981). The Coastal Circulation Off Mauretania, Dt. Hydrogr. Z., 34, 81-118.

Mittelstaedt, E. (1983). The Upwelling Area Off Northwest Africa--A Description of Phenomena Related to Coastal Upwelling, Prog. Oceanogr., 12, 307-331.

Mittelstaedt, E., This volume. On the Subsurface Circulation Along the Moroccan Slope.

Montgomery, R.B. (1938). Circulation in Upper Layers of the Southern North Atlantic Deduced With Use of Isentropic Analysis, Pap. Phys. Oc. Met. MIT, 6, 2.

Montgomery, R.B. (1941). Transport of the Florida Current Off Habana. J. Mar. Res., 4, 198-220.

Mooers, C.N.K. (1975). Several Effects of a Baroclinic Current on the Cross-Stream Propagation of Inertial-Internal Waves. Geophys. Fluid Dyn., 6, 245-275.

Mooers, C.N.K., C.A. Collins and R.L. Smith (1976). The Dynamic Structure of the Frontal Zone in the Coastal Upwelling Region Off Oregon. J. Phys. Oceanogr., 6, 3-21.

Mooers, C.N.K. and A.R. Robinson (1984). Turbulent Jets and Eddies in the California Current and Inferred Cross-Shore Transports. Science, 223, 51-53.

Morales, C., M.L. Argote, A. Amador and E.D. Barton (1978). Mediciones de vientos, corrientes e hidrografia frente a Punta Colonet, B.C. en junio de 1976. Data Report, CICESE, Ensenada. 236 pp.

Munk, W.H. (1950). On the Wind-Driven Ocean Circulation. Journal of Meteorology, 7(2), 79-93.

Nederlandsch Meteorologisch Institut (1949). Seas Around Australia: Oceanographic and Meteorological Data. Koninklijk Nederlands Meteorologisch Institut Publication No. 124, 79 pp. De Bilt.

Nelson C.S. (1977). Wind Stress and Wind Stress Curl Over the California Current. NOAA Technical Report NMFS SSRF-714, U.S. Department of Commerce, Washington, D.C., 89 pp.

Nelson, G. and L. Hutchings (1983). The Benguela Upwelling Area. Prog. in Oceanog., 12(3), 333-356.

Nelson, G. (1985). Notes on the Physical Oceanography of the Cape Peninsula Upwelling System. In: South African Ocean Colour and Upwelling Experiment. Shannon, L.V. (Ed.). Cape Town; Sea Fisheries Research Institute, 63-95.

Nelson G. and A. Polito (1987). Information on Currents in the Cape Peninsula Area, South Africa. In: The Benguela and Comparable Ecosystems. Payne, A.I.L., Gulland, J.A. and K.H. Brink (Eds.). S. Afr. J. Mar. Sci., 5, 287-304.

Neshyba, S. and R. Mendez (1976). Analisis de temperaturas superficiales del mar como indicadores de movimientos de aguas superficiales en el Pacifico Sur-Este. Rev. Com. Perm. Pacifico sur, 5: 139-146.

Newton, C.W. (1971). Global Angular Momentum Balance: Earth Torques and Atmospheric Fluxes. J. Atmos. Sci., 28, 1329-1341.

O'Brien, J.J. and H.E. Hurlburt (1972). A Numerical Model of Coastal Upwelling. J. Phys. Oceanogr., 2, 14-26.

Osborn, T.R. and C.S. Cox, (1972). Oceanic Fine Structure. Geophys. Fluid Dyn., 3, 321-345.

Pace, M.L., G.A. Knauer, D.M. Karl and J.H. Martin (1987). Primary Production, New Production and Vertical Flux in the Eastern Pacific Ocean. Nature, (London), 325, 803-804.

Packard, T.T., P.C. Garfield and L.A. Codispoti (1983). Oxygen Consumption and Denitrification Below the Peruvian Upwelling. In: Coastal Upwelling, Pt. A., E. Suess and J. Theide (Eds.), Plenum Press, N.Y., 147-173.

Pak, H., L.A. Codispoti and R.V. Zaneveld (1980). On the Intermediate Particle Maxima Associated with Oxygen-Poor Water Off Western South America. Deep-Sea Res., 27A, 783-797.

Paquette, R.G., R.H. Bourke, J.L. Newton and W.F. Perdue (1985). The East Greenland Polar Front in Autumn, J. Geophys. Res., 90 (C3), 4866-4882.

Parrish, R.H., A. Bakun, D.M. Husby and C.S. Nelson (1983). Comparative Climatology of Selected Environmental Processes in Relation to Eastern Boundary Current Fish Reproduction. In: Proceedings of the Expert Consultation to Examine Changes in Abundance and Species Composition of Neritic Fish Resources. San Jose, Costa Rica, April 1983. Sharp G. D. and J. Csirke (Eds). F.A.O. Fish Rep. 291 (3), 731-777.

Patzert, W.C., T.J. Cowles and C.S. Ramage (1978). El Nino Watch Atlas of Physical, Chemical, and Biological Oceanographic and Meteorological Data. Ref. Ser. No. 78-7, Scripps Inst. Oceanography, 322 pp.

Pearce, A.F. and B.F. Phillips (1988). ENSO Events, the Leeuwin Current and Larval Recruitment of the Western Rock Lobster. J. Du Conseil., accepted.

Pedlosky, J. (1974). Longshore Currents and the Onset of Upwelling Over Bottom Slope. J. Phys. Oceanogr., 4, 310-320.

Peffley, M.B. and J.J. O'Brien (1976). A Three-Dimensional Simulation of Coastal Upwelling Off Oregon. J. Phys. Oceanogr., 6, 164-180.

Pelaez, J. and J.A. McGowan (1986). Phytoplankton Pigment Patterns in the California Current as Determined by Satellite. Limnol. Oceanogr., 31, 927-950.

Perkin, R.G. and E.L. Lewis (1984). Mixing in the West Spitsbergen Current. J. Phys. Oceanogr., 14, 1315-1325.

Philander, S.G.H. and J-H. Yoon (1982). Eastern Boundary Currents and Coastal Upwelling. J. Phys. Oceanogr., 12, 862-897.

Pierotti, D. and R.A. Rasmussen (1980). Nitrous Oxide Measurements in the Eastern Tropical Pacific Ocean. Tellus, 32, 56-72.

Piper, D.Z. and L.A. Codispoti (1975). Marine Phosphorite Deposits and the Nitrogen Cycle. Science, 188, 15-18.

Poulain, P.M., J.D. Illeman and P.P. Niiler (1987). Drifter Observations in the California Current System (1985-1986). SIO Ref. 87-27, 72 pp.

Preston-Whyte R.A. and P.D. Tyson (1973). Note on Pressure Oscillations Over South Africa. Mon. Wea. Rev. 101, 650-653.

Puls, C. (1895). Oberflachentemperaturen und stromungsverhaltnisse des Aequatorialgurtels des Stillen Ozeans. Doctoral Dissertation, Marburg.

Quadfasel D., J.-C. Gascard and K.-P. Koltermann (1987). Large-Scale Oceanography in Fram Strait During the 1984 Marginal Ice Zone Experiment. J. Geophys. Res., 92 (C7), 6719-6728.

Redfield, A.C., B.H. Ketchum and F.A. Richards (1963). The Influence of Organisms on the Composition of Seawater. In: M.N. Hill (Ed.), The Sea, Vol. 2, Interscience, N.Y., 26-77.

Reed, R.K. and D. Halpern, (1976). Observations of the California Undercurrent Off Washington and Vancouver Island. Limnol. Oceanogr., 21, 389-398.

Reid, B.J. (1987). The Fall Transition of Oregon Shelf Waters. M.S. Thesis, College of Oceanography, Oregon State University, Corvallis, Oregon, 103 pp.

Reid, J.L., G.I. Roden and J.G. Wyllie (1958). Studies of the California Current System. Calif. Coop. Oceanic Ris. Invest. Rep., July 1, 1956-January 1, 1958, 28-56.

Reid, J.L., Jr., G.I. Roden and J.G. Wyllie (1958). Studies of the California Current System. CalCOFI Prog. Report 1, 1956 - 1. 29-57.

Reid, J.L., Jr. (1960). Oceanography of the Northeastern Pacific During the Last Ten Years. CalCOFI Prog. Report 7, 77-90.

Reid, J.L., Jr. (1961). On the Geostrophic Flow at the Surface of the Pacific Ocean with Respect to the 1000-Decibar Surface. Tellus, 13, 489-502.

Reid, J.L., Jr. (1962). Measurements of the California Countercurrent at a Depth of 250 Meters. J. Mar. Res., 20, 134-137.

Reid, J.L., Jr. and R.A. Schwartzlose (1962). Direct Measurements of the Davidson Current Off Central California. J. Geophys. Res., 67, 2491-2497.

Reid, J.L., Jr. (1963). Measurement of the California Countercurrent Off Baja California. J. Geophys. Res., 68, 4819-4822.

Reid, J.L., Jr., R.A. Schwartzlose, and D.M. Brown (1963). Direct Measurements of a Small Surface Eddy Off Northern Baja California. J. Mar. Res., 21, 205-218.

Reid, J.L., Jr. and A.W. Mantyla (1976). The Effect of the Geostrophic Flow Upon Coastal Elevations in the Northern Pacific Ocean. J. Geophys. Res., 81, 3100-3110.

Reid, J.L., Jr. and A.W. Mantyla (1978). On the Mid-Depth Circulation of the North Pacific Ocean. J. Phys. Oceanogr., 8, 946-951.

Reid, R.O. (1948). The Equatorial Currents of the Eastern Pacific as Maintained by the Stress of the Wind. J. Mar. Res., 7, 74-99.

Ridgway, K.R. and R.G. Loch (1987). Mean Temperature-Salinity Relationships in Australian Waters and Their Use Is Water Mass Analysis. Aust. J. Mar. Freshwat. Res., 38, 553-567.

Rienecker, M.M., C.N.K. Mooers, D.E. Hagan and A.R. Robinson (1985). A Cool Anomaly Off Northern California: An Investigation Using IR Imagery and in situ Data. J. Geophys. Res., 90, 4807-4818.

Rienecker, M.M. and C.N.K. Mooers (1989a). A Summary of the OPTOMA Program's Mesoscale Ocean Prediction Studies in the California Current System. In: Three Dimensional Ocean Models of Marine and Estuarine Dynamics (J.C.J. Nihoul and B.M. Jamant, Eds.). Elsevier Scientific Publishing, Amsterdam. In press.

Rienecker, M.M. and C.N.K. Mooers (1989b). Mesoscale Eddies, Jets and Fronts off Pt. Arena, CA, July 1986. J. Geophys. Res. In press.

Robinson, A.R., J.A. Carton, C.N.K. Mooers, L.J. Walstad, E.F. Carter, M.M. Rienecker, J.A. Smith and W.G. Leslie (1984). A Real-time Dynamical Forecast of Ocean Synoptic/Mesoscale Eddies. Nature, 309, 781-783.

Robles, F. (1976). Descripcion general de las condiciones oceanograficas en Aguas Chileans. Instituto de Fomento Pesquero. Santiago de Chile, 90 pp.

Robles, J.M., C. Morales, J. Garcia and C. Flores (1981). Informe de datos hidrograficos de la region de Cabo Colonet, B.C. Crucero Subac 2, Julio de 1980. CICESE, 165 pp.

Rochford, D.J. (1969a). Seasonal Variations in the Indian Ocean along 110°E.I Hydrological Structure of the Upper 500 m. Aust. J. Mar. Freshwat. Res., 20, 1-50.

Rochford, D.J. (1969b). Seasonal Interchange of High and Low Salinity Surface Waters Off South-West Australia. Division of Fisheries and Oceanography Technical Paper No. 29.

Roden, G.I. (1962). Oceanographical Aspects of Eastern Equatorial Pacific. Geofisica Internacional, 2, 77-92.

Roden, G.I. (1970). Aspects of the Mid-Pacific Transition Zone. J. Geophys. Res., 75, 1097-1109.

Roden, G.I. (1971). Aspects of the Transition Zone in the Northeastern Pacific. J. Geophys. Res., 76, 3462-3475.

Roden, G.I. (1972). Thermohaline Structure and Baroclinic Flow Across the Gulf of California Entrance and in the Revilla Gigedo Islands Region. J. Phys. Oceanogr., 2, 177-183.

Roemmich, D. and C. Wunsch (1985). Two Translantic Sections: Meridional Circulation and Heat Flux in the Subtropical North Atlantic Ocean. Deep Sea Res., 32(6), 619-664.

Rogers, J. (1987). Seismic, Bathymetric and Photographic Evidence of Widespread Erosion and a Manganese-Nodule Pavement Along the Continental Rise of the Southeast Cape Basin. Marine Geol., 78, 57-76.

Rojas, O., A. Mujica, G. Ledermann and H. Miles (1983). Estimacion de abundancia relativa de huevos y larvas de peces. CORFO, Gerencia de Desarrollo (AP 83-31). IFOP, Santiago, 98 pp.

Salmon, R., G. Holloway and M.C. Hendershott (1976). The Equilibrium Statistical Mechanics of Simple Quasi-geostrophic Models. J. Fluid Mech., 75, 375-386.

Samelson, R.M. and J.S. Allen (1987). Quasi-geostrophic Topographically Generated Mean Flow over the Continental Margin. J. Phys. Oceanogr., 17, 2043-2064.

Sarmiento, J.L. and J.R. Toggweiler (1984). A New Model for the Role of the Oceans in Determining Atmospheric pCO2. Nature (London), 308, 621-624.

Saville-Kent, W. (1897). The Naturalist in Australia. Chapman & Hall, London, 338 pp.

Schemainda, R., D. Nehrin, S. Schulz (1975). Ozeanologische Untersuchungen zum Produktionspotential der nordwestafrikanischen Wasserauftriebsregion 1970-1973. Geod. Geoph. Veroff R IV, 16, 1-85.

Schwartzlose, R.A. (1963). Nearshore Currents of the Western United States and Baja California as Measured by Drift Bottles. CalCOFI. Reports, 9, 15-22.

Schwartzlose, R.A. and J.L. Reid, (1972). Near-shore Circulation in the California Current. CalCOFI Rept., 16, 57-65.

Send, U., R.C. Beardsley and C.D. Winant (1987). Relaxation from Upwelling in the Coastal Ocean Dynamics Experiment. J. Geophys. Res., 92, 1683-1698.

Shannon, L.V. (1985). The Benguela Ecosystem I. Evolution of the Benguela, Physical Features and Processes. In: Oceanography and Marine Biology. An Annual Review 23. Barnes, M. (Ed.). Aberdeen; University Press, 105-182.

Shannon, L.V. N.M. Walters and S.A. Mostert (1985). Satellite Observations of Surface Temperature and Near-surface Chloroplhyll in the Southern Benguela Region. In: South African Ocean Colour and Upwelling Experiment. Shannon L.V. (Ed.). Sea Fisheries Research Institute, Cape Town, 183-210.

Shannon, L.V. and D. Hunter (1988). Notes on Antarctic Intermediate Water Around Southern Africa. S. Afr. J. Mar. Sci., 6, 107-117.

Shelton, P.A. and L. Hutchings (1982). Transport of Anchovy, Engraulis Capensis Gilchrist, Eggs and Early Larvae by a Frontal Jet Current. J. Cons. Perm. Int. Explor. Mer. 40 (2), 185-198.

Sievers, S., H.A. and N. Silva (1975). Masas de agua y circulacion en el Oceano Pacifico Sudoriental. Latitudes 18°S to 33°S (Operacion Oceanograficas "MARCHILE VIII"). Cienc. y tec. del Mar. Contrib. CONA No 1, 7-67.

Silva, N. and T. Fonseca (1983). Geostrophic Component of the Oceanic Flow Off Northern Chile. Conferencia Internacional Sobre Recursos Marinos del Pacifico. P. Arana (Ed.), 59-70.

Silva, S., N. and S. Neshyba (1979). On the Southernmost Extension of the Peru-Chile Undercurrent. Deep-Sea Res., 26, 1387-1393.

Simons, T.J. (1983). Resonant Topographic Response of Nearshore Currents to Wind Forcing. J. Phys. Oceanogr., 13, 512-523.

Simpson, J.J. (1983). Large-Scale Thermal Anomalies in the California Current During the 1982-1983 El Nino. Geophys. Res. Lett., 10, 937-940.

Simpson, J.J. (1984). El Nino-Induced Onshore Transport in the California Current During 1982-1983. Geophys. Res. Lett., 11, 241-242.

Smith, J.D., B.M. Hickey and J. Beck (1976). Observations from Moored Current Meters on the Washington Continental Shelf from February 1971 to February 1974. Special Report Number 65, University of Washington, Department of Oceanography, 347 pp.

Smith, R.L. (1968). Upwelling. Ocean. Mar. Biol. Ann. Rev., 6, 11-46.

Smith, R.L. (1981). A Comparison of the Structure and Variability of the Flow-Field in Three Coastal Upwelling Regions: Oregon, Northwest Africa, and Peru. In: Coastal Upwelling, F.A. Richard (Ed.), American Geophysical Union, Washington, D.C., 107-118

Smith, R.L. (1983). Peru Coastal Currents During El Nino: 1976 and 1982. Science, 221, 1397-1399.

Sorokin, Y.I. (1978). Description of Primary Production and of the Heterotrophic Microplankton in the Peruvian Upwelling Region. Oceanology, 18, 62-71.

Stabeno, P.J. and R.L. Smith (1987). Deep-Sea Currents Off Northern California, J. Geophys. Res., 92, 755-771.

Stevenson, M.R., J.G. Pattullo and B. Wyatt (1969). Subsurface Currents Off the Oregon Coast as Measured by Parachute Drogues. Deep-Sea Res., 16, 449-461.

Stevenson, M.R., R.W. Garvine and B. Wyatt (1974). Lagrangian Measurements in a Coastal Upwelling Zone Off Oregon. J. Phys. Oceanogr., 4, 321-336.

Stommel, H., E.D. Stroup, J.L. Reid and B.A. Warren (1973). Transpacific Hydrographic Sections at Lats 43°S and 281°S: The Scorpio Expedition..I. Preface. Deep-Sea Res., 20, 1-7.

Stramma, L. (1974). Geostrophic Transport in the Warm Water Sphere of the Eastern Subtropical North Atlantic, J. Mar. Res., 42, 537-558.

Stramma, L. (1984a). Geostrophic Transport in the Warm Water Sphere of the Subtropical North Atlantic. J. Mar. Res. 42, 537-558.

Stramma, L. (1984b). Potential Vorticity and Volume Transport in the Eastern North Atlantic From Two Long CTD Sections. Deutche Hydrographische Zeitschrift, 37, (4), 147-155.

Strub, P.T., J.S. Allen, A. Huyer, R.L. Smith and R.C. Beardsley, (1987a). Seasonal Cycles of Currents, Temperatures, Winds, and Sea Level Over the Northeast Pacific Continental Shelf. J. Geophys. Res., 92, 1507-1526.

Strub, P.T., J.S. Allen, A. Huyer and R.L. Smith (1987b). Large-Scale Structure of the Spring Transition in the Coastal Ocean Off Western North America. J. Geophys. Res., 92, 1527-1544.

Stumpf, H.G. and R. Legeckis (1977). Satellite Observations Mesoscale Eddy Dynamics in the Eastern Tropical Pacific Ocean. J. Phys. Oceanogr., 7, 648-658.

Sturges, W. (1974). Sea Level Slope Along Continental Boundaries. J. Geophys. Res., 79, 825-830.

Suess, E. (1980). Particulate Organic Carbon Flux in the Oceans: Surface Productivity and Oxygen Utilization. Nature (London), 288, 260-263.

Suginohara, N. (1974). Onset of Coastal Upwelling in a Two Layer Ocean by Wind-Stress with Longshore Variation. J. Oceanogr. Soc. Japan, 30, 23-33.

Suginohara, N. (1982). Coastal Upwelling: Onshore-Offshore Circulation, Equatorward Coastal Jet and Poleward Undercurrent over a Continental Shelf Slope. J. Phys. Oceanogr., 12, 272-284.

Suginohara, N. and Y. Kitamura (1984). Long-Term Coastal Upwelling over Continental Shelf-Slope. J. Phys. Oceanogr., 14, 1095-1104.

Sverdrup, H.U., M.W. Johnsen and R.H. Fleming (1942). The Oceans. Prentice-Hall.

Sverdrup, H.V. and R.H. Fleming (1941). The Waters Off the Coast of Southern California, March to July, 1937. Bulletin, Scripps Institution of Oceanography, 4, 261-378.

Swallow, J.C., W.J. Gould and P.M. Saunders (1977). Evidence for a Poleward Eastern Boundary Current in the North Atlantic Ocean, Int. Counc. Explor. Sea, C.M. 1977/C:32, Hydrogr. Comm., 11 pp.

Swift, J.H. (1986). The Arctic Waters, In: The Nordic Seas, ed. B.G. Hurdle, Springer-Verlag, New York, 129-153.

Thompson, R.O.R.Y. and G. Veronis (1983). Poleward Boundary Current Off Western Australia. Aust. J. Mar. Freshwat. Res., 34, 173-185.

Thompson, R.O.R.Y. (1984). Observations of the Leeuwin Current Off Western Australia. J. Phys. Oceanogr., 14, 623-628.

Thompson, R.O.R.Y. (1987). Continental-Shelf-Scale Model of the Leeuwin Current. J. Mar. Res., 45, 813-827.

Thomson, R.E. (1984). A Cyclonic Eddy Over the Continental Margin off Vancouver Island: Evidence for Baroclinic Instability. J. Phys. Oceanogr., 14, 1326-1348.

Thomson, R.E., W.R. Crawford, H.J. Freeland and W.S. Huggett (1985). Low-Pass Filtered Current Meter Records for the West Coast of Vancouver Island: Coastal Oceanic Dynamics Experiment, 1979-81. Can. Data Rep. Hydrogr. Ocean Sci., 40, 102 pp.

Thomson, R.E. and R.E. Wilson (1987). Coastal Countercurrent and Mesoscale Eddy Formation by Tidal Rectification Near an Oceanic Cape. J. Phys. Oceanogr., 17, 2096-2126.

Thorpe, S.A. (1976). Variability of the Mediterranean Undercurrent in the Gulf of Cadiz, Deep-Sea Res., 23, 711-727.

Tibby, R.B. (1941). The Water Masses Off the West Coast of North America. J. Mar. Res., 4, 112-121.

Timofeyev, V.T. (1962). The Movement of Atlantic Water and Heat into the Arctic Basin, Deep-Sea Res., 9, 358-361.

Tomczak, M. (1973). An Investigation into the Occurrence and Development of Cold Water Patches in the Upwelling Region of NW Africa, "Meteor" Forsch.-Ergebn., A, 13, 1-42.

Tomczak, M. 1979. The CINECA Experience. Mar. Policy 3, 59-65.

Tomczak, M. and P. Hughes (1980). Three Dimensional Variability of Water Masses and Currents in the Canary Current Upwelling Region, "Meteor" Forsch.-Ergebn., A, 21, 1-24.

Tomczak, M. (1981). Coastal Upwelling Systems and Eastern Boundary Currents: A Review of Terminology. Geoforum, 12(2), 179-191.

Tont, S.A. (1981). Upwelling: Effects on Air Temperatures and Solar Irradiance. In: Coastal Upwelling, F.A. Richards (Ed.). American Geophysical Union, Washington: 57-62.

Torres Moye, G. and Acosta Ruiz, M.J. (1986). Some Chemical Properties Indicating Coastal Upwelling Events and Subsurface Countercurrent in an Area Punta Colonet, Baja California. Ciencias Marina, 12, 10-25.

Tsuchiya, M. (1974). Variation of the Surface Geostrophic Flow in the Eastern Intertropical Pacific Ocean. Fishery Bulletin, 72, 1075-1086.

Tsuchiya, M. (1975). Subsurface Countercurrents in the Eastern Equatorial Pacific Ocean. J. Mar. Res., 33, (Suppl.), 145-175.

Tsuchiya, M. (1980). Inshore Circulation in the Southern California Bight, 1974-1977. Deep-Sea Res., 27, 99-118.

Tsuchiya, M. 1981. The Origin of the Pacific Equatorial 130°C Water. J. Phys. Oceanogr., 11, 755-770.

Tsuchiya, M. (1983). Use of Nutrient Concentrations for Monitoring the Equatorial Undercurrent. Trop. Ocean-Atmos. Newsl., 20, 3-5.

Tsunogai, S. (1971). Ammonia in the Oceanic Atmosphere and the Cycle of Nitrogen Compounds Through the Atmosphere and the Hydrosphere. Geochem. J., 5, 57-67.

Tucholke, B.E. and R.W. Embley (1984). Cenozoic Regional Erosion of the Abyssal Sea Floor Off South Africa. Memoir American Association. Petroleum Geologists, 36, 145-164.

Tyson, P.D. (1986). Climatic Change and Variability in Southern Africa. Oxford University Press, 220 pp.

Ulrych, T.J., D.E. Smylie, O.G. Jensen and G.K.C. Clarke (1973). Predictive Filtering and Smoothing of Short Records by Using Maximum Entropy. J. Geophys. Res., 78, 4959-4964.

Urrutia-Fucugauchi, J. (1988). Is Mexico Part of North America? EOS, 69, 610.

Wang, D.-P. (1982). Development of a Three-Dimensional, Limited-Area (Island) Shelf Circulation Model. J. Phys. Oceanogr., 12, 605-617.

Warren, B.A. (1981). Transindian Hydrographic Section at Lat. 18°S: Property Distributions and Circulation in the South Indian Ocean. Deep-Sea Res., 28A, 759-788.

Wattenberg, H. (1939). Die Verteilung des Sauerstoffs im Atlantischen Ozean. Atlas. Wissenschaftliche Ergebnisse der Deutschen Atlantischen Exp. "Meteor" 1925-1927, IX.

Webster, I., T.J. Golding and N. Dyson (1979). Hydrological Features of the Near-Shelf Waters Off Fremantle, Western Australia, During 1974. CSIRO Australia Division of Fisheries and Oceanography Report No. 106, 30 pp.

Weichart, G. (1974). Meereschemische Untersuchungen im nordwestafrikanischen Auftriebsgebiet 1968. "Meteor" Forsch.-Ergebn, A, No. 14, 33-70.

Weikert, H. (1986). Occurrence of Calanoides curinatus (Copepoda, Calanoida) Along the Continental Slope Off Morocco-Evidence of a Large-Scale Transport by a Northward Boundary Flow. (Unpublished).

Weiss, R.F. (1981). The Temporal and Spatial Distribution of Trophospheric Nitrous Oxide. J. Geophys. Res., 86, 7185-7195.

Wells, F.E. (1985). Zoogeographical Importance of Tropical Marine Mollusc Species at Rottnest Island, Western Australia. W.A. Naturaliste, 16, 40-45.

Werner, F. and B.M. Hickey (1983). The Role of a Longshore Pressure Gradient in Pacific Northwest Dynamics. J. Phys. Oceanogr., 13, 395-410.

Wickham, J.B. (1975). Observations of the California Countercurrent. J. Mar. Res., 33, 325-340.

Wickham, J.B., A.A. Bird and C.N.K. Mooers (1987). Mean and Variable Flow Over the Central California Continental Margin, 1978 to 1980. Continental Shelf Res., 7, 827-849.

Winant, C.D. (1980). Downwelling over the Southern California Shelf During the Summer. J. Phys. Oceanogr., 10, 791-799.

Winant, C.D. and A.W. Bratkovitch (1981). Temperature and Currents on the Southern California Shelf: A Description of the variability. J. Phys. Oceanogr., 11, 71-86.

Winant, C.D. (1983). Longshore Coherence of Currents on the Southern California Shelf During the Summer. J. Phys. Oceanogr., 13, 54-64.

Winant, C.D., R.C. Beardsley and R.E. Davis (1987). Moored Wind Temperature, and Current Observations Made During Coastal Ocean Dynamics Experiments 1 and 2 over the Northern California Continental Shelf and Upper Slope. J. Geophys. Res., 92, 1569-1604.

Wolf, G., Kaiser, W. (1978). Uber den Jahreszyklus der T-S-Eigenschaften quasipermanenter Wasserarten und Variationen produktionsbiologischer Parameter auf dem Schelf vor Cap Blanc. Geod. Geoph. Veroff, R. IV, 24, 1-81.

Wood, E.J.F. (1954). Dinoflagellates in the Australian Region. Aust. J. Mar. Freshwat. Res., 5, 171-351.

Wooster, W. and M. Gilmartin (1961). The Peru-Chile Undercurrent. J. Mar. Res., 19(3): 97-122.

Wooster, W.S. and J.L. Reid, Jr. (1963). Eastern Boundary Currents, p. 253-280. In: The Sea, Vol. II, M.N. Hill, (Ed.). Interscience Publ., New York, 554 pp.

Wooster, W.S., T.J. Chow and I. Barrett (1965). Nitrite Distribution in the Peru Current Waters. J. Mar. Res., 23, 210-221.

Wooster, W.S. (1970). Eastern Boundary Currents in the South Pacific. In: Scientific Exploration of the South Pacific, W.S. Wooster, (Ed.), National Academy of Sciences, Washington, D.C.

Wooster, W.S. and J.H. Jones (1970). California Undercurrent Off Northern Baja, California. J. Marine Res., 28, 235-250.

Wooster, W., A. Bakun and D.R. McClain (1976). The Seasonal Upwelling Cycle Along the Eastern Boundary of the North Atlantic, J. Mar. Res., 34, 131-141.

Worthington, L.V. (1976). On the North Atlantic Circulation. The Johns Hopkins Oceanographic Studies, No. 6, 110 pp.

Wright, D.G. (1980). On the stability of a fluid with specialized density stratification. Part II. Mixed baroclinic-barotropic instability with application to the Northeast Pacific. J. Phys. Oceanogr., 10, 1307-1322.

Wust, G. (1935). Die Stratosphare des Atlantischen Ozeans. Wissenschaftliche Ergebnisse der Deutschen Atlantischen Expedition "Meteor", 1925-1927, VI, 1, 288 pp.

Wust, G. (1957). Quantitative Untersuchungen zur Statik und Dynamik des Atlantischen Ozeans. Stromgeschwindigkeiten und Strommengen in den Tiefen des Atlantischen Ozeans. Wissenschaftliche Ergebnisse der Deutschen Atlantischen Expedition "Meteor", 1925-1927, VI, 2(6), 420 pp.

Wyllie, J.G. (1966). Geostrophic Flow of the California Current at the Surface and at 200 meters. California Cooperative Oceanic Fisheries Investigations Atlas 4. State of California Marine Research Committee.

Wyrtki, K. (1963). The Horizontal and Vertical Field of Motion in the Peru Current. Bull. Scripps Inst. Oceanogr., 8, 313-346.

Wyrtki, K. (1964). The Thermal Structure of the Eastern Pacific Ocean. Deutsche Hydrographische Zeitschrift. Erganzungsheft, A(6), 84 pp.

Wyrtki, K. (1964). Upwelling in the Costa Rica Dome. U.S. Fish and Wildlife Service, Fisheries Bulletin, 63, 355-372.

Wyrtki, K. (1965). Summary of the Physical Oceanography of the Eastern Pacific Ocean. Institute of Marine Resources, University of California, San Diego, Ref. 65-10. 78 pp.

Wyrtki, K. (1966). Oceanography of the Eastern Equatorial Pacific Ocean. Oceanogr. Mar. Biol. Ann. Rev., 4, 33-68.

Wyrtki, K. (1967). Circulation and Water Masses in the Eastern Pacific Ocean. Journal of Oceanology and Limnology, 1, 117-147.

Wyrtki, K. (1977). Advection in the Peru Current as Observed by Satellite. J. Geophys. Res., 82, 3939-3944.

Yamazaki, H. and R.G. Lueck (1987). Turbulence in the California Undercurrent. J. Phys. Oceanogr., 17, 1378-1396.

Yoon, J.H. and S.G.H. Philander (1982). The Generation of Coastal Undercurrents. J. Ocean. Soc. Japan, 38, 215-224.

Yoshida, K. and H.L. Mao (1957). A Theory of Upwelling of Large Horizontal Extend. J. Marine Res., 16, 40-57.

Yoshida, K. and M. Tsuchiya (1957). Northward Flow in Lower Layers as Indicator of Coastal Upwelling. Records of Oceanographic Works in Japan, 4, 14-22.

Yoshida, K. (1967). Circulation in the Eastern Tropical Oceans With Special Reference to Undercurrents and Upwelling. Japan Journal of Geophysics, 4,1-75.

Yoshinari, T. (1976). Nitrous Oxide in the Sea. Mar. Chem., 4, 189-202.

Zafiriou, O.C. and M.C. True (1979a). Nitrite photolysis in seawater by sunlight. Mar. Chem., 8, 9-32.

Zafiriou, O.C. and M.C. True (1979b). Nitrite Photolysis in Seawater by Sunlight. Mar. Chem., 8, 33-42.

Zenk, W. (1975). On the Mediterranean Outflow West of Gibraltar, "Meteor" Forsch.-Ergebn., A, 16, 23-34.

Zuta, S. and O. Guillen (1970). Oceanografia de las aquas costeras del Peru. Bol. Inst. Mar. Peru, 2, 157-324.

APPENDIX A: MAILING AND PARTICIPANT LIST FOR THE WORKSHOP ON POLEWARD FLOWS ALONG EASTERN OCEAN BOUNDARIES

Monterey, California, December 1986
[*] indicates attendance at the workshop.

David Adamec
Inst. for Naval Oceanography
Bldg. 1103, Rm. 233
Stennis Space Center, MS
39529-5005

Dr. John Allen
College of Oceanography
Oregon State University
Corvallis, OR 97331

W.R.H. Andrews
Wayside
Main Road
Betley, Crewe
Cheshire CW3 9AD, UNITED KINGDOM

Andrew Bakun
Pacific Environmental Group
P.O. Box 831
Monterey, CA 93942

Dr. Richard T. Barber
Monterey Bay Aquarium Res. Ctr.
160 Central Avenue
Pacific Grove, CA 93950

Dr. E.D. Barton *
Dept. of Physical Oceanography
Marine Science Labs.
Menai Bridge
Anglesey
Gwynedd, UNITED KINGDOM

Mary Batteen *
Naval Postgraduate School
Monterey, CA 93940

Dr. Robert Beardsley
Woods Hole Oceanographic Inst.
Woods Hole, MA 02543

Dr. David Behringer
AOML-NOAA
15 Rickenbacker Cswy.
Miami, FL 33149

Dr. Patricio Bernal
Universidad Catolica de Chile
Sede Regional de Talcahuano
4 Casilla 127
Talcahuano, CHILE

Arlene Bird *
Naval Postgraduate School
Dept. Ocean Code 68
Monterey, CA 93940

Dr. Robert Bourke *
Naval Postgraduate School
Monterey, CA 93940

Dr. Alan Bratkovich *
Dept. of Geological Sciences
USC - University Park
Los Angeles, CA 90089-0741

Dr. Larry Breaker *
Moss Landing Marine Lab
Moss Landing, CA 95039

Dr. Kenneth Brink *
Woods Hole Oceanographic Inst.
Woods Hole, MA 02543

Dr. Shenn-yu Chao
Horn Point Laboratory
University of Maryland
Cambridge, MD 21613-0775

Dr. David Chapman
Woods Hole Oceanographic Inst.
Woods Hole, MA 02543

Dr. Dudley Chelton
College of Oceanography
Oregon State University
Corvallis, OR 97331

Dr. John Church *
CSIRO-Marine Laboratories
Division of Oceanography
GPO Box 1538
Hobart, Tasmania 7001, AUSTRALIA

Dr. Allan Clarke *
Dept. of Oceanography
Florida State University
Tallahassee, FL 32306

Dr. Lou Codispoti *
Bigelow Ocean Sciences Lab.
W. Boothbay Harbor, ME 04575

Dr. Curt Collins
Naval Postgraduate School
Monterey, CA 93940

Dr. William Crawford
Institute of Ocean Sciences
Patricia Bay
P.O. Box 6000
9860 W. Saanich Road
Sidney, B.C. V8L 4B2

A. Badan-Dangon *
CICESE
P.O. Box 4844
San Ysidro, CA 92073

Dr. Don Denbo
Ctr. for Earth & Planetary Physics
Harvard University
Cambridge, MA 02138

Dr. Russ Davis
Mail Code A-030
Scripps Inst. of Oceanography
La Jolla, CA 92093

Prof. Richard C. Dugdale
Laboratoire Physique et Chimie Marin
LaDarse, 06230 Villefranche sur Mer
FRANCE

D.J. Ellett
SMBA
P.O. Box 3
Oban, Argyll, PA 34 4AD
UNITED KINGDOM

Felix Espinoza
Inst. Hidrografico de la Armada
Casilla 324
Valparaiso, CHILE

Dr. Armando Fiuza
Oceanography Grp. Physics Dept.
University of Lisbon
Rua Escola Politecnica, 58
1200 Lisbon, PORTUGAL

Tomas Fonseca *
INGEMAR, Condell 1190
Casilla 1443
Valparaiso, CHILE

Dr. Howard Freeland *
Institute of Ocean Sciences
Patricia Bay
P.O. Box 6000
9860 W. Saanich Road
Sidney, B.C. V8L 4B2

Robert Frouis
Mail Code A-021
Scripps Inst. of Oceanography
La Jolla, CA 92093

R.S. Gardiner-Garden
Geophysical Fluid Dynamics Prog.
James Forrestal Campus
P.O. Box 308
Princeton, NJ 08542

Dr. Stuart Godfrey
CSIRO-Marine Laboratories
Division of Oceanography
GPO Box 1538
Hobart, Tasmania 7001, AUSTRALIA

Dr. John Gould
National Institute of Oceanography
Wormley, Godalming
Surrey, ENGLAND

Dr. Dale Haidvogel
The Chesapeake Bay Institute
The Johns Hopkins University
The Rotunda, Suite 340
711 West 40th Street
Baltimore, MD 21211

Dr. David Halpern
J.P.L. M/S 169/236 CAL TECH
4800 Oak Grove Drive
Pasadena, CA 91109

R. Haney *
Naval Postgraduate School
Monterey, CA 93940

Dr. Donald Hansen
NOAA/AOML
15 Rickenbacker Cswy.
Miami, FL 33149

Loren Haury
Scripps Inst. of Oceanography
A-018
La Jolla, CA 92093

Tom Hayward
Scripps Inst. of Oceanography
La Jolla, CA 92093

Dr. Barbara Hickey
Dept. of Oceanography
University of Washington
Seattle, WA 98195

Dr. Greg Holloway *
Institute of Ocean Sciences
P.O. Box 6000
Sidney, B.C. 48L 4B2

Dr. Robert Houghton
Lamont-Doherty Geological Observ.
Columbia University
Palisades, NY 10964

L. Hutchings
Sea Fisheries Branch
Beach Road
Sea Point
Cape Town, SOUTH AFRICA

Dr. John Huthnance
Institute of Ocean Sciences
Bidston, Birkenhead
Merseyside L43 7RA, UNITED KINGDOM

Dr. Adriana Huyer *
College of Oceanography
Oregon State University
Corvallis, OR 97331

Dr. Motoyoshi Ikeda
Bedford Inst. of Oceanography
Dartmouth NS
Vancouver, Nova Scotia
Canada B2Y 4A2

Dr. Donald Johnson
Naval Ocean R&D Activity
Stennis Space Center, MS 39529-5004

Dr. Burton Jones
Dept. of Biological Sciences
Univ. of Southern California
Los Angeles, CA 90007

Dr. P. Michael Kosro *
College of Oceanography
Oregon State University
Corvallis, OR 97331

Dr. Pijush Kundu
Nova Univ., Physical Ocean. Lab
8000 N. Ocean
Dania, FL 33004

Dr. Miguel Lavin
Dept. of Oceanography
CICESE
San Ysidro, CA

Dr. Ants Leetmaa
NOAA/AOML
15 Rickenbacker Cswy
Miami, FL 33149

Steve Lentz *
Woods Hole Oceanographic Inst.
Woods Hole, MA 02543

Rolf G. Lueck
Chesapeake Bay Institute
Johns Hopkins University
315-711 W. 40th Street
Baltimore, MD 21211

Dr. Roger Lukas
Joint Inst. for Marine & Atms. Res.
University of Hawaii at Manoa
1000 Pope Road
Honolulu, HI 96822

R.J. Lynn
Southwest Fisheries Center
P.O. Box 271
La Jolla, CA 92038

T.A. McClimans
Norwegian Hydrotechnical Laboratory
Norwegian Inst. of Technology
Trondheim, Norway

Dr. Julian McCreary *
Nova University
8000 N. Ocean Drive
Dania, FL 33004

Dr. John McGowan
Mail Code A-028
Scripps Inst. of Oceanography
La Jolla, CA 92093

Prof. George L. Mellor
Geophysical Fluid Dynamics Prog.
Princeton University
P.O. Box 308
Princeton, NJ 08540

Dr. Ekkehard Mittelstaedt
Deutsches Hydrographisches Institut
Bernhard-Nocht-Strasse 78
2 Hamburg 4, GERMANY

Dr. C.N.K. Mooers *
Inst. for Naval Oceanography
Bldg. 1100, Rm. 311
Stennis Space Center, MS 39529-5005

Dr. Robin Muench
10533 Ravenna Avenue NE
Seattle, WA 98125

Greville Nelson
Sea Fisheries Research Inst.
Private Bag X2
Roggebaai 8012, SOUTH AFRICA

Dr. Stephen Neshyba *
College of Oceanography
Oregon State University
Corvallis, OR 97331-5503

Dr. H.J. Niebauer
Institute of Marine Science
University of Alaska
Fairbanks, AK 99755

Peter Niiler
Scripps Inst. of Oceanography
Mail Code A-030
La Jolla, CA 92093

Marlene Noble
USGS
Menlo Park, CA

Dr. James J. O'Brien
Dept. of Meteorology Annex
Florida State University
930 Wildwood
Tallahassee, FL 32306

Theresa Paluszkiewicz
Minerals Management Service
U.S. Dept. of Interior
Washington, DC 20240

A. Pares-Kiema
Dept. of Meteorology Annex FSU
930 Wildwood
Tallahassee, FL 32306

Richard Parrish
Pacific Environmental Grp.
P.O. Box 831
Monterey, CA 93942

Dr. Joseph Pedlosky
Dept. of Physical Oceanography
Woods Hole Oceanographic Inst.
Woods Hole, MA 02543

Dr. George Philander
NOAA Geophys. Fluid Dynamics Lab
Box 308
Princeton, NJ 08540

Prof. Steve Ramp *
Naval Postgraduate School
Monterey, CA 93940

Prof. Joseph Reid
Mail Code A-030
Scripps Inst. of Oceanography
La Jolla, CA 92093

Michele Reinicker
Inst. for Naval Oceanography
Bldg. 1103, Rm. 233
Stennis Space Center, MS
39529-5005

Dr. Peter Rhines
School of Oceanography
University of Washington
Seattle, WA 98195

Steve Riser *
School of Oceanography
University of Washington
Seattle, WA 98195

Dr. Lew Rothstein *
School of Oceanography
University of Washington
Seattle, WA 98195

Prof. Tom Royer
Institute of Marine Science
University of Alaska
Fairbanks, AK 99701

Eckart Schumann
Universiteit van Port Elizabeth
Posbus 1600 Port Elizabeth 6000
Republeik van Suid-Afrika

Mr. Nelson Silva
Centro de Investigaciones del Mar
Universidad Catolica de Valparaiso
Casilla 1020
Valparaiso, CHILE

Dr. James Simpson
Mail Code A-030
Scripps Inst. of Oceanography
La Jolla, CA 92093

Prof. D.C. Smith
Naval Postgraduate School
Monterey, CA 93940

Dr. Robert L. Smith *
College of Oceanography
Oregon State University
Corvallis, OR 97331-5503

CPCB Hector SOLDI Soldi
Direccion de Hidrografia y
Navegacion de la Marina
Casilla Postal 80
Chucuito, Callao, PERU

Dr. Raymond K. Steedman
72 Watkins Road
Dalkeith, Western Australia
AUSTRALIA 6009

Lothar Stramma
Institut fur Meereskunde
an der Universitat Kiel
Dusternbrooker Weg 20
D 2300 Kiel 1 FRG

Dr. Rick Thompson
Institute of Ocean Sciences
9860 W. Saanich Road
Sidney, BC V8L 4B2
CANADA

M. Tomczak
Dept. of Geology and Geophysics
University of Sydney
Sydney 2006, NSW AUSTRALIA

Dr. John Whitehead *
Woods Hole Oceanographic Inst.
Woods Hole, MA 02543

Jacob Wickham, Professor Emeritus
Naval Postgraduate School
Monterey, CA 93940

H. Yamazaki
Chesapeake Bay Institute
Johns Hopkins University
315-711 W. 40th Street
Baltimore, MD 21211

Prof. Salvador Zuta
Instituto del Mar IMARPE
Direccion Postal
Casilla 3734
Lima, PERU

Coastal and Estuarine Studies

(formerly Lecture Notes on Coastal and Estuarine Studies)

Vol. 1: J. Sündermann, K.-P. Holz (Eds.), Mathematical Modelling of Estuarine Physics. Proceedings, 1978. 265 pages. 1980.

Vol. 2: D.P. Finn, Managing the Ocean Resources of the United States: The Role of the Federal Marine Sanctuaries Program. 193 pages. 1982.

Vol. 3: M. Tomczak Jr., W. Cuff (Eds.), Synthesis and Modelling of Intermittent Estuaries. 302 pages. 1983.

Vol. 4: H.R. Gordon, A.Y. Morel, Remote Assessment of Ocean Color for Interpretation of Satellite Visible Imagery. 114 pages. 1983.

Vol. 5: D.C.L. Lam, C.R. Murthy, R.B. Simpson, Effluent Transport and Diffusion Models for the Coastal Zone. 168 pages. 1984.

Vol. 6: M.J. Kennish, R.A. Lutz (Eds.), Ecology of Barnegat Bay, New Jersey. 396 pages. 1984.

Vol. 7: W.R. Edeson, J.-F. Pulvenis, The Legal Regime of Fisheries in the Caribbean Region. 204 pages. 1983.

Vol. 8: O. Holm-Hansen, L. Bolis, R. Gilles (Eds.), Marine Phytoplankton and Productivity. 175 pages. 1984.

Vol. 9: A. Pequeux, R. Gilles, L. Bolis (Eds.), Osmoregulation in Estuarine and Marine Animals. 221 pages. 1984.

Vol. 10: J.L. McHugh, Fishery Management. 207 pages. 1984.

Vol. 11: J.D. Davis, D. Merriman (Eds.), Observations on the Ecology and Biology of Western Cape Cod Bay, Massachusetts. 289 pages. 1984.

Vol. 12: P.P.G. Dyke, A.O. Moscardini, E.H. Robson (Eds.), Offshore and Coastal Modelling. 399 pages. 1985.

Vol. 13: J. Rumohr, E. Walger, B. Zeitzschel (Eds.), Seawater-Sediment Interactions in Coastal Waters. An Interdisciplinary Approach. 338 pages. 1987.

Vol. 14: A.J. Mehta (Ed.), Estuarine Cohesive Sediment Dynamics. 473 pages. 1986.

Vol. 15: R.W. Eppley (Ed.), Plankton Dynamics of the Southern California Bight. 373 pages. 1986.

Vol. 16: J. van de Kreeke (Ed.), Physics of Shallow Estuaries and Bays. 280 pages. 1986.

Vol. 17: M.J. Bowman, C.M. Yentsch, W.T. Peterson (Eds.), Tidal Mixing and Plankton Dynamics. 502 pages. 1986.

Vol. 18: F. Bo Pedersen, Environmental Hydraulics: Stratified Flows. 278 pages. 1986.

Vol. 19: K.N. Fedorov, The Physical Nature and Structure of Oceanic Fronts. 333 pages. 1986.

Vol. 20: A. Rieser, J. Spiller, D. VanderZwaag (Eds.), Environmental Decisionmaking in a Transboundary Region. 209 pages. 1986.

Vol. 21: Th. Stocker, K. Hutter, Topographic Waves in Channels and Lakes on the f-Plane. 176 pages. 1987.

Vol. 22: B.-O. Jansson (Ed.), Coastal Offshore Ecosystem Interactions. 367 pages. 1988.

Vol. 23: K. Heck, Jr. (Ed.), Ecological Studies in the Middle Reach of Chesapeake Bay. 287 pages. 1987.

Vol. 24: D.G. Shaw, M.J. Hameedi (Eds.), Environmental Studies in Port Valdez, Alaska. 423 pages. 1988.

Vol. 25: C.M. Yentsch, F.C. Mague, P.K. Horan (Eds.), Immunochemical Approaches to Coastal, Estuarine and Oceanographic Questions. 399 pages. 1988.

Vol. 26: E.H. Schumann (Ed.), Coastal Ocean Studies off Natal, South Africa. 271 pages. 1988.

Vol. 27: E. Gold (Ed.), A Law of the Sea for the Caribbean: An Examination of Marine Law and Policy Issues in the Lesser Antilles. 507 pages. 1988.

Vol. 28: W.S. Wooster (Ed.), Fishery Science and Management. 339 pages. 1988.

Vol.29: D.G. Aubrey, L. Weishar (Eds.), Hydrodynamics and Sediment Dynamics of Tidal Inlets. 456 pages. 1988.

Vol. 30: P.B. Crean, T.S. Murty, J.A. Stronach, Mathematical Modelling of Tides and Estuarine Circulation. 471 pages. 1988.

Vol. 31: G. Lopez, G. Taghon, J. Levinton (Eds.), Ecology of Marine Deposit Feeders. 322 pages. 1989.

Vol. 32: F. Wulff, J.G. Field, K.H. Mann (Eds.), Network Analysis in Marine Ecology. 284 pages. 1989.

Vol. 33: M.L. Khandekar, Operational Analysis and Prediction of Ocean Wind Waves. 214 pages. 1989.

Vol. 34: S.J. Neshyba, Ch.N.K. Mooers, R.L. Smith, R.T. Barber (Eds.), Poleward Flows Along Eastern Ocean Boundaries. 374 pages. 1989.